FIRE INVESTIGATIONS FOR FIRST RESPONDERS

Russell K. Chandler

JONES & BARTLETT
LEARNING

Jones & Bartlett Learning
World Headquarters
5 Wall Street
Burlington, MA 01803
978-443-5000
info@jblearning.com
www.jblearning.com

Jones & Bartlett Learning books and products are available through most bookstores and online booksellers. To contact Jones & Bartlett Learning directly, call 800-832-0034, fax 978-443-8000, or visit our website, www.jblearning.com.

10401-1

Production Credits

General Manager and Executive Publisher: Kimberly Brophy
VP, Product Development: Christine Emerton
Senior Managing Editor: Donna Gridley
Executive Editor: Bill Larkin
Vice President of Sales, Public Safety Group: Phil Charland
Digital Project Specialist: Angela Dooley
Director of Marketing Operations: Brian Rooney

VP, Manufacturing and Inventory Control: Therese Connell
Composition: S4Carlisle Publishing Services
Cover & Text Design: Scott Moden
Rights & Media Specialist: John Rusk
Cover Image (Title Page, Part Opener, Chapter Opener): © Jones & Bartlett Learning. Photographed by Glen E. Ellman.
Printing and Binding: LSC Communications

Library of Congress Cataloging-in-Publication Data
Names: Chandler, Russell K., author.
Title: Fire investigations for first responders / Russell K. Chandler.
Description: Burlington : Jones & Bartlett Learning, 2019. | Includes index.
Identifiers: LCCN 2019033981 | ISBN 9781284180275 (paperback)
Subjects: LCSH: Fire investigation. | Fire prevention--Standards. | Fire fighters.
Classification: LCC TH9180 .C45 2019 | DDC 363.37/65--dc23
LC record available at https://lccn.loc.gov/2019033981

6048

Printed in the United States of America
23 22 21 20 19   10 9 8 7 6 5 4 3 2 1

© Jones & Bartlett Learning. Photographed by Glen E. Ellman.

Brief Contents

Contents

© Jones & Bartlett Learning. Photographed by Glen E. Ellman.

CHAPTER **4**
Science, Methodology, and Fire Behavior   49

CHAPTER **5**
Building Construction for the Fire Investigator   72

CHAPTER **6**
Fire Protection Systems   83

CHAPTER **7**
Electricity and Fire   99

CHAPTER 11
Patterns: Burn and Smoke 177

CHAPTER 12
Examining the Scene and Finding the Origin 199

CHAPTER 13
Fire Cause 213

© Jones & Bartlett Learning. Photographed by Glen E. Ellman.

About the Author

Like so many of the readers, I chose the fire service as my career. As such, it was always a team effort, so it was hard to take individual credit. However, as an author I now need to share my achievements so that you can see why this text is a work of someone who has "been there" and was fortunate to have "done that."

I always teach my students that "we are the sum of our experiences." My experiences started before I got involved in public safety. My grandfather, who taught me my core values, also gave me an opportunity to become a licensed electrician and plumber before I even got my driver's license. In addition to wiring homes and businesses, we did repairs on electrical household appliances. Little did I know at that time how valuable this experience would be in my investigative ventures. A second family member who was an influence in my chosen path was my uncle Carroll Leeman. As a child, I was impressed that he was a volunteer fire fighter, and he had a wealth of common sense and a straightforward way of handling things that I admired.

My employment path has taken me from being a career fire fighter, to a full-time safety manager and fire chief at a major amusement park, to the first fire marshal in Hanover County, Virginia. I was then lured by the fiscal rewards of being a fire investigator in the private sector, primarily serving the insurance industry. This role added to the sum of my experiences, allowing me to see an entirely different side of and motivation for completing the fire investigation. Through this different aspect of the investigative world, I took great pride in our accomplishments and found a whole new world of colleagues and peers.

In 1990, the Virginia Department of Fire Programs (VDFP) hired me as their Manager of Investigations and Inspections. Within a few years, we redesigned that entire division, creating the Virginia Fire Marshal Academy (VFMA). The Academy delivered national certification-level training for Investigations, Inspections, and, eventually, Public Fire & Life Safety Education to Virginia's fire departments. One vital school the Academy provided to Virginia's Fire Marshals was the Law Enforcement School for Fire Marshals. This school enabled local investigators to get their police powers from an academy dedicated to making them investigators, rather than from the other law enforcement academies dedicated to turning out new police patrol officers.

Being promoted to VDFP Branch Chief offered me the opportunity to head up the Technical Services Branch with oversight of six divisions, one of which was the VFMA. I then took over Operations, which allowed me to have oversight of the seven divisions that provided fire training to the entire state. Retiring from VDFP culminated 42 years of career fire service duty.

I was thinking that retirement was finally here! However, the International Association of Fire Chiefs made me a wonderful offer. For almost five years, I was able to add more to my experiences by serving as their subject matter expert (SME) in fire investigations and prevention. I was also able to serve as a program manager, working as staff liaison with the IAFC Fire & Life Safety Board of Directors. I was and remain in awe of this group of individuals who have done so much good in fire prevention at the national and international levels.

In addition to my full- and part-time career jobs, I also served more than 50 years as a volunteer fire fighter. With each job change came a residential change. With each move, I always found time to join the local volunteer department. I was so fortunate to have been afforded the opportunity to serve in every level from fire fighter to Chief.

I had given up my position as an adjunct instructor with J. Sargeant Reynolds Community College after 20 years of teaching investigations, prevention, and other fire service topics. I then took an adjunct position with the Virginia Commonwealth University teaching graduate students at the School of Forensic Science. This is the one job I still hold. I teach a class in fire investigations to future forensic lab personnel. I like to describe it as an opportunity to start their transition from the academic world to the real world. Arson is the perfect topic because it can involve so many other crimes.

I had been fortunate to work with an outstanding team of peers in the validation of IFSTA manuals, including the second and third editions of the *Introduction of Fire Origin and Cause* and the first edition of *Fire Investigator*. I was chosen to work on the Technical Working Group on Fire/Arson Scene Investigation in their creation of the *Fire and Arson Scene Evidence*: *A Guide for Public Safety Personnel*, which was a research report published by the United States Department of Justice, Office of Justice Programs, National Institute of Justice.

One of the greatest rewards—and an unbelievable education—had been the opportunity to serve on the National Fire Protection Association, NFPA 921 Technical Committee, on the *Guide for Fire and Explosion Investigations* text. For more than 20 years,

I served on this committee, working alongside, and in awe of, some of the greatest minds in the fire investigative field, becoming the committee secretary before leaving. These meetings were rewarding, frustrating, and, at times, infuriating but they were always educational.

I was fortunate to be recognized with various awards, including Instructor of the Year for the Virginia Commonwealth University School of Humanities, and I received the Medal of Valor from Hanover County Fire & EMS.

I hope that my experiences and the material in this text will provide the reader with the motivation to conduct the investigation and be a seeker of truth, as well as provide the knowledge and skills necessary to come up with the accurate determination of the fire origin and cause.

Reviewers

Scott E. Avery
Full-Time Faculty
Columbia Southern University
Orange Beach, Alabama

Amy Brooks
Director of Emergency Services
Central Arizona College
Coolidge, Arizona

J. Barry Burnside
Staff Instructor
Mississippi State Fire Academy
Jackson, Mississippi

Paul E. Calderwood
Retired Deputy Fire Chief/Fire Investigator
Everett Fire Department
Everett, Washington

Battalion Chief Jim Campbell
Pike Township Fire Department
Indianapolis, Indiana

Matthew Claflin
Educational Contact and Training Coordinator
    (University of Akron)
Lieutenant, Akron Fire Department
Akron, Ohio

Christopher Easton
Fire Science Technology
Eastern Maine Community College
Bangor, Maine

Stan Fernandez
Deputy Fire Marshal
Alameda County Fire Department
Dublin, California

Robert Goldenberg
Assistant Director, Advanced Fire Program
Technical Education Training Center Osceola
Kissimmee, Florida

Lucas Goodin
Cromwell Wright Area Fire District &
    Hibbing Community College
Hibbing, Minnesota

Kristopher Grod
Associate Dean—School of Public Safety
Northcentral Technical College
Wausau, Wisconsin

Kevin L. Hammons
Metropolitan State University of Denver
Denver, Colorado

James Horton
Fort Worth Fire Department
Fort Worth, Texas

William Justiz
Fire Science Coordinator
Triton College
River Grove, Illinois

Bernard "Sandy" Kromenacker
Senior Analyst/Director of Investigations
Forensic Fire Consultants, Inc.
Durham Tech Community College
Durham, North Carolina

Wes Lail
Director of Fire Protection & Emergency Management Program
Catawba Valley Community College
Hickory, North Carolina

Chad Landis
Training/Safety Officer
Rapides Parish Fire District 3/Alpine FD
Pineville, LA

Danny Nicholson
Battalion Chief/Accreditation Manager
City of Statesville Fire Department
Statesville, North Carolina

Sean F. Peck
Adjunct Faculty of Fire Science
Arizona Western College
Yuma, Arizona

Jared Rouse
Fire Protection Technology Faculty/Advisor
Johnston Community College
Smithfield, North Carolina

Kenneth Staelgraeve
Professor of Fire Science
Macomb Community College
Warren, Michigan

Ian Storm
Program Manager of Emergency Services Administration BAAS
The University of North Texas Dallas
Dallas, Texas

Norcliff W. Wiley
Chief/Fire Marshal—Salinas Fire Department
Salinas, California

Acknowledgments

Cooperative Effort

This book was a cooperative effort, with Chief Bobby Bailey writing the chapter on Vehicle Fires. He has served as a police officer, fire fighter, Battalion Chief of Suppression, and as the Battalion Chief of the Hanover Fire Marshal's Office. He was the chief in charge of the fire-service training academy in his department until I was fortunate to hire him to take over as the chief of the Virginia Fire Marshal Academy. His expertise and energy took the Academy to new heights.

Dedication

Dedicated to Chief Michael G. Harman, Sr.

As we get older and look back upon our careers, we realize that there have been a handful of individuals who have served as mentors and shining examples of what we hope to become in our professions. We would not be where we are today without their guidance.

In my career, I have been fortunate to have encountered numerous individuals who have guided me on my path, but one person shines far above all others: Chief Mike Harman, who hired me as the first Fire Marshal of Hanover County, Virginia. He provided me with the opportunity to start my fire investigative, prevention, and teaching career. He taught me to emphasize prevention and not enforcement unless necessary. He also reinforced my firm belief that all fire investigators should conduct a full investigation to prove innocence as much as guilt. All investigation units should have the same mantra: *We are seekers of truth.*

Mike was the finest leader I encountered in my entire career. He always sought a fair equity in any endeavor for all parties, and he led by example in every part of his professional and personal lives. I am not just a better investigator, instructor, and fire service leader thanks to Mike, but also a better person. Thank you, Chief, for being who you are, not just for me, but for so many others.

Editing Team

A phenomenal editing staff made many suggestions that enhanced this text. They deserve accolades for their suggestions and efforts to educate an old fire fighter. Were it not for Executive Editor Bill Larkin, this text would never have seen the light of day.

Notes to Instructors and Readers

From the very beginning, the primary goal of this text was to provide training to first responders to assure an accurate determination of the fire origin and cause of all fires within their jurisdiction. The second goal was to organize the text so it can be used in a single-semester college environment. Other fire investigator texts concentrate on the intricacies of being a full-time investigator, with all the challenges and requirements to meet those national standards. However, the average community college classroom consists of current or aspiring fire fighters who need the fundamental knowledge of fire origin and cause at the suppression level.

The layout of this text not only meets FESHE requirements but is also designed to easily flow into the average college semester schedule, making it a great tool for instructors to include on their syllabus. It also serves as a perfect reference document for all first responders for ongoing and refresher training, whether it be at the fire station, police academy, EMS facility, or the emergency communications center.

I welcome your feedback on this text. If you have any ideas for improvements, comments on the content, or suggestions for what to include in future editions, please contact me. If you have any questions about the organization or presentation of this material, please let me know.

Russ Chandler
Vafirelaw@aol.com

CHAPTER 1

Role of First Responders

LEARNING OBJECTIVES

Upon completion of this chapter, you should be able to:

- Describe who is a first responder.
- Describe the responsibilities of the fire fighter and the fire department.
- Describe the observations made by the various types of first responders as they approach, upon arrival, and during suppression at the fire or explosion scene.
- Describe the role of the first responder as the potential first investigator on the scene.
- Describe the value of first responders' observations upon arrival and during suppression at the fire or explosion scene.
- Identify incendiary devices.

Case Study

It was 3:00 AM when the alarm came in: A local drinking establishment had flames shooting through a rear window. The site was not far from the local fire station, and in minutes the engine apparatus and the ladder truck arrived on the scene. Both parked in front of the property, blocking any entrance to and exit from the parking lot. The arriving companies had no reason to think this positioning would pose any problem.

The engine company's lieutenant did an assessment, walking around the structure while his crew stretched their attack lines. The first thing he noticed was that the back door to the structure was open and there appeared to be signs of forced entry. He accompanied his crew as they crawled into the structure and quickly put out the fire. Coming back out of the structure, the fire officer took off his self-contained breathing apparatus (SCBA) mask. He immediately noticed a slight trace of a gasoline odor—one that he realized was coming from his turnout gear.

The lieutenant pulled out his crew immediately, ceasing any overhaul. Even though the crew had searched the structure for victims, he did another quick check and ensured all fire fighters had vacated the building. In this process, he made some other observations. The coin boxes for all the video games and pool tables were gone, and the cash register drawer was missing. A wire-bound notebook was laid open in a serving tray and covered in liquid. This finding was unusual because the engine company had not flowed any water in that area of the structure. The fire officer also noticed handwritten entries on the open page of the notebook.

In the meantime, the ladder truck had prepared to set up the ladder pipe in case it was needed; it was not, because the engine crew had quickly extinguished the fire. As the ladder operator stood on his turntable, he looked down and noticed a lone car in the parking lot. He could see that someone was slouched down in the driver's seat but was looking around as the fire fighters went about their business.

Both the engine lieutenant and the truck operator reported their findings to the incident commander, who immediately called for an investigator (the local fire marshal). When the investigator arrived on the scene, he and the lieutenant examined the fire scene. As suspected by the lieutenant, the site appeared to be a crime scene. Indicators included the back door that had been forced open, the missing coin boxes, the cash register that had been forced open, and

the presence of an ignitable liquid. But most revealing was the notebook found in the serving tray. The tray was holding approximately a half inch of what appeared to be gasoline. The submerged book was clearly a bar tab; it was open to a specific patron's tab—a person who owed more than $200 to the establishment. Fortunately, the fire had been set in the kitchen on the counter and had not reached the gasoline vapors in the bar by the time the fire crews arrived.

After documenting the scene, the investigator went to the parked car and found the individual described by the ladder apparatus operator. The person in the vehicle stated that he had come out of the bar and was too drunk to drive, so he fell asleep in his car. This seemed a valid explanation and was a commendable decision. After asking for the driver's credentials, the investigator discovered that the name on the driver's license was the same as that on the bar tab. A police patrol officer was on the scene, and the investigator requested a sobriety test be performed on the vehicle occupant, which the driver clearly passed. This was important to assure the individual would understand the next line of questioning.

As part of the preparation for his role, the fire investigator had received specific training on watching an individual's body language to determine truthfulness. In this particular case, the driver's body language matched the classic description of deceitful behavior, so the investigator asked for permission to look in the trunk of the car. The driver stated that he would like to comply—but it was not his car and he did not have a trunk key. A radio call to the dispatcher, however, revealed that the vehicle was registered to the driver.

The investigator again asked if he could look in the trunk. The driver advised that he had lost the trunk key, but also clearly stated that if the investigator could open the trunk, he was welcome to look in it. The investigator warned the driver that the lock and the trunk itself would be damaged by forcing it open, but the driver stood firm and said to go right ahead.

The truck company lieutenant was standing nearby and had heard the conversation. The lieutenant had already picked out his best fire fighter, who was now standing at the back of the vehicle with his forcible entry tool and a flat-head axe. One strike and the trunk popped open. The trunk contained two shrink-wrapped cases of beer labeled with a delivery sticker for the bar. Next to them were six metal coin boxes (the type from vending machines and games) half-full of quarters. Sitting on top of the coin boxes was a drawer from a cash register, still holding coins and paper money.

As the investigator placed the cuffs on the driver and read him his rights, the driver just stood there shaking his head. It was later confirmed that the serial numbers on the coin boxes matched the video games and pool table in the bar. The bar owner identified the cash drawer; it even contained the owner's business card with some figures written on the back. The owner also confirmed that the beer found in the trunk had been delivered to the bar and had not been sold or given away by the staff. As the suspect left the scene in cuffs, his only comment was that he did not expect such a quick response by the fire department and did not expect the fire trucks to block him from pulling out of the parking lot.

The actions of the lieutenant in recognizing incendiary indicators and getting his crew out of the structure were commendable. His decision to delay overhaul also preserved the scene, allowing the investigator the best possible scenario to examine and document the evidence. This factor, along with the ladder truck operator noticing unusual circumstances, provided everything necessary to get an eventual conviction for arson.

During this fire investigator's entire career, this incident was one of only two investigations where the arsonist was arrested at the scene. If not for the observations and actions of the fire fighters at the scene, however, this case might never have been solved. Overhaul of the interior might have been disturbed the evidence, removing the debt ledger in the tray sitting on the bar. The arsonist could have gone unnoticed, pulling away later with the evidence in the trunk. It is critical that the first responders realize that they are a vital component in the identification of fire origins and causes—information that may prove critical in preventing future fires.

 *Access Navigate for more resources.*

Introduction

A **first responder** is just what this term implies: one of the first public safety providers dispatched to or arriving on the emergency scene. First responders can be patrol law enforcement (police) officers, **emergency medical technicians (EMTs)**, or fire fighters. They are the first public safety officials to observe the scene and the surrounding area, as well as the people in and around the incident. Their observations can be critical to the ultimate disposition of the investigation (**FIGURE 1-1**).

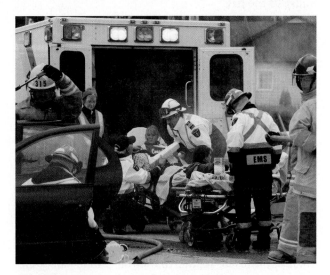

FIGURE 1-1 First responders arriving at the scene make many valuable observations in the course of their duties.
© KellyNelson/Shutterstock.

> ### TIP
>
> First responders can be patrol law enforcement (police) officers, emergency medical technicians (EMTs), or fire fighters.

Responders have their own responsibilities and will do an individual size-up as they approach the scene. As they exit their vehicles, they will assess people present for injuries and look for damage and road blockages that could hamper the arrival of additional resources. They will think about all kinds of safety issues. In particular, if they are arriving at an explosion scene, the thought of a secondary device will not be far from their minds.

While en route, arriving at, and working the scene, responders will also make observations that may prove valuable to the fire investigator. The importance of these observations may not be obvious at first. With training, however, recall of these observations will help the first responder in coming up with potential conclusions as to the fire's origin and classification of the fire's cause.

The **emergency communications dispatcher** is usually the first public safety person to collect any information about the incident. For this reason, these personnel are considered first responders, even though they do not travel to the scene. Dispatchers, like the emergency medical personnel, are not in a position to investigate the fire, but they can be a valuable asset in this process. The patrol officer, too, is not likely to be the primary investigator in most jurisdictions. Nevertheless, owing to the very nature of the occupation, these officers may recognize the value of what they see and will readily present this information to the proper authorities.

The fire fighter is both a first responder and potentially the first investigator on the scene. In this text, we group fire officers and fire fighters within the same category. Their rank does not establish their role so much as their assigned responsibilities. The person who fills out the fire report is the one who should collect the information necessary to complete the fire incident report, which can include the identification of the first material ignited, the heat source, and the cause of the fire.

This chapter focuses on the role of each first responder and how that role can affect the fire investigation. Some first responders, especially fire service personnel, have a tremendous impact on the fire scene. A fire fighter may also be the investigator of record, documenting the fire scene, suppression, and the determination of the fire origin and cause in the National Fire Incident Report System.

Fire Department Responsibility

The well-used phrase "to protect and serve" is more than appropriate to the role of all public safety elements of the government. In particular, the fire department of any locality is responsible for responding to calls for assistance from the public. However, fire departments' exact role is most likely defined in the charter of that locality or defined by the authority having jurisdiction (AHJ), which would be an entity such as the board of supervisors or town council. Essentially, the AHJ is the body of elected officials of that local government entity.

Fire fighters certainly respond to fires, and in most jurisdictions today they handle emergency medical services as well. One key aspect of the firefighting service focuses on the goal of saving lives and

property through fire prevention. This can be accomplished through code enforcement and—most importantly—through fire and life safety education delivered to school children, public groups, and visitors to the fire station.

Of course, you can only prevent fires if you know how they start. The first step is to know and use the principles covered in Chapter 4, *Science, Methodology, and Fire Behavior*—not to suppose, speculate, or guess how fires start, but rather to prove how fires start by applying scientific methodology. The first step in this process is to investigate and identify the area of fire origin. At this location will be the material first ignited and the reliable heat source. then the act that brought them together will need to be identified. These data are then used in fire education and the development of fire codes to prevent any similar incidents. The collection of such data starts with fire fighters filling out the forms in the **National Fire Incident Reporting System (NFIRS)**.

The underlying principle of the NFIRS is for members of the fire service to provide data on the fires they encounter, and to document the type of structure as well as the presence (or absence) of fire protection systems, and if they functioned properly. Perhaps the person doing the report might grumble about how long it takes to complete. However, these data are critical to all fire prevention in the future as well as for justifying future requests for equipment, personnel, and other resources. NFIRS is also the primary source for data on where fires originated, which materials ignited first, and which heat sources ignited those materials. This information is critical to enhance fire prevention efforts.

Full-time fire investigators will also have their own reporting procedure, which covers all aspects of their actions and findings on the scene, plus findings during their more detailed and additional investigation based on interviews as well as lab reports. Sometimes the subsequent investigation reveals that the initial findings were not completely accurate. When this happens, the investigator must amend the NFIRS to assure it accurately reports the final findings. This information could be critical later if the investigation leads to an arrest and court appearance. Both the investigator's report and the NFIRS report should accurately reflect the final findings to prevent any confusion.

Some departments rely on fire operations personnel, such as fire fighters, to fill out the NFIRS report—and that is where any further determination of fire origin and cause stops. Other departments may assign personnel to be either part-time or full-time fire investigators, who have the sole charge of examining the scene and reporting their findings.

If evidence suggests that the fire is incendiary, then these investigators will turn the matter over to the local police, who will then continue the investigation. Still other fire departments have assigned personnel to conduct full in-depth investigations and give them full police powers to do so. Some localities employ a team approach, with both fire and police personnel working as a team to carry out full investigations.

The responsibility of both the first-responder investigator and the assigned investigator will be to find the area of origin and the cause of the fire. The assigned investigator, possibly with the assistance of the first responders, must further examine the reason for any injuries or deaths. Discovering why a victim was injured or why he or she did not escape is just as important as finding the area of fire origin or even identifying an arsonist. The mission of the fire department is not just to save lives during the incident, but also to take steps necessary to prevent future incidents, thereby reducing injuries and deaths.

First Responders' Role in an Investigation

From the fire service's perspective, an incident starts with the first report of that incident. Usually, the first public safety personnel to become involved in an incident are those in the emergency communications center (dispatch). They, in turn, broadcast via radio the information necessary to send the appropriate units for the situation at hand. Those personnel sent to the incident are the first responders, the very first personnel to respond to the scene to handle the emergency situation. They may know information that will be beneficial to the investigation of the incident, including observations they make while en route to the scene or comments they heard as they were packing up to leave.

> **TIP**
>
> Usually, the first public safety personnel involved in an incident are those in the emergency communications center (dispatch).

Although the hierarchy of responders may differ from jurisdiction to jurisdiction, their overall operational tasks remain somewhat consistent across the country. We explore these commonalities in the following subsections.

Emergency Communications Dispatchers

Dispatchers are not typically considered first responders because they do not physically go to the scene. However, when considering the investigation of an incident, dispatchers are usually the first emergency personnel to have information about the fire or explosion scene (**FIGURE 1-2**). Most modern emergency communications centers not only have radios, telephones, and proper protocols, but also have recording equipment that documents all telephone and radio traffic.

The vast majority of emergency communication centers have 24-hour electronic digital recorders. Older systems, which may still be in use in rural areas, use magnetic tape similar to a typical reel to reel tape recorder. In these systems, someone must take the tapes off the machine and rotate the oldest tape back onto the machine for re-recording. The data on the older tape are erased as the newer data are recorded. Most facilities have enough spare tapes to enable them to take a few out of the rotation to be saved for an investigation or court proceedings. Eventually, all of these older systems will likely be replaced with the newer digital formats.

Emergency communications centers with newer systems record on various media with sufficient data memory to capture 30 days' of recordings or more with no need to physically change media devices each day. As would be expected, the new media, such as solid-state devices for digital recordings, have better quality and are capable of discerning which information came off which radio frequency or which phone line. Data can be transferred onto a variety of devices for long-term storage.

FIGURE 1-2 The dispatcher can obtain valuable information from those reporting the emergency.
Courtesy of Russ Chandler.

Whatever type of system is used by the fire department, the assigned fire investigator must be familiar with the dispatch center, its capabilities, and its resources. The investigator must also make sure to request communications' data before they are erased. For its part, the dispatch center should be able to transfer data from the main system to a smaller recording device.

> **TIP**
>
> The assigned fire investigator must be familiar with the dispatch center, its capabilities, and its resources.

The emergency communication center's recorders usually apply a time stamp to the data—that is, recordings are stamped with the exact time of their creation. Time stamps help in searching at a later date and assist with future investigations.

The dispatching staff must also be trained to recognize that their interactions with callers on the phone (911 lines) and first responders on the radio may provide crucial evidence about the incident. Dispatchers are trained to collect the vital information needed to get the proper resources to the scene. In situations where citizens are in harm's way, the dispatcher will advise them to get to safety and terminate the call. In contrast, when witnesses are not in harm's way, the dispatcher may take the initial information and ask the caller to hold as the resources are dispatched. The dispatcher may then have an opportunity to continue the discussion with the caller. The dispatcher may ask if the witness knows where the fire started and where the fire appears in the structure. Other questions include whether the caller saw anything suspicious and, if so, could provide a description of any persons or vehicles leaving the scene. Dispatchers assigned to police phones and radios may already have these skills; those attached solely to a fire department communications center may need to learn these additional skills.

Dispatchers should have developed keen listening skills and can provide the investigator with insights that can prove beneficial before the investigator even listens to the tapes. They can also be a valuable resource regarding what is heard on the tapes. For example, dispatchers may be able to differentiate stray sounds that the recording equipment makes or that are unique to the communications center from sounds that are unusual.

Do not discount a dispatcher's life experiences. Dispatchers have been known to recognize specific sounds, such as coins dropping into a toll-booth basket, the distinctive beeping of a french fry machine in a fast-food restaurant, and the characteristic noise of a water fountain in a park. Call takers in the dispatcher center may have also had years of experience dealing with people reporting emergencies on the phone. They may be able to provide insights into the attitude or demeanor of the person reporting the emergency. Many of these bits of information are not necessarily something that will be used in court for a criminal case, but they can help lead the investigator in the right direction and eventually get the case to court if necessary.

Emergency Medical Personnel

An ambulance or rescue squad is called to the scene of a fire or explosion for two primary reasons. First, there may be a reported injury. Second, they need to be on standby at the scene in case of an injury to a first responder or citizen. The EMT is trained to make assessments, observe patients, and look for the method of injury as a means of diagnosis. The method of injury or any comment made by a patient may be of importance in the subsequent investigation.

> **TIP**
>
> The method of injury or any comment made by a patient may be of importance in the investigation.

For example, the method of injury might involve multiple cuts from flying glass. The incident may have been an explosive scene in the early morning hours in a remote industrial site. The fact that the injuries came from flying glass would be of interest to the investigator: It could indicate that the patient is either an eyewitness or the person who set the explosive device, getting hurt from a premature detonation.

Burn injuries are definitely of interest to the fire investigator. These burns should be thoroughly documented, including the use of a standard system such as the Rule of Nines. This system provides a quick process for determining the percentage of burns on the body based on the location and the extent of burns at each location. If the patient with the burn injuries was inside a business that was closed at the time and supposedly unoccupied, the burns may be a result of setting the fire. The first responder and investigator, however, should never make rash assumptions. The person with burn injuries might also have been a bystander who saw the fire and tried to extinguish the flames.

Even when they arrive at the scene to serve as a standby unit, EMTs may observe suspicious actions of a person in the crowd. When something stands out as being unusual to an EMT, it should be of concern to the fire investigator. For example, consider a bystander who seems overly exuberant in trying to assist the fire fighters and shows an uncanny joy about the scene. Perhaps that individual craves the excitement of a fire incident and was the perpetrator who set the fire. An individual may be hiding in the shadows, trying to avoid being seen or photographed. This, too, could be the actions of someone who set the fire. Recognize, however, that such assumptions can be dangerous: The person might be absolutely innocent and this may be a normal behavior for him or her. But when more than one first responder reports seemingly unusual behavior, this information can be verified or corroborated and could be of value to the investigation.

When EMTs approach the scene, they begin assessing the number of victims and deciding whether they need more resources. If there are no victims, EMTs may be assigned to set up a rehabilitation site (rehab) where the fire fighters can receive fluids, heat, or cooling, depending on the situation, and have their vital signs checked before assuming another suppression task. For this purpose, the EMT crew seeks a location convenient to the scene but out of the way.

On a fire scene, EMTs may observe the location where smoke is originating and take note of the smoke or flame color, but should recognize that their observations are not an accurate determination of the first fuel ignited or a fire cause. EMTs might notice the direction of travel of the smoke, which indicates the wind direction; this may be of interest, for example, if the smoke was blowing from the west to the east yet the fire traveled from the east to the west.

EMTs may also observe victims or bystanders (**FIGURE 1-3**). The attire of the building's occupants should match the time of day. If a residence is burning, for example, it would be unusual to see the family fully dressed at 2 AM. It may also be of interest to see fully dressed bystanders at the scene as responders arrive. Of course, they may have a good reason to be fully dressed: Perhaps they work late shifts or are walking the dog. EMTs might also observe vehicles or pedestrians leaving the scene. Most people are drawn to accident or emergency scenes. In contrast, running from the site or speeding away in a vehicle would not be a normal behavior and would warrant investigation.

On a fire or explosion scene, the EMTs' role in the investigation, in addition to making observations, is

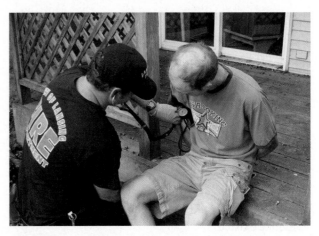

FIGURE 1-3 While assessing patients from the fire or explosion scene, EMTs might hear and observe valuable information.
© Jones and Bartlett Learning. Photographed by Glen E. Ellman.

to preserve any evidence found on victims. Clothing may contain residue of ignitable liquids or chemicals, though this in itself does not indicate criminal behavior. Perhaps an accidental fire started as the result of the victim using gasoline as a cleaning fluid. In all situations, it is critical to make proper and accurate observations but to avoid speculation—just report the facts.

> **TIP**
>
> Clothing may contain residue of ignitable liquids or chemicals, although this in itself does not indicate criminal-related behavior.

If clothing has a noticeable odor, the clothing should be protected from contamination and turned over to a fire investigator as soon as possible. Likewise, any other items that may be potential evidence should be protected and saved. If an investigator is not readily available and no clean metal cans or glass canning jars are available to hold these items, the last resort is to place them in a clean plastic bag that can be sealed. If suspicion arises that this incident may have been more than arson, then other protocols will be followed to preserve any other evidence. EMTs should contact an investigator to arrange for the transfer of all evidence.

> **TIP**
>
> If clothing has a noticeable odor, the clothing should be protected from contamination.

If victims are transported to the emergency department of a hospital, medical personnel there can recover additional evidence. Parts from an explosive device can be embedded in the body of the victim or perpetrator. If the investigator has not arrived at the emergency room, an EMT can stay with the patient and bag that evidence as well. As soon as practical, all evidence should be turned over directly to the investigator; to maintain the chain of custody, the evidence should never leave the EMT's control until this transfer occurs. In years past, investigators' efforts have often been frustrated because no chain of custody for evidence existed.

Emergency medical personnel should be trained and educated on all aspects of their role in assisting in investigations, including collecting, preserving, and protecting evidence. Their help is valuable for all types of investigations. It is just as important to identify the accidental fire scene as it is to solve the criminal case. Thorough investigations of all fire and explosion scenes can go a long way toward preventing similar incidents from occurring in the future.

The Police Patrol Officer

Because police officers are on patrol, they are frequently the first to arrive on the scene of a fire or explosion. On these occasions, law enforcement first responders are in a position to make many important observations should the location turn out to be a crime scene. As trained observers, they can interpret what they see and can even take further actions should they feel it is necessary.

A good example is when a police officer observes someone walking or running away from a fire scene. Most citizens move toward a fire scene out of curiosity. The officer, if time allows, can approach the fleeing individual to investigate (**FIGURE 1-4**). The officer can verify the individual's identification and conduct a brief interview to ascertain why the person was in the area and why the person was leaving. Simply leaving a scene does not make a person guilty of a crime. Indeed, a person may have a good reason for leaving the area. Conversely, the fleeing individual may have been the one to set the fire and may want to distance himself or herself from the scene before being noticed. When a police officer obtains the person's identification information, the scene investigator may be able to interview the person later. If the person is an employee or ex-employee of a business that is burning, the individual would want to stay at the scene and not leave. Quickly departing the scene would make the person's actions suspicious and worthy of future investigation.

FIGURE 1-4 The police officer on the scene can interview or detain suspicious individuals of interest.
Courtesy of Russ Chandler.

Of course, depending on the situation, the police officer may not have time to take any such action. Injured victims might need immediate attention. The officer may need to move a crowd away from the incident scene to protect them and to protect the scene because it might be a crime scene. The scene could be blocked with bystanders' vehicles, and the officer might need to clear traffic and parked vehicles to make room for arriving fire apparatus. Even with all these other duties, a police officer can most likely describe suspicious individuals or even call in the information for another patrol officer to investigate.

Directing traffic away from the scene is an important function of the police officer from a suppression aspect, because it ensures that the fire apparatus can lay lines and get into the best position for pumping and spotting ladders. Sometimes even the police officer's own vehicle can get in the way of fire apparatus placement. As they direct traffic and people away from harm's way, officers might notice an overly enthusiastic citizen who appears eager to assist. It is not unusual to see these individuals go so far as to pull lines off the apparatus or drag a hose behind the fire fighters. If the officer moves such an individual away from the scene, he or she should also obtain the person's identification information so that the investigator can interview this individual at a later time. Obtaining identification can easily be done under the guise of wanting to thank the individual for the assistance at a later date. Again, just because someone appears to be helpful does not make him or her guilty of a crime. At the same time, the motivation of the person setting the fire might be to show the fire department that he or she would be a good candidate for a firefighting job.

In addition to keeping people clear of the scene for their own safety, it is imperative that the officer consider the security of the scene to be a priority. When possible, the police officer should start placing barricade tape around the scene (**FIGURE 1-5**). This tape is usually yellow and carries wording such as *Crime Scene, Fire Line*, or *Police Line* as well as *DO NOT CROSS*—a message usually understood by the public. As additional resources arrive, they can assist in ensuring that only public safety personnel approach and enter the scene. As the fire is placed under control and fewer resources are needed, fire fighters can assist with maintaining the security of the scene.

In the absence of a fire investigator, it can be the responsibility of the police officer to protect the scene or to collect obvious evidence to prevent its destruction or disappearance.

The Fire Fighter

Like the police officer and the EMT, the fire fighter enters strategy-planning mode from the minute the alarm is broadcast over the radio. Considerations about the time of day, weather conditions, location of the fire, and whether it is in a commercial district or residential area are going through fire fighters' minds as they speculate about what they may encounter when they arrive on the scene.

FIGURE 1-5 Police or fire personnel can cordon off and protect the fire scene by using barricade tape.
Courtesy of Russ Chandler.

As they approach the scene, fire fighters will be following through with their strategy of where to lay the lines for water supply, where to place any aerial apparatus and pumpers to enable them to carry out their assigned task, and whether they will need additional resources. During this process, they are also observing the scene.

While carrying out their suppression duties, fire fighters can also make observations that are important to the investigation. For example, the presence of obstacles en route to or at the scene could indicate that someone did not want the fire fighters to put a quick stop to the fire. This aspect of the scene should not be confused with usual problems with traffic or cluttered lawns or hallways. However, should a downed tree block a driveway or road, the fire fighter or officer should note whether it is a natural break or whether saw marks are evident on the trunk. Even if the tree fell from high winds, an arsonist might have waited until a large storm was approaching the area before setting the fire. A word of caution: The mere presence of storm-created obstacles does not mean the fire was incendiary. The culmination of *all* the facts determines the origin and cause of a fire, not the individual circumstances.

Other observations made by fire fighters upon arrival relate to the condition of the doors and windows. Were they locked or unlocked, open or closed? Locked doors are not unusual, but it is unusual to find doors nailed shut. This may be a security measure, or it may be a stalling tactic used by an arsonist to hinder fire fighters from getting to the fire. Likewise, windows nailed shut may be a security measure for the occupants. It is critical that fire fighters ascertain door and window conditions by simply trying to open them before forcing them. Basic firefighting training requires that the fire fighter try the door knob before using forcible entry tools. The point is to prevent unnecessary damage and to not mask prior attempts by perpetrators who may have forced the door before the fire fighter's arrival (**FIGURE 1-6**).

The condition of the windows is important. Were they forced prior to the fire fighter's arrival, indicating an illegal entry? Were they left open on an extremely cold day? This could indicate the arsonist's desire to ensure sufficient air to feed the fire. For example, in one residential fire, the casement windows were cranked open, and then the arsonist removed the handles in an effort to keep anyone from closing the windows. This condition raised a red flag because it was not a typical behavior of the occupants of such a structure. A fire fighter noticed this strange situation and reported it to the officer in charge (OIC) of the scene, who in turn advised the arriving fire investigator.

Fire fighter first responders also usually notice the type, quantity, and arrangement of the structure's contents. Fire fighters should find that the living room contents consist of couches, chairs, and entertainment equipment such as stereos or televisions. Bedrooms should have beds, bureaus, and sufficient clothing to show they were occupied. Kitchens should have cooking and eating utensils along with sufficient quantities of foods, spices, and condiments. Nothing is more interesting than a homeowner who states that he was eating breakfast and just barely escaped with his life, only to have a fire fighter observe that the kitchen cabinets and refrigerator were completely empty.

> **TIP**
>
> Nothing is more interesting than a homeowner who states that he was eating breakfast and just barely escaped with his life, only to have a fire fighter observe that the kitchen cabinets and refrigerator were completely empty.

The arrangement of the structure's contents is another concern. In one instance, fire fighters made their initial entry into a real estate office and found piles of magazines, books, and papers in the middle of the floor, making a path of combustibles leading from one room to another. The obvious intent was to create a trailer to spread the fire. When the fire fighters realized what they had discovered, they radioed the OIC, who immediately started securing the area as a crime scene.

Fire fighters also observe the fire itself. With a few years of firefighting experience, a fire fighter can recognize when a fire has unusual burning characteristics and relate this finding to the fire investigator. Situations in which the fire burns, extinguishes, and then reignites may be an indication of a petroleum accelerant. An unusually colored flame can indicate the presence of a chemical not typical of that type of structure. Of course, obvious facts are of interest as well: Was the structure burning from the outside or was the fire contained in the interior?

> **TIP**
>
> An unusually colored flame can indicate the presence of a chemical not typical of that type of structure.

First Responder Responsibilities

The one ultimate responsibility of first responders is to call for assistance if they cannot make an accurate determination of the origin and cause of the fire. If they suspect for any reason that the fire might be incendiary, intentionally set with malice, or involve any criminal activity, they should call for the assigned investigator.

Before the arrival of the investigator, first responders can take some specific actions to protect any evidence and establish a safer environment. They can also start looking at various systems to ascertain their status, such as fire alarm systems, sprinkler systems, and intrusion alarms. Suppression and individual

FIGURE 1-6 Tool mark impressions must be protected. Unnecessary damage from suppression personnel could mask these imprints.

Courtesy of Russ Chandler.

actions as well as the observations of fire personnel must be documented. The sooner these steps can be taken, the more accurate the results will be.

Securing the Scene

As soon as practical, the scene—which might include both a building and the surrounding area—must be secured to prevent unauthorized entry. Securing the scene is always necessary, regardless of whether the fire is suspected to be incendiary in nature or accidental. Indeed, it is just as important to protect the scene of an accidental fire so that evidence can be secured and preserved in case of a civil trial and to help prevent future fires.

The scene must be secured not just from outsiders, but from other first responders as well. Once the scene is secure, entrance should be limited to those who need to carry out assigned tasks such as overhaul and safety. Unnecessary personnel inside the fire scene can cause unnecessary damage to unknown evidence; create additional damage to the structure, such as by leaning against sheetrock and causing it to crumble, which could destroy a fire pattern (**FIGURE 1-7**); or create further contamination of the scene from fire boots.

> **TIP**
>
> The scene must be secured not just from outsiders, but from other first responders as well.

The area outside may contain evidence such as tire tracks, footprints, or discarded items such as cigarettes and empty or not-so-empty containers that may have

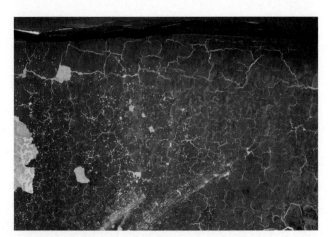

FIGURE 1-7 Sheetrock walls show burn patterns but become fragile as a result of heat. They are easily damaged by fire suppression streams or movement of debris during an initial investigation.
Courtesy of Russ Chandler.

contained an accelerant. Fire fighters can use various methods to protect this evidence. For example, they can place an empty bucket upside down over the evidence to protect it from damage. Unfortunately, this method has a serious drawback: Curious fire fighters may lift the bucket and take a peak underneath. Proper training of the fire crews hopefully will deter this from happening.

The method used to secure the scene needs to be obvious to the general public. Both police and fire apparatus should carry barricade tape to mark the area that is clearly off limits, protecting the scene and any potential evidence. This step also promotes safety by limiting entry into a potentially dangerous area. Natural items around the scene can be used to secure the tape: Trees, signposts, mailboxes, and even bushes may be used to support the tape, keeping it high enough to be seen and sending the message for all to keep out. More information about evidence can be found in Chapter 9, *Evidence*.

Ascertaining the Status of Electrical Service

First responders should note the status of electricity to the structure. Were lights turned on in the structure when they arrived? During the day, the answer to this question can be hard to determine. Does the structure have a working electrical service connection? When conducting size-up, a fire officer may notice that the structure is older and the electric company's meter is missing from the meter base, which now has a blank cover (**FIGURE 1-8**).

Of course, it is always wise to check for secondary sources of electricity. The property could have been upgraded to an underground service, and the new meter may have been relocated to another position on the structure. A thorough examination is important to ascertain the status of electricity for safety and for the eventual investigation of the fire.

An additional note on electricity: Some people steal electricity from their neighbors. Investigators have found extension cords leading to an outside outlet on a neighbor's home and the wire hidden with leaves or dirt. The amount of electricity coming from that electrical branch may be limited, but it can still pose a hazard and represent a source of heat for the ignition sequence. This type of theft typically works on

> **TIP**
>
> Some people steal electricity from their neighbors.

FIGURE 1-8 Just because the electrical company meter is missing, that does not necessarily mean that there is no electricity in the structure. Look for other potential sources.
Courtesy of Russ Chandler.

neighbors who do not usually walk around the outside of their homes, such as the elderly or those who go on frequent or long trips away from home.

First responders should note whether electricity was actually supplied to the structure prior to the fire. No electricity in the structure eliminates most electrical appliances, electrical distribution system, and auxiliary devices from being involved in the ignition sequence. Even so, the first responder conducting the preliminary investigation or the assigned investigator must examine and document all appliances to confirm they were not involved with the fire ignition. For more information on electricity, see Chapter 7, *Electricity and Fire*.

Dealing with Alarm Systems

Both commercial and residential structures can have fire alarms as well as intrusion alarms. First responders can verify from the structure's occupants or owners the status of these alarm systems. Each system will most likely have a local control box within the structure that may provide valuable information. A local alarm system provides an alarm only within or on the outside of the structure. By contrast, a monitored alarm system sends a signal to a remote location staffed by alarm company personnel. Should an activation of a detector or sensor occur in a monitored system, the owner of the system and the local authorities are notified for a response depending on the type

of alarm. Local police or fire may have experienced previous activation of this system that can indicate whether the system was faulty or experienced previous problems potentially related to the emergency at hand. The most obvious fact that should be noted by first responders is whether they heard a siren, horn, or klaxon indicating that the alarm had activated. More information on alarm systems can be found in Chapter 6, *Systems*.

Documenting and Making Sense of Observations

When first responders arrive on the scene and interact with individuals such as occupants, owners, witnesses, or neighbors, they make an effort to remember what they see and hear. Some statements may be crucial to the investigation. The first responder should document any unusual or pertinent comment so that it can more easily be recalled at a later time (**FIGURE 1-9**). Additional information from the witness, suspect, or victim includes the person's full name, date of birth, address, phone numbers, and any other contact information. An investigation is a process of seeking the truth, so documenting statements showing innocence is just as important as noting those indicating guilt.

To help them accurately recall the details of the incident later, first responders should briefly document all activities and actions they took. Such information

TIP

Documenting statements showing innocence is just as important as noting those indicating guilt.

FIGURE 1-9 When the OIC obtains information that may be valuable to the incident, it is always best to take complete notes.
Courtesy of Russ Chandler.

might include when an EMT administered first aid to a victim, a police officer moved around to the rear of the structure to check for activity, or a fire fighter forcibly entered the structure or placed fans for ventilation.

The observations made by different first responders may sometimes differ. For example, the observation of smoke and flame color varies with the stages of the fire. These observations may be helpful to the investigator to provide insight into what was burning, but the first responder needs to document *when* he or she made the specific observation. Although first responders cannot be expected to constantly look at their watches, other methods of documenting time are certainly possible. For example, if the fire company reports black smoke as they pulled onto the scene, their arrival time will be documented by the emergency communications center.

Knowing the time is important to compare the sequence of observations. Consider a barn fire in a rural area. Several first responders report that the smoke was white; another says black; and a third reports that the smoke was brown. After an in-depth interview, the investigators put together a timeline that explains the differences. The local patrol deputy was the first on the scene. As she arrived, she observed brown and gray smoke. Approximately 8 minutes later, fire fighters on the first fire engine arriving on the scene reported black smoke coming from the roof. The rescue squad arrived about 6 minutes after the first engine, and the EMTs reported seeing white smoke.

Each of these observations can be logically explained. The deputy saw ordinary combustibles burning, such as the hay and wood inside the barn, which gave off gray and brown smoke. When the first engine arrived, the fire had progressed from inside the barn to the tar shingle roof, which gave off black smoke when burning. By the time the EMTs arrived, the fire company had started putting water on the fire and, as expected, steam was observed, leading to the EMTs' report of white smoke. The timeline explains the differences in observations. But even better, after knowing the timeline, the observation of the deputy points toward the fire starting in the barn's interior. This simplistic example points to the importance of conducting good in-depth interviews and the value of the observations of first responders.

Documenting Other Suppression Activities

From the beginning of the incident all the way through the overhaul and cleanup, safety issues must be monitored and abated, and these actions should be documented for improved future protocols. This process can include shutting off all utilities; protecting an exposed, downed electrical line; and diverting water from the structure to lessen the weight. Additional resources may be available to determine structural stability, such as the local **building official**. The expertise and experience of the building official are valuable resources for first responders.

TIP

The expertise and experience of the building official are valuable resources for first responders.

Documenting Any Holes in the Floor

During the fire suppression activities, the fire crews may have noticed the existence of holes in the floor that created a serious hazard for the fire fighters and investigators. These holes can be created by flame impingement or over a longer period of time from smoldering embers under the debris. They can also be created by the arsonist as a means to slow down suppression operations.

Such holes can also be created by the fire fighters during suppression activities, such as when they cut holes in the floor to drain off water. Because the weight of water can lead to further unstable conditions, it needs to be removed as soon as practical.

Documenting Fire Suppression Ventilation Activities

Because attack lines and ventilation can change the direction of fire travel, initiation of these activities is important to document and relay to the investigating officer. Fire fighters use ventilation to remove smoke and combustible gases from a structure. Unfortunately, when ventilation is improperly implemented, it can actually change the direction of the fire travel, pulling the fire into areas that would not logically have burned.

TIP

Because attack lines and ventilation can change the direction of fire travel, initiation of these activities is important to document and relay to the investigating officer.

Documenting Delays in Fire Suppression

Investigators are also interested in any problems with the water supply that may have delayed suppression activities and that could explain additional fire damage. When water is flowing, fire fighters need to use caution when directing their fire streams because careless application of fire streams can damage evidence. Of course, preserving evidence should always be a secondary consideration—rescue and suppression are the primary concerns. Nevertheless, fire fighters should recognize that an innocent application of a straight stream of water down an undamaged aisle in a store can knock items off shelves and walls, making it look as though vandals caused the destruction.

The OIC can communicate these issues to the investigator as well as provide details on any water supply problem to assess whether it is important to the investigation. Debris inside a hydrant or a standpipe system may have been the work of juveniles with nothing better to do—but an arsonist could also deliberately plug the water supply to keep the fire fighters from extinguishing an intentionally set fire.

Documenting Hazardous Materials Issues

Hazardous materials can be found in every structure. The discovery of additional hazards from chemical, biological, or other sources must be reported to the investigator for safety reasons. The investigator can then assess whether these hazards are common to the structure and its use. Today, investigators must often address additional concerns about illegal activities that involve hazardous materials, such as drug labs in homes, motel rooms, vehicles, and abandoned structures.

Documenting Potential Devices and Terrorist Indicators

Since the bombing of the Murrah Federal Building in Oklahoma City in 1995 and the terrorist attacks on September 11, 2001, fire investigators have been forced to consider terrorism—whether domestic or foreign—as a motive at fire and explosion scenes. First responders and fire investigators must remain alert for the presence of secondary devices, some of which may be aimed at the first responders to kill or maim them as part of the overall objective.

What does an incendiary or explosive device look like? They can be disguised to look like they belong in any surrounding, as was the case with the explosive device placed in a backpack at the Boston Marathon bombing in 2013. One would expect to see backpacks in that type of environment. Perpetrators may not disguise them very well, however, because they do not believe the devices will be discovered. That carelessness can give investigators an opportunity to identify these devices before or after activation.

Does something look out of the ordinary? Older adults may remember the Molotov cocktail—a large bottle stuffed with a rag. On the Internet and in movies depicting World War II, however, the young may see these devices. One curious youngster saw a Molotov cocktail in a movie and found a case of bottles in his grandfather's garage. He and a friend replicated the devices they saw in the movie. They then went through the neighborhood one evening and threw the devices from a vehicle at neighbors' mailboxes. Reports rolled in about small fires up and down subdivision streets. By the time the engine company arrived, all of the bottles with rags had self-extinguished. The one major flaw in the design was that the youths used antique soda bottles. Owing to the thickness of these bottles, none broke open as required to make the device work properly.

In another incident, an arsonist was making Molotov cocktails openly on his front porch. A neighbor reported his actions after seeing him ladling liquid from a kitchen pot on a small table into bottles, then sticking a rag in the top of the bottle. The arsonist had stopped to take a break and attempted to light a cigarette. He had gasoline residue on his hands and striking the match caused the gasoline vapors to ignite. In trying to extinguish this fire, he shook his hands, which in turn ignited the gasoline in the kitchen pot. The perpetrator attempted to hide the pan inside the structure before emergency services arrived (**FIGURE 1-10**). However, he could not get rid of the burn pattern on the wall, which matched the neighbor's testimony; he also had first-degree burns on his hands and arms. The suspect was caught, booked, and eventually convicted of manufacturing incendiary devices.

Law Enforcement Reporting

Investigative personnel file reports on fires and explosions through the National Incident-Based Reporting System and the Uniform Crime Report (UCR), which are both regulated and maintained by the Federal Bureau of Investigation (FBI). All local, state, and federal law enforcement agencies provide data on incidents and arrests involving certain offenses.

The full-time investigator may also file reports with the Bureau of Alcohol, Tobacco, Firearms, and Explosives (BATF or ATF), which maintains a specialized

FIGURE 1-10 The burn pattern on a wall just above where the suspect had a kitchen pot with gasoline on the table. He was ladling gasoline into old wine bottles to make Molotov cocktails.

Courtesy of Russ Chandler.

reporting system that tracks arsons and explosives data. The Bomb Arson Tracking System (BATS) is actually more than just a reporting process and database collection point. In addition to collecting data from local, state, and federal law enforcement agencies, it offers state-of-the-art case management technology. The database also includes information on explosives and fire cases; information on improvised explosive devices (IED), their components, and incendiary devices; suspect information; and even information on juvenile fire setters.

Every locality must ensure that its fire investigator authority—which could reside within the fire department or the police department—reports fire investigative findings to both the FBI and ATF for these incidents. If the fire department is the investigative

authority, it should also partner with the local police or sheriff's office, with both entities submitting their reports together.

Completing the Fire Incident Report

The OIC has an equal interest in many of the issues that concern the fire investigator. Each department may have different reporting criteria and procedures, but ideally all fire departments will participate in the NFIRS. The data collected from these reports have immense value for many aspects of the fire service. At the conclusion of the event, the OIC or the OIC's designee must fill out the fire report. This report documents items such as the first material ignited, the heat source that provided the energy to start the fire, and what brought the heat source together with the first material ignited.

> **TIP**
>
> Ideally, all fire departments will participate in the NFIRS.

By using the NFIRS, first responders can investigate and report their findings on the fire's area of origin and cause. These data can then be used to educate the public on preventing future fires, to combat arson through education, and to justify fiscal allocations for investigative resources. Such statistics can also be used to justify changes to the fire code or taking a dangerous product off the market through a recall. Another benefit of contributing to the NFIRS is that this step identifies suppression fiscal needs, training, and resource allocations, which all can justify future requests for apparatus, equipment, and increased staffing.

Wrap-Up

SUMMARY

- First responders are public safety personnel who are dispatched to an emergency scene by an emergency communications center on behalf of the government or under contract to the government. First responders include emergency medical personnel, patrol police officers, and fire fighters. In many instances, the fire officer in charge, the OIC's designee, or the patrol police officer on the scene may be the first investigator to work the scene and determine a fire's origin and cause.

- All fire departments put out fires, perform fire prevention activities, and serve the public in a wide range of capacities. Today more than ever, the fire department is providing EMS coverage. The role of the fire department in investigations depends on the department and the AHJ. Some fire services will simply determine the area of origin and cause, entering that information into the NFIRS. Others will investigate the scene in more detail, turning over any evidence to local law enforcement. Others ensure that their fire investigators have full police powers and can conduct a full investigation of any fire incident.

- When the fire scene is more complex or the fire potentially appears to be incendiary in nature, the OIC contacts a full-time fire investigator to conduct a detailed investigation. In many situations, the OIC, or the OIC's designee, is also responsible for ensuring that the fire report is accurately and completely filled out.

- Incendiary devices are a fact of life for today's fire fighters. The fire service must be aware of the potential presence of both primary and secondary devices. This is important not just for addressing the active device but also for identifying the remnants of an expended device. Such a device may have any sort of appearance, so the first responder must take note of anything that looks out of the ordinary and report that information to the full-time investigator.

- First responders are critical to the overall success of fire investigations. Their observations upon arrival and in handling the emergency provide a wealth of knowledge that can be put to use in the final analysis of the scene. They may be the first to recognize evidence and to take steps to protect that evidence. They hear comments and see the actions of individuals that may give insight into what happened on the scene prior to the arrival of first responders. They might even observe unusual behavior of individuals at the scene that could point to the perpetrator, should the incident turn into a crime scene.

KEY TERMS

Barricade tape Wide, brightly colored tape with a clear message that will prevent entry onto the fire scene. The wording can be anything from *Fire Line—Do Not Cross* to *Crime Scene—Keep Out*.

Building official The person responsible for the enforcement of the local and state building codes in that jurisdiction. The building official's office is responsible for issuing construction permits and certificates of occupancy as a means of ensuring compliance with the code.

Casement windows Windows with hinges on the side that usually open with a rotating crank assembly.

Emergency communications center (dispatch) Sometimes referred to as the E 911 center, the place where calls from the public for emergency assistance are received and whose staff (dispatchers) alert and send the appropriate emergency unit(s) to handle the situation.

Emergency communications dispatcher A person who works in an emergency communications center, whose duties are to receive calls from the public for emergency assistance, and then to send the appropriate emergency unit(s) to handle the situation. Commonly referred to as the dispatcher.

Emergency medical technicians (EMTs) Individuals with specialized medical skills, trained to offer basic first aid in the field under emergency circumstances, and then to assist in transporting the patient to a medical facility if necessary.

First responder Any public safety responder who may be or may have the potential of being first on the scene as the result of being sent there by the emergency communications center.

Incendiary fire A fire set willfully and intentionally, with malice.

National Fire Incident Reporting System (NFIRS) A national computer-based reporting system that fire departments use to report information on fires and other incidents to which they respond. This uniform system collects data for both local and national use.

Self-contained breathing apparatus (SCBA) designed equipment that fire fighters wear when entering an unsafe atmosphere. The bottle in the apparatus is filled with compressed air.

Sheetrock Often called drywall, gypsum, or wallboard; a crumbly material called gypsum sandwiched between two layers of thick paper. Gypsum is fire resistive.

REVIEW QUESTIONS

1. List the roles and responsibilities of EMT first responders, and explain how these first responders can contribute to the fire investigation.

2. List the roles and responsibilities of patrol police officer first responders, and explain how they may contribute to the fire investigation.

3. What is the importance of finding all windows open in a burning structure in freezing weather?

4. Why should the first responder observe the contents of the structure? Provide an example of a situation that would be considered unusual or suspicious.

5. When the fire is under control and essentially put out, why is it important to keep fire fighters from entering the structure, unless necessary, to accomplish an assigned task or mission? Provide examples.

6. Why is it important for first responders to write down, as soon as practical, any comments they heard made by witnesses, victims, neighbors, and others on the scene?

7. Why does the fire investigator want to know about the use of attack lines and ventilation?

8. Why might a fire investigator consider the dispatcher to be a first responder?

9. Why is it important that the fire investigator be advised of any delay in applying water to the fire when first responders arrived on the scene?

10. Does the fact that windows are nailed shut prove that the fire was incendiary in nature? Explain your answer.

DISCUSSION QUESTIONS

1. A building on fire has an audible fire alarm system, but on arrival the first responders hear no audible alarm. What is the significance of this finding?

2. If you have prior fire or police experience, what have you observed that you now recognize may be of interest to a fire investigator?

3. A fire is clearly incendiary in nature, and the homeowner has been mean and behaved badly toward the first responders. Even though the first responder may suspect that the homeowner may have been involved in setting the fire, the first responder hears the homeowner utter a comment that may indicate his innocence. Should the first responder report this statement to the investigator? Why or why not?

CHAPTER 2

Safety

LEARNING OBJECTIVES

Upon completion of this chapter, you should be able to:

- Describe the various hazards found at the fire or explosion scene.
- Discuss the importance of federal, state, and local regulations to a fire scene investigation.
- Describe the protective equipment available to the assigned fire investigator and regulations requiring the use of such equipment.
- Describe the protective measures that must be taken at the fire/explosion scene using accepted safety practices.
- Describe the necessity of safety training for each person involved in the investigation of a fire or explosion.

Case Study

After spending several hours at the fire scene of a residential fire, the company officer of Engine 6 determined that the fire's cause was accidental. However, he wanted an assigned fire investigator to look at the wiring in the kitchen. Following safety protocols, the investigator put on full turnout gear and grabbed a flashlight before entering the residence. Seeing the engine company officer, he walked toward him, forgetting that he had left his portable radio in the charger in his vehicle—his first error. The investigator had to chuckle when he looked at the scene: The roof was gone, and the debris in the rooms had been almost completely removed. This state made a complete investigation more difficult, but the investigator was confident that the engine company officer had done his job well and had narrowed down the area of origin prior to doing the full overhaul. After all, the request was to just look at some wiring.

The engine company officer passed on his findings and the reason for wanting a second set of eyes on the electrical system. He had no doubt that the fire started from the electrical wiring, but he was concerned about whether the structural wiring was up to code. He had no sooner relayed his concerns than he was asked by dispatch if he was available for another call. The assigned investigator decided that he could handle the scene and recommended that the officer respond with his crew to the new call.

The floor showed a lot of charring, so the investigator took a tentative step inside. It felt stable; he knew that fire fighters had walked on this same floor only moments earlier. Had they thought the floor was unsafe, he believed that they would have advised him of this suspicion. Assuming that the floor was safe was his second error. Like so many others in his field, he felt sure there would be telltale signs if the floor was about to give way—his third error in judgment.

Only four steps into the structure, the investigator stopped to take a photo of the wiring in the hall. With no warning whatsoever, no warning crack, no telltale softness in the flooring, the floor gave way, and down the investigator. Along with poor judgment came a fateful placement of the feet—one foot on one side of a floor joist and one foot on the other side of the floor joist. The floor joist had no damage, and it gave no indication of instability as the investigator came in contact with it. Although there was only a 3-ft crawl space under the floor, such a drop can cause quite a jarring and even some minor injuries such as nail punctures, sprains, or even broken bones. Fortunately, the investigator suffered only minor physical injuries. Even so, the potential of serious injury is always present when any floor suffers fire damage.

Investigators must always follow the guidelines and policies pertaining to post-fire scene examination. In the situation just described, it was appropriate for the engine company to leave and take the new call. Unfortunately, investigative staffing rarely allows for two investigators to work every scene. When possible, though, it is always beneficial for the first responder investigator to remain on the scene for the most important reason: safety.

Access Navigate for more resources.

Introduction

For years, the fire service has stressed that the most important thing to remember at every fire or other scene is fire fighter safety. Not thinking of your own safety may very well put others in danger unnecessarily. Taking care of yourself also allows you to be around to take care of your family. The first step toward safety is being aware of all federal, state, and local regulations as they pertain to fire or explosion scenes. In addition, the assigned fire investigator must be aware of and follow the safety protocols established by the fire department. Those policies and procedures should cover safety during the post-fire scene examination. Your agency or department should be diligent in creating policies that address the safety of all personnel, including the first responder investigator and the assigned full-time investigators.

Even before arrival at the scene, members of fire suppression teams will be thinking of many things while en route to an incident. Many of these thoughts will reflect the dispatch information given at the time of the assignment. The very first concern, though, must be for the safety of the team. If the alarm was reported as a structure fire, the crew will assure they are ready by donning their appropriate turnout gear. If it was reported as hazardous material incident, they will think about the potential exposures and ways to safely work in such an environment.

The full-time investigator will be doing the same thing. Based on the dispatch information, the investigator will be considering which safety steps need to be taken, which gear or safety clothing and breathing apparatus will be needed, and in the case of an explosive scene, which types of exposure issues might be present.

The full-time assigned investigator will most likely have the same skill set and equipment as the fire department first responders. These resources will include protective clothing, monitoring devices, and other safety equipment. The first responder who conducts preliminary investigations also needs a complete understanding of the safety aspects of a full investigation. This chapter covers the various safety topics considered by the entire team.

Every fire investigator must have full knowledge of federal and state regulations dealing with labor laws and the proper use of safety equipment. This information includes the fit testing policy for **self-contained breathing apparatus (SCBA)** masks, which most—if not in all—fire departments have adopted. This process needs to be expanded to include testing and use of filter masks not just for the assigned investigator, but also the first responder investigator.

TIP

The procedure for fit testing filter masks needs to be added to the department safety policy.

TIP

SCBA or filter masks must be worn to protect the first responder or investigator from carcinogens and toxins off-gassing from recently burned material.

Both suppression and investigative personnel must be proficient with the use and operation of atmosphere multi-gas meters, which provide immediate readings of oxygen in the atmosphere and the presence of unwanted gases. If turnout gear is not used, then wearing, maintaining, and caring for alternate protective clothing is important for the investigator's safety, health, and well-being.

Proper building construction knowledge is essential in determining structural stability. More information on building construction is available in Chapter 5, *Building Construction for the Fire Investigator*. Investigators must be cognizant of special hazards such as slippery conditions, sharp objects, and electrical and water hazards. Even unusual hazards such as territorial pets and the occasional snake can create a challenge for the investigator.

Federal, State, and Local Regulations

All investigators must be aware of the many fire, construction, and industrial safety rules and regulations and follow them diligently. These range from the department's rules on safety to the requirements outlined in state and federal regulations. In 1970, the U.S. Congress enacted the **Occupational Safety and Health Act**, also known as the OSH Act or simply "OSHA." This legislation empowered the U.S. Department of Labor to create a new agency, the **Occupational Safety and Health Administration (OSHA)**, to administer and enforce the provisions of the act. One stipulation of OSHA allows states (and territories) to come up with their own job safety and health plans; thus, the state, rather than federal authorities, enforces these regulations. The state plan must incorporate the requirements under OSHA, but may also include more stringent standards or delve into areas not covered. To date, approximately half of the states have successfully presented their own plans. You must know which agency acts as the enforcing entity within your state.[1]

TIP

All first responders and investigators must be properly fit tested with the respirator they will use in the conduct of their duties.

Part of the federal requirements deal with the minimum training necessary before an individual is allowed to work in a hazardous atmosphere. Without such training, an investigator may not legally enter a fire scene to conduct a fire investigation. Contact your state or federal enforcer of OSHA regulations for more information about the minimum safety training requirements and ways to obtain this training. For the most part, training already received by the investigator from the fire department or state fire training agency will meet the OSHA requirements.

Your own department may have locally mandated safety standards that are enforced by your department's safety officer or by a quality assurance officer (or other title). Other agencies within your local jurisdiction, such as the **Building Officials Office**, may establish additional safety requirements. Building officials include the individuals who approve building permits and review plans—valuable skills for determining whether a structure is safe. The Building Officials Office also includes electrical and plumbing inspectors: Plumbing inspectors can assist with water issues, and electrical inspectors can be an added resource for safety. Additional resources may be found in the Department of Personnel, which may have a division titled Loss Prevention. *Loss prevention* is exactly what it implies: Personnel in this office seek means to prevent accidents and, as such, could be beneficial in a fire investigation.

Other state and federal agencies establish rules and regulations specific to a certain industry or situation. For example, the Bureau of Mines has its own enforcement regulations separate from the OSH Act's requirements. A state may have its own enforcement authority that works much like OSHA. As a fire investigator, you must do your research on all safety requirements well in advance so that you can be in compliance prior to pulling onto the scene of the emergency incident.

Protecting All Investigators

First responder investigators from the fire service must be well versed in most safety issues that might potentially be encountered at the fire scene. In addition, they will already be wearing the best protection possible as the result of conducting fire-ground operations. When suppression and overhaul operations are complete and they begin the search for evidence of the fire's origin and cause, first responder investigators should continue to use any breathing apparatus worn during the operational stage until the atmosphere is safe. Only then can the SCBA be removed. If the assigned fire investigator is dispatched, he or she will also need to don breathing apparatus to do the job.

At any incident, fire services personnel must consider many safety issues. Fire fighter training encompasses electric hazards, problems with standing water, and myriad other issues. This safety training must expand as the fire fighter transitions into the role of first responder fire investigator; the same is true for the assigned investigator. As investigators take a more in-depth look into the structure, they may encounter more hazards.

Like terrorism-related incidents, illegal activities may create hazardous environments. Just walking into a motel room could be a hazard if it was used as a temporary methamphetamine lab. Vehicles have been used for these labs, only to then catch fire; fire personnel responding to such a fire scene may not be fully aware of the hazards present. In some cases, the lab's operator might set up explosive devices that are automatically triggered when the lab is discovered; the goal is to destroy evidence of the criminal activity. Thus, when the fire team members recognize the presence of a lab, their first action should be to retreat

and call for law enforcement, the bomb squad, and the hazardous materials team. Investigators face the same hazard when conducting an investigation at these locations. All are encouraged to seek additional training on safety issues to better prepare them to recognize these hazards and call for assistance.

Terrorist IEDs and methamphetamine chemicals, all of which can be explosive, may also be found together. When first responders see these items, they need to secure the scene and turn it over directly to specialists to clear the explosives and/or drug manufacturing materials. Only then can the fire investigation proceed.

Protective Clothing

Many regulations require that investigators wear protective clothing appropriate for the environment in which they will be working. Depending on the hazard, protective clothing may range from attire as simple as a Tyvek coverall and filter mask (**FIGURE 2-1**) to fully encapsulated suits with SCBA. At an active fire scene, they must don the same full protective clothing used by suppression personnel, as specified in the most recent issue of National Fire Protection Association (NFPA) 1971, Standard on Protective Ensemble for Structural Fire Fighting.[2] This clothing, which protects and covers the investigator from head to toe, consists of boots, pants, coat, hood, helmet, and gloves (**FIGURE 2-2**). If the department has more stringent requirements, the fire investigator should follow them instead.

To be considered completely extinguished, a scene should have no active fire or hot spots. In this situation, the investigator may switch from suppression coat and pants to wearing disposable Tyvek coveralls, especially during the warmer months. If a first responder is assisting the investigator, most investigators carry extra sets of coveralls to accommodate

FIGURE 2-1 Investigators may be required to wear protective clothing, ranging from basic Tyvek coveralls to fully encapsulated suits with self-contained breathing apparatus.
Courtesy of Russ Chandler.

FIGURE 2-2 Fire investigator suited out in full turnout gear with self-contained breathing apparatus.
Courtesy of Russ Chandler.

those assisting. The coveralls are intended to keep out the cancerous carbon along with any other dry contaminates potentially found at the extinguished fire scene. A key word in the description of the coveralls is *disposable*: Each pair of coveralls should be discarded after use, with the investigator always putting on a new set at the next event.

> **TIP**
>
> Is carbon a carcinogen? Carbon from smoke or the carbon residue found in a structure after a fire has most likely absorbed polyaromatic hydrocarbons. Thus, this carbon is a carcinogen when inhaled. The inhaled particles will most likely be deposited in the lungs and in this form can be cancerous. In contrast, activated carbon or activated charcoal is not a carcinogen.

One major reason for donning respirators and coveralls is to protect the investigators from fire-scene carbon—a known carcinogenic. Fire-scene carbon dust on clothing can be a further hazard when those clothes are worn home, then placed in a washer and dryer. Residue carcinogens and other contaminates can then remain in the washer or dryer, attaching themselves to other family members' clothing and putting them at risk.

To date, several studies have focused on cancer risks in fire fighter, and in the future even more research may be carried out on this crucial safety issue. The Centers for Disease Control and Prevention (CDC), through its subsidiary, the National Institute for Occupational Safety and Health (NIOSH),[3] conducted a multiyear study of nearly 30,000 fire fighters to better understand the link between fire fighters and cancer. The findings showed that fire fighters had a higher rate of cancer diagnoses and cancer-related deaths. Most of these cancers occurred in the digestive, oral, respiratory, and urinary systems. Compared to the general population, nearly twice as many fire fighters had malignant mesothelioma, a rare type of cancer caused by exposure to asbestos. They also had higher incidence of bladder and prostate cancers.

Turnout gear and other clothing exposed to carbon and other contaminants should be properly laundered in machines designed to handle such materials. Most departments have at least one washer and dryer installed at each fire station. If no machine is available, the gear or clothing should be cleaned in an area such as the station apron, where it can be scrubbed with brushes and rinsed off. Care should be taken to assure runoff does not spread the contamination.

Respiratory Protection

While working in a hazardous atmosphere, investigators (like all fire suppression personnel) should wear suppression-quality SCBA. Before using any SCBA, investigators must undergo a physical examination to ensure that they are physically fit enough to don and work in such equipment. Before wearing any face piece, they must be fit tested to ensure a proper seal can be maintained while wearing the mask. Finally, they should receive documented training on the proper use and care of such equipment before being allowed to wear the SCBA in a hazardous atmosphere.

When the fire is out and the structure has been properly ventilated, fire fighters should test the atmosphere throughout the structure to ensure an adequate oxygen supply is available. The atmosphere should also be tested for **carbon monoxide (CO)**, hydrogen cyanide (HCN), and other potential toxic gases, as well as for the presence of any explosive gases by deploying a **multi-gas detector** (**FIGURE 2-3**). Even though the atmosphere may indicate proper oxygen levels exist and CO and HCN are not present, the scene will not necessarily remain safe. Constant monitoring may be necessary.

> **TIP**
>
> When the fire is out and the structure has been properly ventilated, fire fighters should test the atmosphere throughout the structure to ensure an adequate oxygen supply is available.

FIGURE 2-3 Fire investigator using a multi-gas detector. The most important test is for a sufficient supply of oxygen in the atmosphere and the absence of dangerous gases such as carbon monoxide.
Courtesy of Russ Chandler.

FIGURE 2-4 An investigator may wear either a full-face or half-face canister mask. If wearing a half-face mask, the investigator must wear properly fitting non-vented goggles.
Courtesy of Russ Chandler.

Airborne particulates stirred up during the act of conducting the fire scene examination include bits of dust that can be toxic and carcinogenic. First responder investigators will need to remain on SCBA air to rule out the possibility of inhaling carbon dust or any other particulates floating in the air. In contrast, assigned investigators may have other forms of protection that can be used in these situations. For example, they may pass up the SCBA in favor of a lightweight, easy-to-use respirator. It is essential that continued checks on oxygen levels and the presence of toxic gases continue if anyone is working at the scene.

Respirators must be of the **canister** type, as paper mask filters do not effectively block toxins or create an adequate seal. For this reason, paper mask filters are not acceptable at a fire scene. The canister masks are sized so that the wearer can attain a proper fit, which ensures that all breathed-in air passes through the filter canister. No mask should be worn until the wearer has been fit tested for an accurate fit. Because carbon accounts for the vast majority of airborne particulates, investigators should check with the respirator manufacturer to ascertain which filter is the safest to use in that atmosphere. Other contaminants, such as asbestos, may also be present. When deciding which canister to wear, investigators should choose the filter that provides the most protection.

The choice of half- or full-face mask is up to the investigator (**FIGURE 2-4**). If wearing a half-face mask, the investigator must also wear properly fitted **non-vented goggles** to prevent the absorption of contaminants through the sclera of the eye. Both types of equipment must be worn at all times while within the hazardous atmosphere. When using any eye protection, fire fighters may experience fogging up of the lenses of the goggles or the face piece of full masks.

Fogging reduces visibility, creating an additional safety issue by limiting the ability to see clearly. Numerous chemicals are marketed as coatings for the lenses that can prevent fogging. Failure to take steps necessary to stop fogging can further endanger the wearer from unseen hazards.

Hand Protection

In the search for evidence, the investigator may handle fire debris to a greater extent than do the fire suppression personnel on the scene. Indeed, the investigation will likely include much sifting and collecting of evidence. When handling this debris, the investigator's gloves may absorb some chemicals, thereby transferring contaminants, such as flammable liquids, to the inner surface of the glove and allowing them to be absorbed by the skin. To prevent this possibility, investigators must wear latex gloves under their work gloves (**FIGURE 2-5**). Some investigators even take the extra precaution of wearing two or more layers of latex gloves under their leather gloves. After the fire is under control, suppression forces have completed all overhaul, and the fire is completely out including embers, the fire scene is considered a **cold scene** and the investigator can wear lighter leather gloves that provide for more dexterity.

> **TIP**
>
> Wear latex gloves under work gloves.

Foot Protection

Nails, glass, and metal shards create puncture hazards. While first responder investigators will conduct their investigation in suppression boots, the assigned

FIGURE 2-5 To protect against the risk of absorbing chemicals through the skin, investigators can wear latex or plastic gloves under their leather work gloves.
Courtesy of Russ Chandler.

FIGURE 2-6 Switching from a heavier fire suppression helmet to a lighter hard hat will not only be more comfortable but also reduce the investigator's fatigue.
Courtesy of Russ Chandler.

investigator may wear alternative footwear. Assigned investigators should invest in proper footwear. **Steel-toe boots** protect the wearer from falling items that could crush the toes. **Steel-sole boots** protect the wearer from puncture wounds with a thin steel plate, usually stainless steel, embedded in the sole of the boot. This structure should not to be confused with a steel shank, which just provides better support. Manufacturers are now producing woven cloth protection for fire fighters' boots that is similar to the materials used in ballistic vests. This new material provides more puncture protection compared to the standard boot, but does not yet provide as much protection as the steel sole. Of course, the first line of defense is to watch where you step. Even steel-sole boots are capable of being punctured under certain conditions.

Properly fitting boots are important as well. Uncomfortable boots can be a considerable distraction while examining and digging at a fire scene. Well-fitting boots allow the investigator to concentrate on the job at hand. Take the time to be fitted by the boot retailer to assure a fit that will not create irritation to the feet and allow the wearer to endure long hours at the fire scene.

Head Protection

All fire personnel should always wear head protection. When at a fire scene for any appreciable amount of time, assigned investigators may consider switching from a suppression-type helmet to an industrial hard hat (**FIGURE 2-6**). The lighter hard hat can both protect the head and reduce fatigue and stress on the neck. Nevertheless, investigators should wear this type

of lighter headgear only when the hazards on the scene do not dictate the wearing of heavier protection such as a suppression helmet.

Site Safety Assessment

Fire investigators must perform a site safety survey assessment at every fire or explosion scene. Recognition of structural stability is a skill that every investigator must acquire, and it is recommended that all investigators become proficient in the subject of building construction. This knowledge is just as crucial in scene assessment as it is when making a proper determination of the fire's area of origin and cause. Should the investigator not possess this knowledge, he or she should seek out classes in building construction available from state agency training institutions as well as through local colleges and universities. Other ways to become proficient in this knowledge include having a background or experience in building construction or engineering.

Construction Features and Hazards

When they know about construction features, investigators can recognize hazards that could make the structure unsound. Suppression water and weather can affect both a fire scene and the stability of the structure. The investigator must take into account the added weight of water when doing the initial and ongoing scene assessments. For example, a flat roof may readily handle standing water under normal circumstances. But if the structure becomes weakened by a fire or explosion, the additional weight of water could push the stability of the remaining structural elements beyond their capability, causing failure. Likewise, snow and ice on any roof can add enough weight to cause structural failure.

TIP

When they know about construction features, investigators can recognize hazards that could make the structure unsound.

When a roof is not present, rain or snow can collect on the floor and be absorbed by the contents and debris, creating more weight and causing collapse. In winter conditions with below-freezing temperatures, suppression water that would normally flow back out of the structure may potentially freeze inside, adding more weight than would be experienced in the spring or summer.

The presence of any chimney, whether free standing or still attached to structural elements, is a potential hazard. Even those with mass that gives the appearance of stability can become unstable enough to topple, putting anyone in proximity in danger (**FIGURE 2-7**).

FIGURE 2-7 A free-standing chimney may give the appearance of stability, but the mortar may become damaged to the point that a gust of wind can cause it to topple.
Courtesy of Russ Chandler.

Unsafe Chimney Leads to a Loss of Life

A scene assumed to be in a safe condition may not always live up to that assumption. In the state of New York, a fire department investigator was looking at a fire scene that was 5 days old. The fire involved an older residence that had been converted into three apartments. The fire was attacked from the exterior with deck guns and brought under control in only an hour. However, the investigation was delayed because of subzero temperatures.

On the day of the investigation, the assigned fire investigator was accompanied by a private investigator, an insurance adjuster, an electrical consultant, and the homeowner. The insurance adjuster was downstairs assessing damages while the other four stayed in the attic to find the origin and cause of the fire. The roof had suffered considerable damage, leaving the chimney as a free-standing structure rising about 13 ft above the attic floor. This chimney had direct fire exposure, had been struck by water streams from the deck guns operating on the day of the fire, and had been exposed to 54-mi/h wind gusts. These assaults, along with freezing and thawing temperatures, had eroded the structural integrity of the chimney.

About two and a half hours into the investigation, the group had worked their way closer to the chimney. Those working the investigation had commented to each other that they had seen the chimney swaying slightly in the wind gusts. Although the private investigator had

actually pushed and pulled on the stabilizing bar connected to the chimney, the chimney did not move—that point, along with the fact that it was still standing after 5 days, gave the group a false sense of safety.

The department investigator was 8 ft from the base of the chimney, directly in line with the chimney. The other members of the group were off to one side or the other. No one knew there was a problem until they heard the chimney hit the attic floor, crushing a fire department investigator. Those still on the scene tried in vain to move the chimney off of the victim, but to no avail. One of them had to run next door to call 911. At the same time, another investigator arrived on the scene and two fire chiefs arrived shortly thereafter. Through the combined efforts of these five individuals, they were able to move the chimney off the victim. They initiated cardiopulmonary resuscitation (CPR), and he was transported to the hospital. The investigator died later at the hospital from his injuries.

The section of the chimney that landed on the victim measured 98 in. high, 38 in. wide, and 17 in. deep, and weighed between 2500 and 3000 lb. At the time of the incident, the winds were between 33 and 36 mi/h. The construction of the chimney was bricks and mortar only, with no flue liner. Two types of brick were observed near the roofline, indicating that the chimney had been extended at some point in the past. All of these issues contributed to the failure of the chimney.[4]

All chimneys should be evaluated for their potential as a hazard. If necessary, fire fighters should have the chimney toppled to reduce the likelihood of it presenting a danger to the investigator or those visiting the scene well after the emergency is over.

NIOSH is a federal agency responsible for conducting research and coming up with prevention recommendations to prevent future work-related accidents. Its May 14, 1999, report, number F99-06, made the following recommendation:

> Fire departments should conduct an assessment of the stability and safety of the structure, e.g., roofs, ceilings, partitions, load-bearing walls, floors, and chimneys before entering damaged, e.g., by fire or water, structures for the purpose of investigation.

Not all fire investigators and first responders will have the necessary skills to make a fair assessment of the scene, so calling in qualified experts may be necessary. These individuals could include staff from the Building Officials Office or other appropriate individuals identified by the investigator. Structural engineers may also be willing to make such assessments for a fee.

To make a scene safe, heavy equipment may sometimes need to be brought in to remove walls and overhangs. When this is done, make sure to photograph the area adequately prior to disturbing the scene; the photos can be used in the future when making your determination of origin and cause. Photographs are also beneficial for others who may be working at the scene after the structure has been altered.

The use of heavy equipment is a last resort to make the structure safe because, as a result of its actions, some evidence may be inadvertently contaminated. The investigator must take steps to avoid such contamination:

- Choose the location for equipment placement based on the minimum exposure to the debris and the scene.
- Prior to the arrival of the heavy equipment, thoroughly examine the placement area for potential evidence.
- Clean any part of the equipment that will be entering the scene, either in direct contact or hovering over the scene, and test it using a hydrocarbon detector to ensure that contaminants are not being brought into the scene. Lubricating grease and oils cannot be completely removed from the parts of the heavy equipment operating within the scene, but samples of these oils and grease can be taken for elimination purposes should that be necessary.

FIGURE 2-8 Damaged structural components may be unstable, capable of failure at any time.
Courtesy of Russ Chandler.

The stability of the structure is an ongoing concern and should be monitored constantly (**FIGURE 2-8**). If resources are available, maintain the safety officer position used during the suppression phase of the event.

Entering the Structure

Before an investigator enters the structure, the fire department must have established its legal right to be on the scene. See Chapter 3, *Legal Issues and the Right to Be There*, for more information. Upon arrival at the scene, investigators must follow their department's directive on the **Incident Command System (ICS)**. Most departments have a standardized procedure for command, control, and coordination that covers all types of emergency situations. It provides a hierarchy for not just one agency, but rather the coordinated efforts of multiple agencies. Investigators should report to the incident commander (IC) or the IC aide and request permission to enter the building, advising where they need to go and who will be accompanying them on their initial survey of the scene. In some jurisdictions, fire investigators have concurrent jurisdiction with the IC. Regardless of statutory authority, safety issues dictate that investigators follow the existing ICS. Likewise, being accompanied by a fellow investigator or a fire fighter is important to ensure assistance is available should a dangerous situation arise.

> **NOTE**
>
> Before entering the scene, investigators must follow their department's directive on the ICS.

All ICSs should include accountability tags of some type. Two examples are shown in **FIGURE 2-9**. Leaving your first accountability tag with the IC serves to

FIGURE 2-9 Many versions of accountability tags exist. A common type is the plastic tag with the name and unit number attached with Velcro so that the tag can be attached to the ICS accountability board.

Courtesy of Russ Chandler.

remind command that the investigator is still on the scene. If the scene is large enough or tall enough to require separate command sectors, additional accountability tags should be left with sector commanders. The investigator must be sure to pick up those tags when leaving that sector and when leaving the scene.

When entering a burning building, investigators should always keep track of their location and establish at least two routes of escape at any given time. They should always carry a strong, reliable hand light, whether it is day or night. This light may be necessary to observe items such as burn patterns or evidence, but it is just as valuable a tool to observe tripping hazards and sharp objects that could be a risk. The light can also be used to discover other hazards such as exposed—and energized—electrical wires. Other items to carry would be a radio and an assortment of hand tools, including cutters, pliers, and a screwdriver. Many investigators use a multi-tool in lieu of carrying multiple hand tools.

Electricity and Gas Lines

Before entering a building, the investigators should ascertain the status of all utilities. They should look around the structure to find the location where the utilities enter the building. The gas lines will have a cutoff, which should be checked and then shut off. Many firefighting tools have appropriate notches or openings that fit the national standard gas valves. Even a simple pair of pliers can be used to shut off the gas valve.

When the structure has only minor damage and there is no damage to the electrical panel or the site

where the electrical line enters the structure, fire fighters can shut off the electricity at the electrical panel. If structural damage has even remotely compromised the panel or feed from the power company, then the source of electricity to the structure should be completely disconnected—not just at the breaker or fuse panel, but at the street, whether it is overhead service or underground. In such a case, the power company should disconnect the electrical service because its personnel have the tools and necessary equipment to perform this operation safely. In the past, it may have been common practice to pull meters; today, however, most departments avoid such actions because doing so creates a potential hazard. Investigators should also examine overhead power lines, especially when they are near the damaged structure. Have the lines been damaged, or are they hanging closer to the structure now, creating a potential hazard?

Even if electricity has been disconnected, all lines should be considered "hot" until each circuit can be verified to be not energized. Electrical meters or disconnects should be tagged and locked out using a lockout device like that shown in **FIGURE 2-10** to ensure they stay de-energized. Each individual working in the structure will put his or her own personal padlock on the device. When leaving the scene, they should take their padlock with them. This way no other person can energize that line by accident.

When using ladders in or around overhead power lines, use extreme caution so as not to come into contact with the wires. Even if the wires appear to be

FIGURE 2-10 A lockout device can accommodate many different locks for each individual working on or near that circuit. The disconnect device will have a tag advising that the circuit is locked out and de-energized.

Courtesy of Russ Chandler.

insulated, that does not negate the potential for energy leakage, which could cause electrocution should a ladder come in contact or get near these wires.

When doing a safety assessment of electrical hazards, investigators must consider alternative sources of electricity other than that provided by the power company. Check whether extension cords or heavier wires are leading from a neighboring structure into the fire scene. Borrowing or stealing electricity through such means is not common, but certainly possible.

> **TIP**
>
> Check whether extension cords or heavier wires are leading from a neighboring structure into the fire scene.

As a final measure, even if investigators are sure that the electricity has been disconnected, they should be familiar with the operation of a simple electric **multi-meter** (**FIGURE 2-11**). This device can verify the presence of electricity or lack thereof when properly used. Alternatively, the simple devices known as **voltage detectors** give an audible or visual signal when the device comes in close proximity to an energized electrical cord or wire. Although great devices and good tools, voltage detectors should not serve as a replacement or be used solely instead of a voltmeter. As with all tools, investigators must properly maintain and test their equipment on a regular basis.

Slippery Surfaces and Sharp Objects

At any fire or explosive incident, slippery surfaces can be present—not just because of ice, but owing to the presence of water, chemicals, or even soot on certain surfaces. In attempting to make these areas safe, the investigator must not introduce additional materials to the scene that could cause contamination or the appearance of contamination. Instead of adding salt or sand, consider marking the area so that others are aware of the hazard and finding an alternative path to avoid the slippery area whenever possible.

Sharp objects are of particular concern. Debris may contain a multitude of objects such as metal shards, glass, nails, and staples that can easily penetrate most clothing if the wearer does not take precautions. Wearing protective equipment such as gloves and boots is just one measure to prevent injuries from sharp objects, but is not enough. Thus, investigators must also be cautious about where they step and where they reach with their hands.

Rehabilitation and Switching to Respirators

Investigators must follow all local policies regarding the amount of time they spend breathing compressed air or using respirators and the necessary time to spend in rehabilitation. Rest and rehydration can be important at an emergency scene even for the fire investigator, not just for physical well-being but to refresh the mind and facilitate clear thinking. Monitoring vital signs and maintaining proper body temperature are also part of the rehabilitation process. NFPA 1584, *Standard on the Rehabilitation Process for Members During Emergency Operations and Training Exercises,* describes the recommended procedures for incident scene rehabilitation. Even if investigating a cold scene, investigators should always have sufficient water for rehydration and snacks for energy. They should also clean up or at least wash their hands and face before handling these foods to prevent contamination.

Before removing their SCBA, investigators should test the atmosphere for proper oxygen levels of at least 19.5 percent, preferably 20 percent, and CO readings that are under the recommended levels established by their department. Headaches can develop from CO levels as low as 40 parts per million, but the threshold for safety should be much lower.

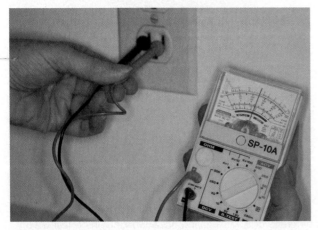

FIGURE 2-11 Every investigator must be familiar with the use of a typical multi-meter.

Courtesy of Russ Chandler.

> **TIP**
>
> Before removing the SCBA, the investigator should test the atmosphere for proper oxygen levels of at least 19.5 percent and for CO readings at a level established as permissible by OSHA (less than 50 ppm for an 8-hour exposure).

Carbon monoxide is a colorless, odorless gas that acts as an anesthetic. In higher concentrations, it becomes an asphyxiant. CO combines with hemoglobin in the blood to form carboxyhemoglobin. Hemoglobin carries oxygen throughout the body; essentially, the hemoglobin prefers to bind with CO 250 more times than it binds with oxygen.[5] When the hemoglobin carries no or limited oxygen, insufficient oxygen reaches the brain and muscles, including the heart. The end effect is narcosis, a condition in which the individual enters a stupor and eventually loses consciousness. Unfortunately, the effects of CO poisoning are often not recognized in time.

Every investigator should have access to a multi-gas detector. These units can be small, constantly operating monitoring units that are attached to your clothing while you are working on the scene, or they can be larger models that you use to check the atmosphere on a regular basis.

If a scene safety officer is on site, this person may be monitoring atmospheric gas levels and will give the all clear for removal of SCBA when appropriate. However, once the fire suppression personnel have completed their tasks at the scene and left the premises, investigators should continue to monitor the atmosphere to ensure that they are working in a safe environment.

If the scene is an industrial or mercantile property, fire fighters should take extra precautions to avoid exposure to dangerous chemicals. A hazardous response team can assess for the presence of such hazards and can advise you as to when it is safe to enter the scene.

Other Hazards

Holes in floors are not always obvious, as they can be masked by loose debris or standing water. The investigator should always use caution when stepping into any standing water because it may be much deeper than expected and may hide sharp edges that could cause injuries. Holes or weakened floors could lead to a fast entry into the floor, basement, or crawl space below, with serious consequences. Holes in floors need to be identified and marked or rendered safe.

Fire fighters should never enter standing water in a basement. Unknown electrical sources there could lead to electrocution hazards, and hidden debris under the water could present tripping hazards. Even basement floors could have multiple levels or contain cisterns, sump pump wells, or other deeper openings. To remove the water and enable a safer examination of the basement area, fire fighters can use portable pumps—available in most departments or through a local equipment rental company. It is important to test any devices before they are put into the water to ensure that they do not have any trace residue of hydrocarbons that could contaminate the scene.

At no time should any device powered with hydrocarbon fuels be used in the structure, as accidental spill or dripping of the fuel may contaminate any evidence in the area. Even the presence of such devices presents the probability of contamination, potentially rendering any evidence collected there useless in court. Another concern with equipment such as generators running in a structure is the risk of CO production, endangering the personnel.

Animals can present special hazards. Even when they appear timid, a family pet can turn and cause serious injuries. Cats rarely pose a threat unless you decide to detain them when they do not want to be handled. Even thick firefighting gloves and gear may not be full protection from a bite or scratch. Scratches need to be taken seriously; they must be immediately cleaned to prevent infections and other problems such as cat scratch fever. Dogs, regardless of their size, can also be a hazard and should be treated with caution (**FIGURE 2-12**). When a loose dog appears in a threatening manner, it may be defending its domain and perceiving the investigator as an intruder. Fire fighters who think that they can "calm the savage beast" may painfully come to realize the flaw in their plan. Even pay careful attention to distraught dogs that appear to be safely tethered.

> **TIP**
>
> Animals are especially dangerous when they appear timid or frightened, as they are more likely to suddenly lash out. The animal that appears aggressive is a known quantity and can be avoided or approached with a fair expectation of what it is about to do. If available, an animal control officer could be summoned to assist.

FIGURE 2-12 A large, aggressive dog can give obvious signs that cause you to approach cautiously, but even a scared, small, timid dog can cause a serious injury.
Courtesy of Russ Chandler.

If family members are not available to assist with managing a pet, the local animal control officers may be able to assist with both the investigators' protection and the protection of the animal. Another tool that may work from time to time is offering a dog or cat snacks as appropriate. A snack may not necessarily make friends with worried or aggressive animals, but a barking dog can be quite distracting, and snacks have been known to satisfy dogs long enough so that the barking will stop or at least lessen.

Dogged Determination

An investigator pulled onto a property where a fire had occurred the night before. No one was at the property, but the investigator could hear a barking dog in the backyard, which was the location of most of the fire damage. The investigator's systematic process of checking the exterior of the structure first revealed that the backyard was not fenced in. However, a dog—specifically, a Rottweiler mix—was chained to a doghouse in the center of the yard. For the record, Rottweilers can weigh between 80 and 100 lb, or more. This particular dog was clearly at the high end of that scale.

What appeared comical at first was then seen as reassuring. The doghouse was made from 2- by 6-in. treated boards and looked large enough to house six dogs. Around the dog's neck was a heavy, thick, 2-in.-wide collar with chrome spikes. In the D-ring of the collar was a climbing carabiner that attached a chain to the collar. It was a short chain, only about 8 ft long, but looked heavy enough to use with a boat anchor. Then, a large eyebolt and base attached the chain to the doghouse using four large bolts. As large as the chain seemed, the dog had no problem pulling it taut as the investigator approached the house. Even so, there seemed no way this dog could escape its tether.

As with any repetitive noise, the investigator eventually blocked out the barking and went to work sifting through the scene. After finding the fire area of origin, he was kneeling down and looking at different items while searching for the fire cause. The investigator eventually realized that the barking was louder and looked up. There was the dog, only 2 ft away, snarling and snapping, looking the investigator right in the eye. What originally was a 20-ft barrier between the dog and the fire scene was now only a few feet. The dog had dragged the doghouse across the yard, leaving a furrow in its wake. It did not take long for the investigator to decide that the final cause of the fire could be determined at a later date.

Depending on where one works, snakes, spiders, and insects can be a hazard. Knowledge and recognition of poisonous snakes known to the area may be beneficial for fire investigators. Spider bites, although not usually deadly, can cause serious sickness and damage. If poisonous spiders are known to inhabit the area, investigators should take precautions to prevent exposure to them, such as covering all exposed skin, sealing clothing at the ankles and wrists, and ensuring that shirts stay tucked in. The same precautions can be taken in southern states where fire ants or other swarming insects may be present.

A can of wasp/hornet spray can be a valuable tool to remove nests that pose a potential hazard. However, because these products may contain petroleum distillate, it is very important to use extreme caution; do not spray them where collecting evidence. You would not want to risk contamination of the scene anywhere around the potential area of fire origin, as this could jeopardize the likelihood of an accurate result from lab testing of the debris sample.

Explosives and Terrorism

No safety chapter is complete without mention of explosive scenes and terrorism. Investigators should be well aware of all safety concerns at any emergency scene involving explosives. Issues with residues, hazardous materials, and incendiary devices are all quite real and potentially deadly. Terrorist acts are no longer a distant concern, but represent a threat on the home front. The risk of encountering an improvised explosive device (IED) is not that remote. A multitude of training events are available from local, state, and federal investigative units that address explosives, their detection, and the handling of explosive debris. The most important step in suspected explosive/terrorism scenes is to contact the local bomb squad to have them clear the area. Even if no devices are readily apparent, if there is even the slightest suspicion of the potential for a device, call for the bomb squad to clear the area, possibly with its specialty K-9.

How can you identify a threatening device? The clichéd TV and movie device of sticks of dynamite with a timer will not necessarily be lying about the scene. Instead, the perpetrator's goal will dictate the device's appearance. If attackers want to prevent the device from being found, they will make an attempt to have it blend in with the occupancy and thereby keep it concealed. In contrast, if attackers want to send a big message, they may not worry about concealment to prevent discovery. Sometimes attackers have insufficient time or resources to make a device blend in; in that case, the device may be as small as a briefcase, backpack, or brown-wrapped package. Investigators' best course is look for something out of the ordinary. They must be careful when touching or picking up items, while keeping an open mind and a cautious outlook. If something out of the ordinary is discovered, the investigator can examine it only to the extent of the investigator's level of training. If any doubt or concern arises about an item's potential for being an explosive device, report it to the local bomb disposal unit.

As the saying goes, if we do not know our history, we are doomed to repeat it. All public safety responders should be familiar with major incidents that have occurred, especially within their own country. When thinking of terrorism, many will envision the planes hitting the World Trade Center. The government and the airline industry have taken great strides to prevent such future events. But we must also remember that the Oklahoma City bombing, which involved a large rental truck, was the first major home-grown terrorism event. In that case, the perpetrators created their bomb by following the directions in a book. More recently, the Boston Marathon bombing showed that the location of the device is critical—and that even a simple device constructed from readily available items can create a devastating event. All students of public safety should learn the history of major terrorist events.

The federal government, in particular the Bureau of Alcohol, Tobacco, Firearms and Explosives (ATF), provides outstanding training events on terrorism, explosives, and the post-blast scene. These trainings usually include hands-on experiences with primary and secondary devices.

TIP

The federal government, in particular the Bureau of Alcohol, Tobacco, Firearms and Explosives (ATF), provides outstanding training events on terrorism, explosives, and the post-blast scene.

Wrap-Up

SUMMARY

- The first responder investigator needs to be aware of all safety aspects of not just the fire suppression scene but also the investigation of a fire scene. The first responder investigator must also be familiar with the aspects of a full investigation.
- First responder investigators will already be wearing the assigned turnout gear. It would also be wise for them to be trained and fitted for filter respiratory masks, as these devices enable them to work longer on the scene assisting the assigned investigator.
- The assigned investigator will begin thinking about safety aspects upon receiving the assignment, and safety will always be the primary concern. Investigators tend to start their size-up of the situation on their way to the scene. Considering safety conditions and steps to make a scene safe can be part of the thought process while responding, but safety must actually start long before the assignment is received.
- Terrorism is a concern in today's society. All investigators should be concerned when they find items out of the ordinary or where they would not expect to find such items. These items can be examined at the investigator's specific level of training. When in doubt, contact a bomb disposal expert.
- All investigators should have strong knowledge of all laws, regulations, and standards to be met while carrying out their assignments.
- Knowledge of building construction and structural integrity is essential for carrying out the initial safety assessment. In addition, investigators must be familiar with hazards they may encounter and all the tools at their disposal to work within these hazardous environments.
- Most of all, investigators must have a proper frame of mind to keep safety in the forefront while completing their assessment, carrying out the fire scene examination, and cleaning up after the scene examination. They must always remember that they are the most important person, and that by taking care of themselves, they will be able to take care of others.

KEY TERMS

Building Officials Office An agency empowered by the authority having jurisdiction to enforce the provisions of a state or local building code, electrical code, plumbing code, and in some instances, the fire code.

Canister A manufactured metal or plastic cylindrical device filled with filtering material. Canisters are designed to be attached to a filter mask with an airtight seal, forcing all air entering the mask to pass through the filtering medium, which removes hazardous airborne particulates.

Carbon monoxide (CO) A colorless, odorless gas usually produced as a by-product of a fire. It can also be created from faulty appliances such as a furnace with a leaking exhaust flue.

Cold scene The remains of a fire scene after the fire has been completely extinguished. Although the debris may still be warm, there is little or no chance of being reignited. Cold scenes may be days old.

Incident Command System (ICS) An on-scene, standardized, all hazard management process.

Lockout A situation in which an electrical disconnect box is in the open position with no current flowing and a lock is in place to keep the circuit from being energized. Some lockouts provide space for multiple locks to provide safety for all personnel working on the scene.

Multi-gas detector A portable device for reading the levels of oxygen and hazardous gases in the atmosphere.

Multi-meter An electronic device used to measure AC/DC voltage, current, resistance, capacitance, and frequency. Also referred to as a multi-tester.

Non-vented goggles Eyewear that makes an airtight seal with the face, preventing the introduction of any airborne particulates.

Occupational Safety and Health Act Also referred to as the OSH Act. Federal law that governs the safety and health practices for all industries, both government and the private sector. The purpose is to guide and require employers to provide a safe working environment for their employees.

Occupational Safety and Health Administration (OSHA) A federal agency created by the Occupational Safety and Health Act whose function is to ensure safety in the workplace.

Respirator A mask (full- or half-face) designed to cover the mouth and nose, allowing the wearer to breathe through filters attached to the mask and thereby prevent the inhalation of dangerous substances, usually in the form of dust or airborne particulates.

Self-contained breathing apparatus (SCBA) A delivery system of breathable air from a compressed cylinder.

Steel-sole boots Boots with a lightweight steel plate that provides a barrier between the foot and the surface being walked on. The intent is to keep the wearer safe from punctures through the boot.

Steel-toe boots Work boots designed with a steel protective cap covering the toes to prevent crushing blows that would otherwise injure the wearer's toes.

Tagged The action of placing a tag on an electrical disconnect switch warning others not to turn the handle to the on position due to a safety concern.

Tyvek coveralls Coveralls made out of Tyvek, which the manufacturer DuPont describes as lightweight, strong, vapor-permeable, water-resistant, and chemical-resistant material that resists tears, punctures, and abrasions. These coveralls are essential to protect the investigator working the extinguished fire scene against carbon and other contaminants and should be properly disposed after each incident.

Voltage detector A tester or probe that either glows or emits a sound when placed in proximity of an energized electric circuit (electrical wires or cords).

REVIEW QUESTIONS

1. List and describe the safety protective clothing used by the assigned fire investigator.

2. When clothing is exposed to fire scene debris, where should it be washed, and why?

3. What needs to be done to ensure that all electrical circuits are not energized?

4. Why is standing water a concern on a floor or in a basement?

5. What type of protective clothing should the fire investigator wear when entering a structure still under suppression activities?

6. Why is it important for the first responder fire investigator and the assigned fire investigator to follow national guidelines on incident scene rehabilitation?

7. Why should proper eye protection be non-vented?

8. It is recommended that the investigator wear latex gloves under leather gloves. Why?

9. The presence of some animals may constitute a hazard to the investigator. What are some measures that could be taken to render the condition safe?

10. What is the benefit of the assigned investigator reporting to the officer in charge (OIC) upon arrival at the emergency scene?

DISCUSSION QUESTIONS

1. What resources are available to assist the investigator in making the scene safety assessment?

2. How could the investigator render slippery surfaces safe without introducing any foreign material that may contaminate the scene?

3. Holes in floors constitute one of the most common hazards encountered by the fire investigator. How could the investigator render such holes safe without introducing something onto the scene that would contaminate the scene and affect any evidence collected?

4. List and explain the reason for using all the safety equipment that should either be carried by the fire investigator or made available to investigators at the fire scene.

5. Discuss who might be available in your jurisdiction to give expert opinions regarding the stability of a structure.

REFERENCES

1. Occupational Safety and Health Administration. *OSH Act of 1970*. Retrieved from https://www.osha.gov/laws-regs/oshact/toc

2. National Fire Protection Association. (2018). *NFPA 1971: Standard on protective ensembles for structural fire fighting and proximity fire fighting*. Retrieved from https://www.nfpa.org/codes-and-standards/all-codes-and-standards/list-of-codes-and-standards/detail?code=1971

3. National Institute of Occupational Safety and Health. (2016, July). *Findings from a study of cancer among U.S. fire fighters.* Retrieved from https://www.cdc.gov/niosh/firefighters/health.html

4. Braddee, R. W. (1999, May 14). *Fire investigator dies after being struck by a chimney that collapsed during an origin and cause fire investigation—New York* (NIOSH Report F99-06). Retrieved from http://www.cdc.gov/niosh/fire/reports/face9906.html

5. Cote, A. E. (Ed.). (2008). *Fire protection handbook* (20th ed., pp. 6–14). Quincy, MA: National Fire Protection Association.

CHAPTER 3

Legal Issues and the Right to Be There

LEARNING OBJECTIVES

Upon completion of this chapter, you should be able to:

- Describe legal issues as they pertain to the right of the fire service to be present and stay at the fire or explosion scene.
- Describe the difference between legislative laws and case law.
- Describe how the U.S. Constitution and the Bill of Rights impact an investigation.
- Describe criminal law and its potential impact on the fire service.
- Describe and understand U.S. Supreme Court decisions that have affected fire investigators and the ways that they carry out their duties.
- Describe civil law and its potential impact on the fire service.
- Describe national standards and guidelines, and explain how they can help the investigator stay within the legal boundaries.

Case Study

Michigan v. Clifford

The fire occurred on October 18, 1980, at approximately 5:40 AM at the home of Raymond and Emma Jean Clifford. At the time, the couple was away on a camping trip. The Detroit Fire Department responded and extinguished the fire, clearing the scene at 7:04 AM, after which the responders left. The fire department had reported the suspicious nature of the fire, but the assigned investigator, Lieutenant Beyer, was not notified until 8:00 AM. Because he had other appointments, he and another investigator did not arrive until 1:00 PM, 5 hours after being notified.

In the meantime, a neighbor notified the Cliffords about the fire, and the Cliffords asked the neighbor to contact their insurance company to have it secure the building. When the investigators arrived, they found a work crew at the scene. They were pumping out the basement and putting plywood on the windows and doors. The neighbor told the investigator about notifying the owners and receiving their instructions about getting the house secured by their insurance company.

The Detroit Fire Department policy stated that as long as the owner was not present and the premises were open to trespass, fire personnel could search the scene within a reasonable amount of time after the fire. The investigators waited for the basement to be pumped out, an operation that was completed at 1:30 PM. In the meantime, they discovered a Coleman lantern in the driveway and marked it as evidence.

The investigators entered the Cliffords' house without consent or a warrant. They determined that the fire started under the basement stairs from a slow cooker plugged into a timer that was set to go off at 3:45 AM and stop at 4:30 AM. With this evidence and a strong smell of a flammable liquid in the basement, the investigators determined the fire was incendiary in nature. Evidence was secured and properly marked.

The investigators did not stop there: They continued searching the remainder of the house. They called in a photographer to document the contents of the structure. They found the drawers full of old clothing and nails on the wall where pictures used to hang. There were even wires where a cassette player would have been placed, but the player was missing.

The Cliffords were charged with arson. Their attorney moved for suppression of the evidence on the basis of the investigators' warrantless search. The trial court denied the request to suppress the evidence, ruling that the investigators acted under exigent circumstances. In interlocutory (occurring between the start and the end of a trial) appeal, the Michigan Court of Appeals held that no exigent circumstance existed. It reversed the findings of the lower court, which made the evidence obtained by the investigators during their interior search of the premises inadmissible.

 *Access Navigate for more resources.*

Introduction

First responders responsible for conducting an investigation to determine a fire's origin and cause must be aware of the various legal issues that pertain to their task. The first and most obvious concern is to understand their legal right to be on the scene. The more detailed the investigation is, the more potential there is for legal issues to arise. To understand the legal requirements pertaining to fire investigations, the investigator must know where laws come from, how they are created, and how laws can change.

As a rule, laws, regulations, codes, and ordinances are not created because we believe something *might* happen, but usually because something *did* happen. In other words, most laws exist for a reason. Knowing why something is legal or illegal can ensure better compliance. Knowing how laws are created, who adjudicates the law, and how laws are interpreted can be vital to investigators whether they are first responders or full-time assigned investigators.

A full discussion of laws that impact emergency services can be found in the text *Legal Aspects of Emergency Services* (Jones & Bartlett Learning). This chapter addresses the differences between legislated law and case law, and considers those points that are specific to both accidental and intentionally set fires. In particular, it concentrates on two U.S. Supreme Court cases that define the rights that the fire investigator has to be on the scene, to collect evidence, and to make a determination of the fire origin and cause.

Laws

Laws are the system of rules created and enforced by society to ensure the safety, health, and welfare of the general public. These laws must be fair, ethical, and equal to all. They teach us how to behave properly, based on moral and ethical standards. Failure to obey laws could result in fines, imprisonment, or both. Courts can also mandate payment of damages. This chapter concentrates on those laws that are relevant to conducting an initial fire investigation.

Constitutional Laws

The Declaration of Independence states that Americans' inalienable rights are "life, liberty and the pursuit of happiness." The U.S. Constitution was written to further define these rights, these freedoms.

Constitutional law comprises the laws established by the language of the Constitution. As the supreme law of the land, it is the standard against what all other laws are measured. It lays out the relationship between different branches of government and—more importantly in the fire services context—it provides for the rights of the individual.

The first 10 amendments to the Constitution establish the basic rights guaranteeing certain freedoms; collectively, they form the Bill of Rights. These rights are critical in the handling of any fire investigation. All investigator must take great efforts to assure that they do not abridge these rights in the course of an investigation. These rights include, but are not limited to, the following:

- The right to free speech
- The right to freedom of religion
- The right to freedom of the press
- The right to a speedy public trial
- The right to bear arms
- The right of protection against unreasonable search and seizures

Legislated Laws

The U.S. Constitution establishes a balance of power under which bills (future laws) are developed by the legislative branch of government. These bills are sent to the executive branch (the president), who signs them into law or rejects them through a veto, sending them back to Congress. There, either the bill dies or, with two-thirds votes in both the Senate and the House of Representatives, the veto is overridden and the bill becomes law. Should the bill go forward to the executive branch and the president fails to take action within 10 days, the bill automatically becomes law.

Case Laws

As the result of criminal or civil action, a point of law may eventually become an issue addressed in a court of law. Some laws are ambiguous, complicated, or difficult to understand. Furthermore, not all circumstances involving laws are the same each and every time. For this reason, a judge interprets the law in a court proceeding. When a court decision is made, that interpretation relevant to that case under those circumstances becomes a precedent. When similar circumstances come up in other courts on the same point of law, judges will apply that precedent in making a decision—a process known as case law. Knowing case law can prevent future problems in court; in the investigative context, that means being familiar with previous cases and decisions relevant to fire investigation.

As cases are adjudicated, parties aggrieved with the decisions may sometimes take their cases to higher courts. When that happens, the entire case is not tried again, but rather points of law are discussed to ensure a fair hearing occurred and all rules were followed.

Supreme Court Decisions

The freedoms and rights of U.S. citizens are spelled out in the Constitution of the United States. The U.S. Supreme Court has the responsibility to interpret the Constitution and its amendments by reviewing the proceedings from the lower courts. These decisions are not always unanimous, but rather reflect the majority position on the Court.

The Supreme Court hears only certain types of cases, including cases dealing with the following topics:

- Constitutional law
- Issues between parties from different states
- Conflicts between states
- Issues between U.S. citizens and foreigners
- Cases involving both state and federal laws
- Cases involving maritime law or admiralty
- Cases in which the United States is a party[1]

The Supreme Court justices decide which cases will be heard based on the need to help further define issues that may not be totally clear in our court system today. Several Supreme Court decisions exist that affect fire investigators and their right to be on the scene collecting evidence as well as other factors dealing with criminal law such as *Miranda* warnings.

Assigned fire investigators with full police powers may seize property, conduct searches, detain and arrest suspects, and testify in court as to their expert opinion on the fire. Each of these activities entails abridging the rights and freedoms of some citizens. Investigators should do this only when necessary to carry out their assigned duties. To perform this role properly, investigators rely on the rulings of the Supreme Court for guidance. Although a multitude of court decisions address investigator duties and responsibilities, the discussion here is limited to those dealing with the right to be on the fire scene and when investigators can collect evidence properly. Also, we will discuss court cases where decisions have affected assigned investigators' ability to give expert testimony.

Crimes

The government has set up a series of statutory laws to define what is not acceptable in society today and tomorrow. Violating one of these laws through an action or an omission is a crime. These laws were not created because legislators thought someone would do such an act; instead, they were created because someone did commit such an offense and legislators wanted to prevent similar acts in the future. Thus, laws are intended to serve as a deterrent to committing particular crimes.

Crimes against legislated laws are put into two categories: Misdemeanors represent lesser offenses, and felonies are more serious crimes. By definition, felonies are crimes from which the sentence can range from a year in prison to the death penalty. Both misdemeanors and felonies are broken down into subclassifications, such as a class 1 misdemeanor or a class 4 felony. Class 1 felonies and misdemeanors are the most serious classifications.

Being found guilty of a misdemeanor could result in a fine and/or brief imprisonment. The length of imprisonment varies from state to state, but predominantly ranges from 6 months up to a year. Imprisonment, if imposed as punishment for a misdemeanor, is usually less than the maximum time possible. If sentenced to serve time for a misdemeanor, that incarceration will usually take place in a county, city, or local jail. In some localities, the fine for a misdemeanor could be only $500 and the sentence no more than 60 days in jail. Fines for misdemeanors vary from state to state, however, and usually range from $2500 to $5000. The highest amounts are usually reserved for more serious and violent misdemeanor crimes.

Felonies, by definition, are serious crimes. The minimum prison time imposed for a felony is more than one year. Federal and most state laws allow imprisonment for life—and in some states, the death penalty—in the most severe cases. Felonies are also referred to as "high crimes" in the U.S. Constitution.

Michigan v. Tyler

On January 21, 1970, shortly before midnight, a fire occurred at Mr. Loren Tyler's Auction, a furniture store in Oakland County, Michigan (**FIGURE 3-1**). Around 2 AM, as the final flames were being extinguished, fire fighters discovered plastic containers with what appeared to be a flammable liquid. Chief See, thinking this incident might involve an arson fire, notified the local police. Detective Webb arrived and took photographs but had to leave the building because of smoke. By 4 AM, the fire was extinguished and the fire department left the scene. Because there was still steam and it was dark, Chief See and Detective Webb also left, taking the containers with them.

Around 8 AM, Chief See and Assistant Chief Somerville returned to check the building; an hour later, Detective Webb and Assistant Chief Somerville did another examination of the scene, collecting even more evidence: a piece of carpet and taped sections of the stairway.

On February 16, 1970 (26 days after the incident), the state police arson investigator arrived, took photographs, and examined the scene. He returned several more times, taking evidence and collecting information. Mr. Tyler and his partner Robert Tompkins were charged with conspiracy to burn real property as well as other offenses. They were convicted, but the case was appealed based on the evidence being obtained without a proper search warrant. The Michigan Supreme Court agreed and reversed the convictions.

The U.S. Supreme Court agreed to hear arguments on admissibility of evidence in this case. It ultimately held that the two plastic containers with flammable liquid were admissible as evidence; because of the exigencies of the situation, the fire chief had the right to secure them as evidence. Because the fire scene was still hazardous with smoke and with problems of seeing resulting from steam and darkness (it was nighttime), the Supreme Court agreed that the evidence

Michigan v. Tyler

436 U.S. 499 (1978)

Supreme Court of the United States

MR. JUSTICE STEWART delivered the opinion of the Court. . . . Shortly before midnight on January 21, 1970, a fire broke out at Tyler's Auction, a furniture store in Oakland County, Mich. The building was leased to respondent Loren Tyler, who conducted the business in association with respondent Robert Tompkins. According to the trial testimony of various witnesses, the fire department responded to the fire and was "just watering down smoldering embers" when Fire Chief See arrived on the scene around 2 AM. It was Chief See's responsibility "to determine the cause and make out all reports." Chief See was met by Lt. Lawson, who informed him that two plastic containers of flammable liquid had been found in the building. Using portable lights, they entered the gutted store, which was filled with smoke and steam, to examine the containers. Concluding that the fire "could possibly have been an arson," Chief See called Police Detective Webb, who arrived around 3:30 AM. Detective Webb took several pictures of the containers and of the interior of the store, but finally abandoned his efforts because of the smoke and steam. Chief See briefly "[l]ooked throughout the rest of the building to see if there was any further evidence, to determine what the cause of the fire was." By 4 AM, the fire had been extinguished and the firefighters departed. See and Webb took the two containers to the fire station, where they were turned over to Webb for safekeeping. There was neither consent nor a warrant for any of these entries into the building, nor for the removal of the containers. . . . .

Four hours after he had left Tyler's Auction, Chief See returned with Assistant Chief Somerville, whose job was to determine the "origin of all fires that occur within the Township." The fire had been extinguished and the building was empty. After a cursory examination, they left, and Somerville returned with Detective Webb around 9 AM. In Webb's words, they discovered suspicious "burn marks in the carpet, which [Webb] could not see earlier that morning, because of the heat, steam, and the darkness." They also found "pieces of tape, with burn marks, on the stairway." After leaving the building to obtain tools, they returned and removed pieces of the carpet and sections of the stairs to preserve these bits of evidence suggestive of a fuse trail. Somerville also searched through the rubble "looking for any other signs or evidence that showed how this fire was caused." Again, there was neither consent nor a warrant for these entries and seizures. Both at trial and on appeal, the respondents objected to the introduction of evidence thereby obtained.

On February 16, Sergeant Hoffman of the Michigan State Police Arson Section returned to Tyler's Auction to take photographs. During this visit or during another at about the same time, he checked the circuit breakers, had someone inspect the furnace, and had a television repairman examine the remains of several television sets found in the ashes. He also found a piece of fuse. Over the course of his several visits, Hoffman secured physical evidence and formed opinions that played a substantial role at trial in establishing arson as the cause of the fire and in refuting the respondents' testimony about what furniture had been lost. His entries into the building were without warrants or Tyler's consent, and were for the sole purpose "of making an investigation and seizing evidence." At the trial, respondents' attorney objected to the admission of physical evidence obtained during these visits, and also moved to strike all of Hoffman's testimony "because it was got in an illegal manner." . . .

The decisions of this Court firmly establish that the Fourth Amendment extends beyond the paradigmatic entry into a private dwelling by a law enforcement officer in search of the fruits or instrumentalities of crime. As this Court stated in Camara . . . the "basic purpose of this Amendment . . . is to safeguard the privacy and security of individuals against arbitrary invasions by governmental officials." . . .

[T]here is no diminution in a person's reasonable expectation of privacy nor in the protection of the Fourth Amendment simply because the official conducting the search wears the uniform of a firefighter, rather than a policeman, or because his purpose is to ascertain the cause of a fire, rather than to look for evidence of a crime, or because the fire might have been started deliberately. Searches for administrative purposes, like searches for evidence of crime, are encompassed by the Fourth Amendment. . . .

The petitioner argues that no purpose would be served by requiring warrants to investigate the cause of a fire. This argument is grounded on the premise that the only fact that need be shown to justify an investigatory search is that a fire of undetermined origin has occurred on those premises. . . . In short, where the justification for the search is as simple and as obvious to everyone as the fact of a recent fire, a magistrate's review would be a time-consuming formality of negligible protection to the occupant.

FIGURE 3-1 *Michigan v. Tyler*, 436 U.S. 499, full text.

The petitioner's argument fails primarily because it is built on a faulty premise. To secure a warrant to investigate the cause of a fire, an official must show more than the bare fact that a fire has occurred. The magistrate's duty is to assure that the proposed search will be reasonable, a determination that requires inquiry into the need for the intrusion on the one hand, and the threat of disruption to the occupant on the other. . . . Thus, a major function of the warrant is to provide the property owner with sufficient information to reassure him of the entry's legality. . . .

In short, the warrant requirement provides significant protection for fire victims in this context, just as it does for property owners faced with routine building inspections. As a general matter, then, official entries to investigate the cause of a fire must adhere to the warrant procedures of the Fourth Amendment. . . . Since all the entries in this case were "without proper consent" and were not "authorized by a valid search warrant," each one is illegal unless it falls within one of the "certain carefully defined classes of cases" for which warrants are not mandatory. . . .

Our decisions have recognized that a warrantless entry by criminal law enforcement officials may be legal when there is compelling need for official action and no time to secure a warrant. . . . A burning building clearly presents an exigency of sufficient proportions to render a warrantless entry "reasonable." Indeed, it would defy reason to suppose that firemen must secure a warrant or consent before entering a burning structure to put out the blaze. And once in a building for this purpose, firefighters may seize evidence of arson that is in plain view. . . . Thus, the Fourth and Fourteenth Amendments were not violated by the entry of the firemen to extinguish the fire at Tyler's Auction, nor by Chief See's removal of the two plastic containers of flammable liquid found on the floor of one of the showrooms.

Although the Michigan Supreme Court appears to have accepted this principle, its opinion may be read as holding that the exigency justifying a warrantless entry to fight a fire ends, and the need to get a warrant begins, with the dousing of the last flame. . . . We think this view of the firefighting function is unrealistically narrow, however. Fire officials are charged not only with extinguishing fires, but with finding their causes. Prompt determination of the fire's origin may be necessary to prevent its recurrence, as through the detection of continuing dangers such as faulty wiring or a defective furnace. Immediate investigation may also be necessary to preserve evidence from intentional or accidental destruction. And, of course, the sooner the officials complete their duties, the less will be their subsequent interference with the privacy and the recovery efforts of the victims. For these reasons, officials need no warrant to remain in a building for a reasonable time to investigate the cause of a blaze after it has been extinguished. And if the warrantless entry to put out the fire and determine its cause is constitutional, the warrantless seizure of evidence while inspecting the premises for these purposes also is constitutional.

The respondents argue, however, that the Michigan Supreme Court was correct in holding that the departure by the fire officials from Tyler's Auction at 4 AM ended any license they might have had to conduct a warrantless search. Hence, they say that even if the firemen might have been entitled to remain in the building without a warrant to investigate the cause of the fire, their reentry four hours after their departure required a warrant.

On the facts of this case, we do not believe that a warrant was necessary for the early morning reentries on January 22. As the fire was being extinguished, Chief See and his assistants began their investigation, but visibility was severely hindered by darkness, steam, and smoke. Thus they departed at 4 AM and returned shortly after daylight to continue their investigation. Little purpose would have been served by their remaining in the building, except to remove any doubt about the legality of the warrantless search and seizure later that same morning. Under these circumstances, we find that the morning entries were no more than an actual continuation of the first, and the lack of a warrant thus did not invalidate the resulting seizure of evidence.

The entries occurring after January 22, however, were clearly detached from the initial exigency and warrantless entry. Since all of these searches were conducted without valid warrants and without consent, they were invalid under the Fourth and Fourteenth Amendments, and any evidence obtained as a result of those entries must, therefore, be excluded at the respondents' retrial.

In summation, we hold that an entry to fight a fire requires no warrant, and that, once in the building, officials may remain there for a reasonable time to investigate the cause of the blaze. Thereafter, additional entries to investigate the cause of the fire must be made pursuant to the warrant procedures governing administrative searches. . . . Evidence of arson discovered in the course of such investigations is admissible at trial, but if the investigating officials find probable cause to believe that arson has occurred and require further access to gather evidence for a possible prosecution, they may obtain a warrant only upon a traditional showing of probable cause applicable to searches for evidence of crime. . . .

These principles require that we affirm the judgment of the Michigan Supreme Court ordering a new trial. Affirmed.

FIGURE 3-1 (Continued).

taken the next morning was reasonable, as an extension of the previous evening's search. However, the Court also ruled that the subsequent visits after that point were improper and all evidence collected then was inadmissible in court.

This decision makes it clear that investigators—whether first responders or assigned investigators—can do a search for the origin and cause, and under the exigencies of the circumstances can seize evidence in plain sight. However, if investigators want to return, they must obtain an administrative search warrant if the fire cause has yet to be determined. If any evidence suggests that the fire is incendiary in nature or evidence of any other criminal activity, they must obtain a criminal search warrant before continuing the search.

> **TIP**
>
> If investigators want to return to a fire scene, they must obtain an administrative search warrant if the fire cause has yet to be determined.

Of course, no search warrant needs to be obtained if the building owner gives consent to search or the renter gives consent to search the leased area (**FIGURE 3-2**). If this permission is rescinded, then the scene must be secured, preferably by someone guarding the property, and the investigator should go immediately to the magistrate to secure a warrant, administrative or criminal, depending on the circumstances.

Other Court Decisions

For the first responder investigator, the court case that most commonly comes into play would be *Michigan*

FIGURE 3-2 The assigned fire investigator should have the building owner sign a consent agreement to search before proceeding with the scene search.
Courtesy of Russ Chandler.

v. Tyler. In some situations, the fire company officer has returned to the scene later, which could have been an issue as described in the *Michigan v. Clifford* case. However, other court decisions or standards must be considered as the first responder investigator increases his or her involvement in determining the origin and cause of fires.

Courts and Scientific Methodology

The scientific method is used in all investigations, even though investigators likely will not explicitly describe their process using this terminology in their reports. (Don't be surprised when you realize one day that you use the scientific method in making many decisions in your life.) However, when more complicated fire cases end up in court, investigators will need to explain in detail how they came to their conclusions. They will need to show that their findings were scientifically based and were not junk science. Two court rulings—*Frye* and *Daubert*—have become the standard for determining whether evidence being submitted in a trial reflects a true scientific perspective.

Frye v. United States. In 1923, the *Frye v. United States* case addressed the question of the admissibility of systolic blood pressure as a means to prove deception as evidence in court. The court held that expert testimony must be based on scientific methods that are sufficiently established and accepted—and that blood pressure measurement did not meet this criteria. The resulting *Frye* standard remains in effect in many jurisdictions today.

Daubert v. Merrell Dow Pharmaceuticals. In 1993, the *Daubert v. Merrell Dow Pharmaceuticals* case set a new precedent regarding the admissibility of non-scientific-based evidence in court. Jason Daubert and Eric Schuller, along with their parents, sued Merrell Dow Pharmaceuticals. Their claim was that the antinausea drug Bendectin, which both of their mothers had taken during their pregnancies, resulted in serious birth defects in both Jason and Eric. The pharmaceutical company called on an expert who relied on unpublished reports and studies to refute the claim that Bendectin caused birth defects.

This case rose to the Supreme Court. Seven members (a majority) of the Supreme Court agreed on future guidelines for expert testimony; these guidelines were subsequently amended in 2000. The Federal Rules of Evidence, Article VII, Rule 702 (Testimony by Expert Witness), set up the trial judge as the gatekeeper who may exclude unreliable expert testimony. Some states still hold to the *Frye* standard, but the majority now follow the *Daubert* ruling.

The *Daubert* decision set up a nonexclusive checklist for trial courts to assure that:

- Scientific testimony is based on scientific knowledge.

- Testimony is relevant to the task at hand and based on a reliable foundation.

- Experts' testimony is reliably applied to the facts at hand.

- The proponent can demonstrate that the scientific knowledge is based on sound scientific methodology using the scientific method. Courts have defined "scientific methodology" as the process of formulating a hypothesis, then conducting experiments to prove or disprove the hypothesis, using the following criteria:
  - Is the theory or technique used by the expert generally accepted in the scientific community?
  - Has it been subject to peer review and publication?
  - Can it be or has it been tested? Can it be challenged in some objective sense?
  - Is the known or potential rate of error acceptable?
  - Was the research conducted independent of the litigation at hand or was it dependent on an intention to provide the proposed testimony?

Kuhmo Tire Co., Ltd. v. Carmichael. In 1999, in *Kuhmo Tire Company, Ltd. v. Carmichael,* the courts held that, depending on the circumstances of the particular case, these factors might also be applicable in assessing the reliability of nonscientific expert testimony. In this case, the car Patrick Carmichael was driving blew a tire, the vehicle overturned, and one passenger died and others were injured. The survivors and decedents' representatives sued the tire maker, Kuhmo Tire Company, claiming that the tire was defective.

The plaintiffs were depending on testimony from their tire failure analyst expert, who planned to testify that the tire had a defect which caused the blowout. The expert's opinion was based on visual and tactile inspection of tire. The defendant (Kuhmo Tire) objected, stating that the plaintiff's expert's proposed testimony was not sound and ignored substantial evidence of other causes of the blowout. The judge agreed to not let the plaintiff's expert testify, noting that the expert visual analysis was not scientific. Without the expert's testimony, the judge ruled in favor of the defendant.

The plaintiff appealed to the Eleventh Circuit Court, which found that the trial judge erred by applying the *Daubert* standard to nonscientific evidence and remanded the case back to the trial court. It included instructions to reevaluate the expert's testimony, but this time not using *Daubert* factors. The defendant appealed, and the U.S. Supreme Court agreed to hear the case.

Kumho Tire: Applying the Daubert Factors to Experts

In its *Daubert* ruling, the U.S. Supreme Court set the standards for admitting testimony about scientific evidence under the Federal Rules of Evidence. Specifically, the Court upheld the trial court's exclusion of evidence in the *Daubert* case. As part of its opinion, the Court set out four factors that the trial judge should consider in determining whether evidence should be admitted: (1) a theory's testability, (2) whether it "has been a subject of peer review or publication," (3) the "known or potential rate of error," and (4) the "degree of acceptance . . . within the relevant scientific community." The *Daubert* ruling stressed that this analysis was intended to be flexible, with the trial judge determining how and whether the factors applied in a given case.

The *Kumho Tire* case looks at whether the *Daubert* factors may also be applied to nonscientific expert testimony. In its review of *Kumho Tire*, the Supreme Court's primary ruling was that trial judges should review evidence to assure that it was reliable and not overly prejudicial. It stressed that the trial courts have great latitude in how they do this review, and that they should be flexible in setting their standards to assure that different types of evidence receive appropriate reviews.

The Court held that it was not an error for the trial judge to use the *Daubert* standards to evaluate the expert's testimony on the nature of the defect in the defendant's tire. It stressed that the trial judge has looked at not just the theory behind the expert's methods, but also the extensive discovery record questioning these methods and the problems the expert had in determining basic information about the tire. The Court made it clear that the trial judge did not reject all methods of tire analysis, but rather looked specifically at this expert's own variant of accepted methods and found them inadequate.

Justice Antonin Scalia's brief concurrence is a good summary of this case:

I join the opinion of the Court, which makes clear that the discretion it endorses—trial-court discretion in choosing the manner of testing expert reliability—is not discretion to abandon the gatekeeping function. I think it worth adding that it is not discretion to perform the function inadequately. Rather, it is discretion to choose among reasonable means of excluding expertise that is false and science that is junky. Though, as the Court makes clear today, the Daubert factors are not holy writ, in a particular case the failure to apply one or another of them may be unreasonable, and hence an abuse of discretion.

Adapted with permission from Edward P. Richards, JD, MPH, Director of LSU Law Center.[2]

Michigan Millers Mutual Insurance Corporation v. Janelle R. Benfield.

The *Benfield* decision, although not a Supreme Court ruling, is case law that the fire investigator should understand. At time of this writing, it has not been cited or used in any other courts.

In 1988, Mr. and Mrs. Benfield secured an insurance policy from Michigan Millers Mutual Insurance Company for their home in Sarasota, Florida. Part of this homeowner's policy included the provision to provide protection for loss for personal property destroyed by fire, but most policies cited exclusions and conditions that limited this protection.

The Benfields separated, but were still married and both were living in the house for economic reasons. On the weekend prior to the fire, Mrs. Benfield had been staying with a friend but came back to the house to get some items when she discovered a fire.

The local fire investigator did not come up with an exact cause of the fire. Even though he labeled it accidental, he still believed the fire was suspicious, but found no physical proof of arson. When the insurance investigator examined the scene three days later, he also found no absolute proof of the fire being incendiary. However, he could find no possibility that the fire started accidentally, so he labeled the fire as incendiary in nature.

The insurance company indicated that it had some concerns about the incident. The claim was considered to be significantly inflated; Mrs. Benfield had tried to sell the house but could not do so; and she was trying to convince her estranged husband to do a quitclaim deed, but he refused. Furthermore, during interviews, she gave conflicting and contradictory accounts of her activities immediately prior to the fire. In addition, she gave a statement indicating that her husband was the one who set the fire.

Millers Mutual denied the claim based on the fact that under Florida law, "the insured" refers to all parties insured under that policy. Based on interpretation, the insured have joint rights and joint obligations. Even though the Benfields eventually divorced during this time, they were still joint holders of the insurance policy. Mrs. Benfield could not collect from the insurance that belonged to both her and her ex-husband, whom the insurance company believed set the fire.

Millers Mutual petitioned the court for summary judgment. In doing so, the company asked the court to hear both sides of the issue and decide whether the denial was appropriate even though the local fire department listed the fire as accidental based on a lack of proof of an incendiary nature. The fire investigator hired by Millers Mutual stated that the fire was incendiary in nature due to the fact that there was no indication of an accidental fire.

The court's decision was succinctly stated: "Because this Court has determined that a genuine issue of material fact exists as to whether the fire, giving rise to the damage, was the result of an intentional act, Plaintiff Michigan Millers' Motion for Summary Judgment is denied."

For the first responder investigator, this type of concern may not be an issue. In contrast, for the assigned investigator, who is charged with determining a fire's cause, it is critical that any fire be proven to be incendiary based on facts and through application of scientific methodology, and that this evidence be properly documented before ever going to court. Without proof of an incendiary nature, all fires are considered accidental unless proven otherwise.

TIP

All fires are considered accidental unless proven otherwise.

Right to Be There

For the assigned investigator, there are several ways to legally be on the scene of the incident. For the average first responder, in case of an emergency, he or she can enter wherever necessary. Once the emergency is over, however, the property owner or tenant has certain rights that investigators must respect. Even though first responders, for the most part, deal only with the exigencies of the situation, they should be aware of how assigned investigators and others can enter the scene after the emergency is over.

As Supreme Court decisions have emphasized, once the exigencies of the incident have ended, public investigators must have permission to be on the scene and must have the right to be there. There are four ways to be on a property legally: exigent circumstances, consent, administrative search warrant, and criminal search warrant.

> **TIP**
>
> There are four ways to be on a property legally: exigent circumstances, consent, administrative search warrant, and criminal search warrant.

Exigent Circumstances

The fire department must have the right to enter private properties without the necessity of obtaining consent or a search warrant. Put simply, delaying entry in an emergency may put lives and property at risk. Any delay may also allow the fire to extend, endangering other properties and other lives. Thus, for the good of the public, fire departments are allowed to enter property under exigent circumstances.

> **TIP**
>
> For the good of the public, fire departments are allowed to enter property under exigent circumstances.

This right is extended to making a determination of the area of origin and cause of the fire, within reason. This activity, too, is undertaken for the good of the people—that is, in hopes of preventing future fires. This right does not encompass an unlimited amount of time; it is not an open-ended invitation to revisit the property at will. If any delay in conducting the scene examination occurs, it must be justified. Waiting for daylight is acceptable, as shown in *Michigan v. Tyler*. However, beyond that point, using exigent circumstances to justify a search for the cause of the fire beyond a certain time frame may be questionable.

Consent

Once the emergency is over and the suppression forces are leaving the scene, the assigned investigator, for the good of the public, may be able to stay on the scene for a bit more time to conduct an origin and cause investigation. The precise amount of time depends on local policies. The assigned investigator will need to respect the property owners and protect himself or herself and the collection of any future evidence to assure he or she can legally stay on the property to complete the investigation.

The person who owns or has lawful control of the property can give consent to search. Control of the property is the issue: The owner of an apartment building can give consent to a search of the common areas, but not the leased apartments. The tenant can give consent to search the apartment or leased area that this person controls. Case law shows that 16-year-olds (or other minors) can give consent to the general areas of the apartment where they live with their parents, but not to their parents' bedroom.

> **TIP**
>
> No search warrant need be obtained if the property owner gives consent.

It is important that the consent to search be documented with each individual. Even when documented, such consent can be rescinded. When this happens, the investigator must seek a search warrant to stay on the scene.

> **TIP**
>
> It is important that the consent to search be documented with each individual.

> **TIP**
>
> Even when documented, consent to search can be rescinded.

Administrative Search Warrants

All search warrants must be justified. Unlike a criminal search warrant, the only probable cause necessary is a government interest in the fire investigator completing the investigation of the origin and cause of the fire. In most states, an administrative warrant is issued only in the following circumstances:

- There is proof that the investigator has the authority to conduct fire investigations in that jurisdiction.
- There is proof that a fire has occurred.

- The investigator cannot lawfully be on the property either because consent has been denied or because the exigencies of the incident are over as identified in the *Michigan v. Tyler* decision.[3]

Once there is any indication that the fire is criminal in nature, the administrative search is over and the investigator must obtain a criminal search warrant.

Criminal Search Warrants

The provisions of the U.S. Constitution as well as the laws of all states require the issuance of a search warrant by a magistrate or judge to be based on an independent determination that probable cause exists. Magistrates and judges are considered to be neutral and detached when deciding the merits of the information provided on the search warrant application.[4]

TIP

Magistrates and judges are considered to be neutral and detached when deciding the merits of the information provided on the search warrant application.

The assigned investigator must swear to the facts written in the application for a warrant, which is commonly referred to as an affidavit. The request must be specific as to what is sought and who or what is to be searched. The application must also state when the property will be searched as well as whether the search will be made without giving notice to the property owner. In the course of carrying out the search, the investigator must stay within the boundaries identified in the search warrant. In the course of the investigation, if it becomes apparent that other areas of the structure must be searched, the investigator must seek another search warrant and outline the probable cause justifying the extended search. To do otherwise may violate someone's civil rights as well as lead to any evidence obtained as a result of that illegal search being inadmissible in court. The best course for investigators is to consult in advance with the prosecuting attorney to establish a policy and procedure to ensure that there is full compliance with all applicable laws.

Private investigators or engineers generally do not have any problem gaining entry into the insured property. They can be on the scene based on the consent of the owner or, if representing the insurance company, they can enter based on the entry provisions of the insurance policy. Should the policy owner refuse entry to the private investigator or engineer assigned by the insurance company, the policy holder risks the chance that the insurance company will deny the claim due to the language in the insurance contract requiring the policy holder to cooperate with the insurance company. In essence, "no scene inspection, no money."

As long as the fire (or police) department investigator maintains the scene, investigators from the private sector do not have a right of entry. In some instances, however, the public and private investigators do work together. They may and should share information, but that cooperation may be limited if the incident turns out to be criminal in nature. In such a case, the fire department investigators will need to maintain some confidentiality on certain items as they build their criminal case.

TIP

The course for investigators is to consult in advance with the prosecuting attorney to establish a policy and procedure.

Insurance Company Investigations

The insurance industry has two ways of handling the investigation of fires based on claims submitted by their policy holders. First, some have their own investigative staff who perform the fire scene investigation and background work, and make the final report on the fire origin, cause, and responsible party. Second, insurance companies may hire independent licensed private fire investigators or licensed engineers to complete either a scene report or a full investigation on incidents reported by their policy holders. This concept can be less expensive for the insurance company. It also incorporates the notion that outside independent investigators are more likely to present unbiased findings.

Private fire investigators and engineers are also hired by self-insured entities and sometimes by individuals. In this text, the term *private investigator and/or engineer* include those individuals employed by the insurance industry as well as those representing any other entity or person having a vested interest in the fire-damaged property.

National Standards and Guides

Those in the fire service are usually aware of the national consensus standards established for fire operations and fire services management. In particular, they are familiar with the professional qualifications standards for fire fighters and fire officers. Likewise, standards have been established for full-time fire investigators.

There is no absolute way to ensure that the fire investigator's actions stay within the various rules established by the courts, but the creation of the national consensus standard is the first step toward addressing those concerns as well as many other fire investigator–related issues. Today the National Fire Protection Association (NFPA) provides a specific standard on fire investigator competencies called NFPA 1033, *Standard for Professional Qualifications for Fire Investigator*.

A professional qualifications standard is a document designed to give the minimum qualifications for an individual to meet a job description such as fire fighter, fire officer, fire instructor, or fire investigator. An authoritative body whose expertise includes both technical and educational experts promulgates these documents. Professional qualifications documents contain two primary features: prerequisite knowledge and prerequisite skills. These items are measurable, enabling the student or a training agency to evaluate training progress.

Note that the professional qualifications are minimum standards. It is the responsibility of individuals to constantly improve their knowledge and skills, taking them beyond these minimum expectations.

NFPA 921, *Guide for Fire and Explosion Investigations*

Many documents, publications, and texts on the market today are good resources for the investigator but are not necessarily standards. One such document is NFPA 921, *Guide for Fire and Explosion Investigations*. Notice that NFPA 921 is called a guide and not a standard. This document is intended to provide recommendations for the safe and systematic investigation of a fire or explosion incident; it is not intended to be adopted as a mandate as are NFPA standards.

NFPA 921 is a vital tool for ensuring the credibility of investigators who will be rendering an opinion in a fire case. As it is a guide, not every aspect of the NFPA 921 text needs to be followed in each and every case. However, when a process used runs contrary to the guide, the investigator should ensure that another authoritative document can substantiate that this process is scientifically valid and proper. This foundation is important to establish the fact that the investigator is adhering to a standard of care, and is being reasonable and prudent in the application of science to the findings of the fire's origin and cause and the identification of the person or thing that was responsible for the incident.

NFPA 921 is continually updated as technology, environmental, and legal issues change. The current schedule calls for the publication of an updated document every three years.

NFPA 1033, *Standard for Professional Qualifications for Fire Investigator*

NFPA 1033, *Standard for Professional Qualifications for Fire Investigator*, addresses performance requirements and the qualifications necessary to meet the requirements for the fire investigator. An investigator must be able to testify in court and the individual must be credible if he or she is to sway a jury to see testimony as accurate and factual.

The authority having jurisdiction (AHJ) is that entity that has the legal authority to employ and empower individuals to conduct fire investigations within the political boundaries of that entity. The AHJ can be the elected officials of a city, county, or town, or it could be those appointed and given this authority by a political board or council, such as the county administrator or the fire chief.

A fire investigator must be of good character and cannot have any convictions of moral turpitude—that is, crimes contrary to justice, honesty, and good morals. This is an important requirement when you consider that this individual will take an oath in court to give expert testimony. If anything in the person's past might give a jury reason not to believe the testimony, an arsonist could win the case and return to the street.

Continuing Education

Completing a basic course in fire investigation that meets the NFPA 1033 requirements is only the first step toward becoming a full-time fire investigator. The NFPA 1033 standard states that an individual must continually keep current on new court decisions and changes in codes, science, and methodology by attending additional training courses throughout the investigator's entire career.

Groups such as the **International Association of Arson Investigators (IAAI)** host a variety of seminars that address almost all aspects of the fire investigation field. State or province associations such as chapters of the IAAI also put on training events. These events are usually good at identifying training topics that address local problems or issues. In many areas, local and regional associations sponsor training as well. Some private entities such as Sirchie host training to address the proper use of their evidence collection kits, specialized equipment, and supplies. NFPA puts on seminars on systems and code applications that can also be of benefit to the fire investigator who is pursuing continuing education.

By attending organized training events, especially those put on locally and regionally, the investigator complies with another NFPA 1033 mandate—namely, maintaining a liaison with fellow investigators and investigative associations. Thus, this standard clearly recognizes that interactions between professionals create an atmosphere in which the exchange of ideas and concepts can take place.

The first responder investigator is encouraged to take part in the same training as is provided for the full-time fire investigator. These events provide a wealth of knowledge and will add to the individual's experience and knowledge base. If becoming a full-time investigator is the first responder's goal at some point in the future, adding these certifications will aid in that endeavor.

Ethics

Every fire investigator, whether employed by a public or private agency, must possess strong ethics. The IAAI is a private, nonprofit organization dedicated to the professional development of fire and explosion investigators through training and research of new technology. Its code of ethics succinctly describes what investigators must know and do to be ethical in carrying out their duties (**FIGURE 3-3**).

I.	I will, as an arson investigator, regard myself as a member of an important and honorable profession.
II.	I will conduct both my personal and official lives so as to inspire the confidence of the public.
III.	I will not use my profession and my position of trust for personal advantage or profit.
IV.	I will regard my fellow investigators with the same standards as I hold for myself. I will never betray a confidence or otherwise jeopardize their investigation.
V.	I will regard it as my duty to know my work thoroughly. It is my further duty to avail myself of every opportunity to learn more about my profession.
VI.	I will avoid alliances with those whose goals are inconsistent with an honest and unbiased investigation.
VII.	I will make no claim to professional qualifications which I do not possess.
VIII.	I will share all publicity equally with my fellow investigators, whether such publicity is favorable or unfavorable.
IX.	I will be loyal to my superiors, to my subordinates, and to the organization I represent.
X.	I will bear in mind always that I am a truth-seeker, not a case maker; that it is more important to protect the innocent than to convict the guilty.

FIGURE 3-3 The International Association of Arson Investigator's (IAAI) Code of Ethics is an excellent reminder on why we investigate fires.
Courtesy of International Association of Arson Investigators.

The last article of the IAAI Code of Ethics sums up the most important task assigned to fire investigators: They must be truth seekers and not case makers. In other words, investigators must keep an open mind and not enter into the investigation with any preconceived ideas about the case.

Wrap-Up

SUMMARY

- Investigators must first and foremost know and understand constitutional law, which provides the basis of assuring the protection of the rights of all involved. They must also understand and distinguish between legislated law versus case law. Case law includes decisions and interpretations from the Supreme Court as well as lower courts.

- Understanding federal and state laws is essential to knowing your rights and the rights of others. Through case law, such as *Michigan v. Tyler* and *Michigan v. Clifford*, fire investigators have been given direction as to what is acceptable and what is not. Even so, there is no clear, definitive line to show you everything you need to know under all circumstances.

- Court rulings mandate that there must be substantial proof to label a fire incendiary. Without such proof, fires are considered to be accidental in nature. Furthermore, the judge in a court is the gatekeeper who assures that expert testimony meets Federal Rule of Evidence 702 and that all testimony being considered is based on proper science.

- Courts often rule differently from one state to another. To ensure that as an investigator you are a seeker of truth and never deny anyone their constitutionally guaranteed rights, it is best to seek advice from the prosecuting attorney, who will assist you, in advance, to establish rules, guidelines, approved forms, and policies on how to handle legal situations.

- Standards and guides from entities such as the NFPA provide a framework of knowledge and skills that provide individual fire investigators with an opportunity to acquire the knowledge necessary to carry out their duties effectively and efficiently.

KEY TERMS

Administrative search warrant Warrant issued by a magistrate or judge that allows the investigator to be on the scene to determine the cause of the fire for the good of the people.

Authority having jurisdiction (AHJ) A term used by code- and standard-writing organizations to indicate the organization, office, or individual who is responsible for approving equipment, installations, procedures, and other resources and policies; an entity responsible for developing, implementing, and maintaining a qualification process.

Bill of Rights The first 10 amendments to the U.S. Constitution, which include such rights as freedom of speech, due process, and a speedy public trial.

Case law Laws that are established by judicial decisions in the courtroom, rather than being promulgated by legislation.

Consent Permission given by the person responsible for or controlling a property, which allows the investigator to search the property.

Constitutional law Laws that establish the relationship between the executive branch, the legislative branch, and the judiciary branch of government.

Crime An illegal act that is an action or omission as defined in legislative laws, allowing for prosecution by the state or federal courts.

Criminal search warrant A warrant issued by a magistrate or judge based on the sworn testimony (and written affidavit) of an investigator that probable cause exists that a crime has been committed and the person or place to search and the items being sought.

Exigent circumstances An emergency; for the good of the people, without permission, public safety personnel can enter property on fire to save lives and control the fire in such a situation.

Felony A crime that most likely involves violence and is more serious than misdemeanors. Punishment for such is usually more than a year in jail or even death.

Guide A document that is advisory or informative in nature and that contains only nonmandatory provisions. It may contain mandatory statements, such as when a guide can be used, but the document as a whole is not suitable for adoption into law.

International Association of Arson Investigators (IAAI) A private, nonprofit organization dedicated to the professional development of fire and explosion investigators through training and research of new technology.

Junk science Findings that are untested or unproven theories that are inappropriately presented as scientifically based.

Misdemeanor Under federal law, an offense that is less serious than a felony. Punishment is usually a year or less in jail and/or a fine.

National Fire Protection Association (NFPA) A private, nonprofit organization dedicated to reducing the occurrence of fires and other hazards.

NFPA 921, *Guide for Fire and Explosion Investigations* A document designed to assist individuals investigate fire and explosion incidents in a systematic and efficient manner.

NFPA 1033, *Standard for Professional Qualifications for Fire Investigator* Performance requirements and the qualifications to meet the requirements for the fire investigator.

Statutory laws Written laws usually enacted by a legislative body.

U.S. Supreme Court The highest court in the United States.

REVIEW QUESTIONS

1. In regard to creating laws, how does the Constitution ensure a balance of power?

2. The U.S. Constitution establishes the types of cases to be heard by the U.S. Supreme Court. List five types of cases that the Supreme Court can hear.

3. Are U.S. Supreme Court decisions required to be unanimous?

4. What evidence was allowed in the *Michigan v. Tyler* case as per the U.S. Supreme Court?

5. Why was the search deemed unconstitutional in the *Michigan v. Clifford* Supreme Court ruling?

6. What right does the private investigator working for the insurance company have to be on the scene?

7. Who can give consent to search a house?

8. What is an exigent circumstance?

9. What is the basis for a magistrate to issue an administrative search warrant that allows a fire investigator to conduct a preliminary fire origin and cause?

10. When an investigator swears out an application for a criminal search warrant before a magistrate, what is allowed to be searched?

DISCUSSION QUESTIONS

1. In the *Michigan v. Tyler* case, some evidence implied that the defendants were guilty, as did the verdict in the lower court decision. Whose fault was it that their conviction was overturned?

2. On television and in the papers, you see when criminals are set free because evidence was not legally obtained. Potential felons are back on the street. It has been suggested that instead of letting the felons go free, we should allow the evidence and also punish the law enforcement agencies. How would this affect our society? How would it affect those law enforcement agencies? How would it affect the actions of the individual officers?

REFERENCES

1. Supreme Court of the United States. *About the Court*. Retrieved from https://www.supremecourt.gov/about/about.aspx
2. Richards, E. (2019, April 4). Expert Witnesses—Kumho Tire: Applying Daubert to Experts. In *LSU Law Center Public Health Law Map—Beta 5.7*. Retrieved from https://biotech.law.lsu.edu/map/index.htm
3. Varone, J. C. (2007). *Legal considerations for fire and emergency services* (pp. 159–160). Clifton Park, NY: Delmar, Cengage Learning.
4. Varone, J. C. (2007). *Legal considerations for fire and emergency services* (pp. 149–153). Clifton Park, NY: Delmar, Cengage Learning.

CHAPTER 4

Science, Methodology, and Fire Behavior

LEARNING OBJECTIVES

Upon completion of this chapter, you should be able to:

- Describe the concept of scientific methodology.
- Describe the process of a systematic search.
- Describe the presumption of fire cause.
- Describe the aspects of fire behavior as they relate to the fire investigator.
- Describe and understand the concept of heating at the molecular level.
- Describe and understand the concept of heat transfer as it relates to the fire investigator.
- Describe the concept and use of the heat release rate.
- Describe and understand the concept of flame spread and complications associated with flashover in an investigation into the fire's area of origin and cause.

Case Study

Engine 33 was returning from a false alarm at the local grammar school when the tones went off, reporting a fire that just happened to be less than a mile away. As the engine neared the scene, black smoke could be seen just above the horizon. As they pulled in front of the house, the crew could see that the garage door was open. Flames stretched from floor to ceiling inside the garage, with more flames lapping out from the top edge of the garage door. The homeowner was in the yard; he told the crew that he was the only one home and had some burns on his hands and arms. One fire fighter started treating the homeowner for his burns while the rest of the crew pulled an attack line, accomplishing a quick knockdown of the fire.

Fortunately, an ambulance is an automatic assignment for any structure fire and arrived on the scene shortly after the engine company, taking over the treatment of the homeowner. The temperature was just above freezing, so these first responders took the patient into the ambulance to treat his burns. Soon thereafter, they left the scene with the patient, taking him to the local hospital for treatment of first- and second-degree burns on his hands. The fire officer was busy with the suppression, and did not get a chance to talk with the victim.

Because of the reported burn injury, the full-time fire investigator was dispatched. While on the scene, the engine company officer and fire fighters made several observations that they relayed to the investigator when she arrived. They noticed an odor similar to gasoline on the clothing of the homeowner while they were treating his injuries. Further, the burn patterns clearly indicated that the fire started and burned at floor level.

Although the crew conducted an overhaul, making holes in the wall and ceiling to ensure they had extinguished all extensions of the fire, they kept it to a minimum. In doing so, they took precautions not to disturb the fire scene any more than absolutely necessary.

The fire marshal, who was the full-time assigned fire investigator, arrived on the scene, talked to the engine crew, and then conducted her investigation by working from the least damaged area to the most damaged area. She photographed the scene, and took notes along the way. She confirmed what the fire fighters had noticed: The fire occurred at floor level across a large percentage of the garage. Burn patterns indicated that an ignitable liquid might be present. Although containers of ignitable liquids are often found in any garage, no patterns were evident at this site that would indicate a leak of any kind. Further examination showed no ignition source; the garage had neither heat nor appliances of any kind. Electricity was present, but it consisted of light switches, light fixtures, and outlets mounted at and above waist height. None of the wiring or devices showed any involvement in the ignition. The fire marshal collected evidence from the floor, placing it in 1-gal evidence cans for processing by the state forensic lab. The fire officer—the first investigator on the scene—reported the observations and facts that he and his crew gathered. The fire marshal instructed the fire officer to complete the fire incident report (NFIRS) but leave the section on fire origin and fire cause blank. She promised to update that section once she completed her investigation. Before the engine crew left the scene, the investigator thanked them for their great observations and careful efforts not to disturb the scene.

Back at the office, the investigator completed her report, including amending the engine company officer's fire incident report. In this particular jurisdiction, the fire department investigated the scene only and the local sheriff's department then took over the investigation if a fire was suspected to be incendiary in nature. The investigator shared her information with the sheriff's office, whose personnel then assumed responsibility for the case.

Sharing an investigation is not a bad thing, provided that the law enforcement detective is a trained fire investigator with full knowledge of fire operations, building construction, fire chemistry, and fire behavior. Regretfully, in this scenario, this was not the case. At the fire scene, a sheriff's detective caught up with the homeowner, who had been treated and released from the hospital and was standing in the yard when the detective arrived.

The two went into the garage, where the homeowner explained that he was using diesel fuel to clean some engine parts; they were soaking in a 5-gal container. He further explained that when he went to move the bucket, he bent over and the cigarette in his mouth fell into the bucket, causing an explosion that knocked him off his feet. In his haste, he accidentally tipped over the bucket of diesel fuel, spilling its contents across the entire floor. When asked where the bucket was now, the homeowner explained that he had grabbed the bucket, which was how he received his burns. He said the bucket was now in the back of his pickup, where he had placed it after the fire.

Sure enough, an empty metal bucket was seen in the bed of the truck, but the detective neither examined the container nor took it for evidence to be tested for the type of hydrocarbon. (Note: The detective should not have attempted to smell for the odor of gasoline or diesel fuel, as this is a violation of safety practices. However, had he taken the container to the laboratory, testing could have confirmed the presence of diesel fuel or, more likely, gasoline.) The detective thanked the homeowner for his time and wished him well in the healing of his wounds. He returned to his office, where he closed his investigation, noting that the fire as described by the homeowner was plausible and was accidental in nature.

What was the real issue? Diesel fuel is combustible, but it is difficult to ignite it with a cigarette, especially given the volume described. Under most circumstances, the cigarette would have dropped in the liquid and extinguished. At the very least, the detective should have looked further into the case rather than accepting an unlikely scenario.

The only saving grace was a thorough investigation by an insurance fire investigator, who uncovered ample evidence that allowed the insurance company to deny the claim based on the fire being intentionally set by the insured to collect insurance proceeds. The evidence used to support this decision were the fire report, the insurance fire investigator's report, and a private lab report showing the presence of gasoline in the sample provided by the insurance company fire investigator. Although the individual was never charged with arson, at least he did not profit from his actions.

The fire investigator must know the properties of the fuels involved in a fire situation. Being able to differentiate a plausible explanation from an implausible one can make or break the case.

 *Access Navigate for more resources.*

Introduction

Several types of investigations take place at any fire scene. The first investigator may be the fire officer or a fire fighter; he or she may call in a full-time investigator to take on the task of investigating more complicated fires. In addition, the insurance company may assign its own investigators (private fire investigators or staff investigators). In rare situations, the property owner may hire an investigator, though this step is more frequently taken by large, self-insured corporations. In this chapter, the "investigator" means all of these types of investigators, including the first responder investigator.

To fully appreciate and understand the science of fire as it pertains to fire investigations, investigators need a basic understanding of the behavior of fire. If they want to be more proficient or eventually move on to conduct detailed fire investigations, they will need additional training in topics such as chemistry and physics. They will learn about heat and molecular movement as well as heat transfer, and master basic concepts in electricity and mechanics. Most are surprised to hear that training in English composition is quite beneficial in writing fire reports. Investigators might conduct excellent investigations, but if they cannot put the results down in writing, their efforts will be of little value.

However, continuing education is just the beginning: Books, research papers, and professional articles can add to the investigator's knowledge. Continual studies of such documents are a necessity for every individual conducting fire investigations. It does not matter if the person is young or old, new or experienced. In this education endeavor, investigators need to ensure that they are studying good science—that is, information that can be used in the fire scene evaluation process. Good science is based on proven and reproducible scientific principles, whereas junk science is unproven or founded on speculation, conjecture, and outdated concepts and principles.

> **TIP**
>
> Good science is based on proven and reproducible scientific principles, whereas junk science is unproven or founded on speculation, conjecture, and outdated concepts and principles.

To ensure a more accurate determination of the area of origin and the cause of the fire, investigators must use the process called the **scientific method**, which provides a systematic framework to assist the investigator. The scientific method is a technique used to acquire new knowledge by gathering evidence and evaluating that evidence using reasoning.

Scientific Methodology

Walking onto a fire scene where massive destruction has occurred can be a daunting prospect. Even a single-family dwelling fire can give the investigator pause, just

> **TIP**
>
> To conduct detailed fire investigations effectively, the investigator will need additional training in topics such as chemistry, physics, and English composition.

as anyone would stand in the front yard and ponder what happened.

One way to approach the investigatory process is to use a systematic approach to examine the fire scene—namely, the scientific method. This method was not invented by the fire service, but takes fundamental principles from the scientific community and adapts them to the process of conducting fire investigations. The steps in the process can vary, but NFPA 921, *Guide for Fire and Explosion Investigations*, outlines the scientific method as including the following steps: (1) recognize the need, (2) define the problem, (3) collect data, (4) analyze the data, (5) develop a hypothesis, (6) test the hypothesis, and (6) select a final hypothesis (**FIGURE 4-1**).[1]

Recognize the Need

The first step is to *recognize the need*—in other words, identify the problem. The problem is that a fire or explosion has occurred, creating the need for

an investigation. More definitively, one can state that there is a need to investigate the fire's area of origin and the cause of the fire. Armed with these findings, the investigator may be able to identify the need for an education program to prevent future incidents. The total body of data related to a fire's origin and cause provides the basis for statistical analysis aimed at preventing future events.

> **TIP**
>
> The total body of data related to a fire's origin and cause provides the basis for statistical analysis aimed at preventing future events.

For a municipality, the need can be established well in advance. The very reason a fire investigator position may exist is that the jurisdiction recognizes that fires will occur and need to be investigated. Recognizing the need can take on a legal tone as well: For example, did the fire occur within the investigator's jurisdiction or area of responsibility? Even if a suppression mutual aid agreement exists with a neighboring jurisdiction and the fire apparatus respond within that other jurisdiction, this may not give the fire investigator the legal right to cross that jurisdictional line and investigate the fire in the neighboring jurisdiction.

Define the Problem

The next step is to *define the problem*. In the first step, the investigator identified the existence of a problem; in this step, the investigator understands the problem and must determine how to come up with a solution. The answer seems quite simple: conduct a detailed investigation into the fire's origin and cause. Most likely, the locality has already created a policy that will dictate how fires will be investigated. Most fire scenes receive an initial investigation by the fire officer in charge or a person assigned by the fire officer. This investigation is usually limited to a brief scene examination, along with possible brief interviews of witnesses or the property owners or tenants. Depending on the fire officer's findings, a duty investigator may be called to the scene to perform a more detailed investigation, perhaps because the fire officer cannot find the origin or cause, or the circumstances discovered are unusual or suspicious. In some instances, such as a multiple-alarm fire or unusual circumstances, the fire investigator may go to the scene immediately and conduct the investigation in lieu of the fire officer.

Collect the Data

Next, the investigator must *collect the data*. Every aspect of the incident must be recorded in one form or

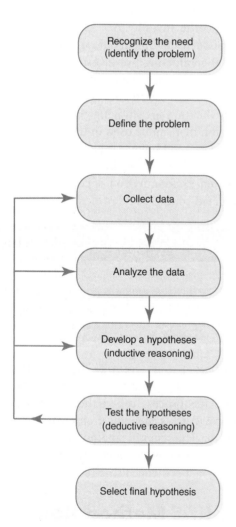

FIGURE 4-1 The scientific method shows the steps to follow and to retrace when a proposed hypothesis cannot withstand the test.

another—from photographing the fire scene to recording interviews, the investigator must collect all evidence about the scene. The data to be collected include everything and anything that can, or may, be used in determining the fire origin and the fire cause. What is collected is considered empirical data, which are individual pieces of information based on observations, experiments, or experience.

> **TIP**
>
> The data to be collected include everything and anything that can, or may, be used in determining the fire origin and the fire cause.

Analyze the Data

After all the available data are collected, the investigator must *analyze the data* using logic and reason. The only data that can be analyzed are those obtained from observation or experiments. In doing this analysis, the investigator uses his or her training, education, knowledge, and experiences. An investigator's level of expertise is based on both knowledge and experiences. Knowledge is a culmination of the individual's formal education, such as obtaining a degree, along with additional education from seminars, conferences, and other specialty schools. Just as important are the investigator's experiences, those both directly and indirectly connected to the investigation of fires. Someone with a strong background as an electrician may have a better understanding of an electrical fire's cause. Likewise, someone with experience in military ordinance may understand explosive devices better than other investigators. Regardless, *we are the sum of our experiences.* All experiences add to an investigator's expertise.

> **TIP**
>
> *We are the sum of our experiences.* All experiences add to an investigator's expertise.

Develop a Hypothesis

Based on the data collected and analyzed, the investigator can *develop a hypothesis* as to the fire's origin and cause. A hypothesis is an assumption made based on the empirical data collected by the investigator. This hypothesis is not the final determination until all other steps in the scientific method are completed.

Test the Hypothesis

Before establishing a final determination of the fire's origin and cause, the investigator must *test the hypothesis.*

This is done using deductive reasoning—that is, taking the facts of the case and thoroughly and meticulously challenging all of those facts. Unless the hypothesis can stand up to all reasonable challenges, it is an unacceptable theory and must be discarded. When this happens, the investigator must reexamine and reanalyze the data to develop a new hypothesis; the investigator must subject this new hypothesis to a thorough and detailed test to see whether it can withstand deductive reasoning.

> **TIP**
>
> Unless the hypothesis can stand up to all reasonable challenges, it is an unacceptable theory and must be discarded.

If no hypothesis withstands the tests, the investigator must collect new data to enable a proper determination of the fire's origin and cause. New data may be gathered by conducting additional interviews, following new leads from those interviews, or conducting additional scientific testing. This process continues until a hypothesis can withstand all tests, showing itself to be the final determination as to the fire's origin and cause. The testing of the hypothesis is dynamic, and it must be retested whenever new data are presented.

Of course, not all fires have a proven hypothesis. In fact, the cause and origin of many fires may remain undetermined. This ambiguity is hard to accept sometimes, especially when a fire leads to injuries or a loss of life. Nevertheless, the investigator cannot solve every fire puzzle. Sometimes not enough data are available to support a plausible explanation as to what happened. Sometimes the jurisdiction has inadequate resources or time to carry out an in-depth investigation. Regardless, the investigator must resist the temptation to attach a cause based on limited data. In these situations, when an investigator has no success in identifying the fire's origin or cause, the fire cause and origin must be labeled "undetermined."

> **TIP**
>
> When an investigator has no success in identifying the fire's origin or cause, the fire cause and origin must be labeled "undetermined."

Presumption of the Fire's Cause

Just as dangerous as labeling a fire based on limited knowledge after an extensive investigation is presuming the fire's cause before even starting a fire investigation. All too often individuals make presumptions about the fire's cause based on false impressions, frequency of certain types of fires, or a piece of loose information heard when arriving on the scene.

The fundamental problem with making a presumption is that an individual may look only for evidence that supports the presumption, rather than keeping an open mind and collecting all data. For this reason, many fire investigators want to look at the fire scene before they conduct any interviews. This approach ensures that the investigator will not be influenced by something someone said in an interview, such as the homeowner indicating she had ongoing problems with the furnace. Even though the investigator is more than capable of keeping an open mind following interviews, others, such as a jury, may believe the fire investigator could have been influenced during the interviews.

As an example, consider a homeowner who indicates that the fire had to be arson because of an argument he had with the neighbor only hours before the fire began. Physical evidence does indicate the neighbor was involved, an arrest is made, and a trial date is set. During the investigator's testimony in the trial, the defense counsel asks whether the investigator had any indication that the neighbor was involved before the fire scene investigation was conducted. Counsel could paint a picture in which the investigator looked for incendiary indicators based only on the information gathered in the interview, and not on the actual scene. Although this argument might sound like a desperate maneuver for the defense, it can put the investigator in a bad light. But imagine the investigator was able to say, "No, Counselor, I looked at the fire scene and found evidence of an incendiary fire before conducting my interview with the homeowner." That testimony would be much more convincing.

The bottom line is that investigators should keep themselves beyond reproach as much as possible—and one way to do this is to perform the fire scene examination first and then conduct the interviews. If something emerges from an interview that needs to be verified, the investigator can always go back to the scene.

Systematic Search

Doing a thorough search of the fire scene is essential to the successful conclusion of an investigation. Along with using the systematic approach outlined by the scientific method, the investigator should do a systematic search of the fire scene, working from the least damaged area to the most damaged area. This method may allow the investigator to follow the path of the fire back to its origin, and it can work extremely well for most fire scenes.

> **TIP**
>
> The investigator should do a systematic search of the fire scene, working from the least damaged area to the most damaged area.

However, this is not the only method for conducting a search. Working from the highest point of the structure to the lowest can work as well, as long as the investigator uses that strategy consistently in each and every investigation. This process is critical to aid in a thorough and complete investigation. A secondary benefit relates to the investigator's courtroom testimony. If investigators systematically conduct each and every fire search in the same way, day after day, month after month, or even year after year, they will have no problem answering in the affirmative if and when challenged in court about their procedures for conducting an investigation.

Of course, the investigator must have somewhere to start to conduct a systematic approach. With a fire in a typical single-family residence, the starting point could be on the street in front of the house. Starting from this point, the investigator can work in a clockwise direction, observing the house and property from all sides, not only examining burn patterns but also looking for evidence. Working toward the structure, the investigator will look at the damage, observing the least damaged area and eventually working into the most damaged area. This process can enable the investigator to follow the path of fire travel right back to where the fire started.

One note of caution regarding recommended procedures: No single procedure or process will fit each and every situation. For this reason, investigators should always be prepared with alternative, viable plans. When conducting a systematic search, they may have a logical reason to work from the bottom up or the top down in a particular fire. Investigators need only be ready to provide a logical explanation as to why they varied from their usual procedure should it be challenged in court.

Chemistry of Fire for the Fire Investigator

When considering the chemistry of fire, an assigned investigator must be able to bring everything down to the molecular level. Sharing electrons or the excitement of a molecule that causes rapid movement can explain how a fire starts, reacts, grows, and is extinguished. Any investigator without this knowledge is limited in understanding the entire realm of fire science and will be handicapped in defending a proper hypothesis.

When going to court, the investigator will have to explain fire in the simplest possible terms to a judge and jury when testifying how a small fire can eventually burn down a large structure. Explaining how fire behaves at the molecular level can help others understand fire and its behavior. Likewise, a fire fighter or fire officer may be able to use this information to explain how fires start to public groups visiting the fire station or to anyone in the general public who wonders how a fire can start on their property.

> **TIP**
>
> To truly understand fire, the investigator must be able to bring everything down to the molecular level.

This text assumes that readers have a fluent knowledge of basic fire behavior and fire chemistry. A good refresher in this area is available in the Fire Behavior chapter of *Fundamentals of Fire Fighter Skills* (Jones & Bartlett Learning). For example, one basic principle that fire fighters learn is that to extinguish a fire, they must remove one of the sides of the fire triangle or **fire tetrahedron** (**FIGURE 4-2**): Remove the oxygen, and the fire smothers; remove the fuel, and there is nothing to burn; remove the heat or the uninhibited chain reaction, and the fire goes out.

Fire investigators must also consider the importance of heat, fuel, and oxygen. Specifically, they are interested in learning what action or event brought these items together to start the fire. Generally speaking, oxygen is always available, so that leaves heat and fuel as question marks. To determine how a fire started, investigators must know how the first item ignited (fuel) came in contact with a capable ignition source (heat). If they cannot explain this relationship, they do not have a true hypothesis on how the fire started.

Fire Tetrahedron

The fire service also uses a model called the fire tetrahedron, which is a four-sided solid object with four triangular faces, a pyramid. The first three sides are the same as those in the fire triangle: heat, fuel, and oxygen. The fourth side is the self-sustained chain reaction; this side explains why the fire continues to grow rather than staying contained in one place (Figure 4-2).

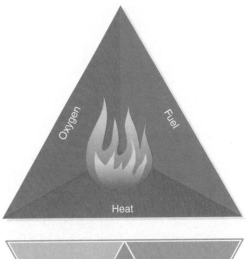
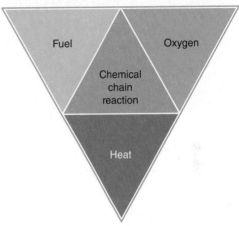

FIGURE 4-2 The fire triangle and the fire tetrahedron are the basic models that explain how to extinguish a fire as well as how a fire started.

Self-Sustained Chain Reaction

Combustion is a complex process that results in the rapid oxidation of the fuel, producing heat and light. When this fuel continues to burn on its own in a self-sustained manner through an **exothermic reaction**,[2] the fire will grow. An exothermic reaction is a chemical reaction created when the bonds of molecules are broken, which releases heat and light. Another term for this chain reaction is "uninhibited chain reaction," and it represents one side of the fire tetrahedron.

> **TIP**
>
> An exothermic reaction is a chemical reaction created when the bonds of molecules are broken, which releases heat and light.

For example, an exothermic reaction occurs when you hold a piece of paper horizontally and expose the underside to the heat from a match (**FIGURE 4-3**). Even without the flame of the match touching the paper,

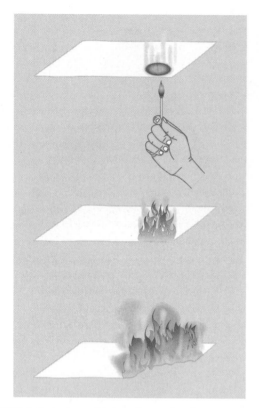

FIGURE 4-3 A piece of paper shows the chain reaction of fire.

the heat is sufficient to start the ignition process. This heat begins to break down the paper, the paper starts to turn brown, and the decomposing paper gives off vapors. These vapors are combustible, and when they are heated to their **autoignition temperature** or come in contact with the flame, they ignite. Now the paper is burning. The leading edge of the flames heats up more of the paper's surface area, releasing more vapors that in turn ignite, contributing to the growth of the fire. This self-sustained chain reaction continues across the surface of the paper until the fire is extinguished. If disturbed by no other outside factors, the fire will eventually go out as a result of a lack of fuel because the paper is completely consumed, which leaves behind carbon and other compounds in the form of ash. If the paper is placed next to other combustibles, the fire could continue to grow as the nearby materials are exposed to the heat and those items too begin to give off vapors and eventually ignite, creating a larger fire.[3]

Oxygen

Fires cannot occur without an **oxidizer**. In most situations, the atmosphere, which contains approximately 21 percent oxygen, is the primary oxidizer that completes the combustion process. However, fires can occur in the absence of atmospheric oxygen. Many compounds, when mixed with other chemicals or when heated, can give off oxygen in sufficient amounts to allow combustion to occur or to continue. One such chemical, ammonium nitrate, was the fuel for one of the United States' largest accidental disasters—in Texas City, Texas—and also has been used to make explosive devices on multiple occasions.

> **TIP**
>
> Fires cannot occur without an oxidizer.

In the Texas City incident, the chemical ammonium nitrate fertilizer (NH_4NO_3) was stored in the hold of the *S.S. Grandcamp*, a French liberty ship that was tied up at the Monsanto plant on April 16, 1947. Smoke was seen coming from the hold at approximately 0800 hours. The ship already held 2300 tons of ammonium nitrate fertilizer in 100-lb bags. The master of the ship decided to have the hatches battened down and steam pumped into the hold to extinguish the fire.

Around 0830 hours, the hatches blew off from pressure building in the hold. A thick column of orange smoke was seen coming from the hold. At 0900 hours, flames were coming from the hold. Then, at 0912 hours, the ship disintegrated in an unbelievable explosion that was heard 150 miles away. One of the ship's anchors, weighing approximately 3200 lb, was thrown 1.62 mi from the ship and buried itself about 10 ft into the ground where it landed. The end result of that explosion and the other explosions and fires that resulted from the initial blast was 568 dead or missing persons and more than 3500 injured (**FIGURE 4-4** and **FIGURE 4-5**). The key issue in this incident is that ammonium nitrate, when broken down, acts as its own oxidizer.[4]

> **TIP**
>
> The explosive reaction from a mixture containing ammonium nitrate can be quite violent. One of the S. S Grandcamp anchors, weighing approximately 3200 lb, was thrown 1.62 mi from the ship and buried itself about 10 ft into the ground where it landed.

Many years later, the same chemical, ammonium nitrate fertilizer, was used in a different manner but also with a devastating effect: It was the key component of a bomb that was loaded into the back of a rental truck, with the truck then being parked just outside the Murrah Federal Building in Oklahoma City, Oklahoma. When detonated, the 2.5 tons of fertilizer mixed with fuel oil tore off the face of the nine-story building, resulting in 168 deaths and more than 800

FIGURE 4-4 The remains of one of the three engines from the Texas City Volunteer Fire Department responding to the initial alarm.
Photo courtesy of Moore Memorial Public Library, Texas City, Texas.

FIGURE 4-5 Smoke from the fires created from the explosion of the S.S. Grandcamp, as seen from Galveston about 8 mi away.
Photo courtesy of Moore Memorial Public Library, Texas City, Texas.

FIGURE 4-6 At 0902 hours on April 19, 1995, an explosion rocked downtown Oklahoma City, destroying the Alfred P. Murrah Federal Building and damaging more than 300 other buildings in the vicinity. Fatalities occurred in 14 separate buildings, taking 168 lives and injuring more than 800 people.
© AP/Shutterstock.

injuries (**FIGURE 4-6**). Nineteen of the fatalities were children younger than the age of six.

Many chemicals are capable of being oxidizers; they include nitrates, nitrites, peroxides, and chlorates. The key component of their makeup is oxygen, which under certain circumstances can be released, enabling combustion. Investigators should be aware that an atmosphere of chlorine could also support combustion.

> **TIP**
>
> Many chemicals are capable of being oxidizers; they include nitrates, nitrites, peroxides, and chlorates.

> **TIP**
>
> Investigators should be aware that an atmosphere of chlorine could support combustion.

Another factor that the investigator must be aware of is the effect of the amount of oxygen present. Chemically speaking, the oxidation process occurs when oxygen bonds to other elements at the molecular level. This process can be slow or fast—slow in the form of rusting or fast during combustion. The rate of combustion reflects the amount of oxygen available. Not all compounds containing oxygen in their molecular makeup are capable of providing enough oxygen for combustion. Typical combustion (if there is such a thing) can occur in the ambient atmosphere, which usually contains 21 percent oxygen. However, in settings with a higher concentration of oxygen, the fire can be more intense. This is evident from a blacksmith's forge, where blowing air across the coals creates more intense heat. Today, high concentrations of oxygen can be mixed with combustible gases to produce an intense flame that is capable of cutting through iron.

If an investigation reveals reports of white, intense heat, far beyond ordinary combustion, the investigator should consider the proximity of an oxidizer as a force creating such a situation.

Fuels

Simply stated, a fuel is anything that can burn. Fuels can be in any physical state: solids, liquids, or gases

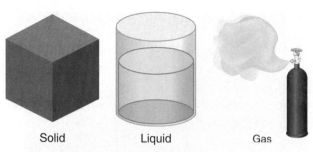

FIGURE 4-7 Status of matter.

(**FIGURE 4-7**). The physical state of the fuel affects the combustion process. Most fuels are required to be in the gaseous state before combustion can occur, but this is not always the case.

Solids. Solids have a physical size and shape. To ignite, they must be heated to the point of decomposing, such that they produce vapors. Sometimes this transition can be easy to see, such as in a fireplace where a log is burning: A space can be seen between the surface of the log and the visible flame. This space is the layer where the solid log is being vaporized and the vapors above the log are burning. The same phenomenon can be seen even more clearly on a burning candle. Although the wick is burning, a solid—wax, in this case—is also being heated and turned into a liquid. Capillary action allows the liquid to be drawn up the wick, where the liquid is vaporized. Once these vapors reach the correct proportion of vapor/air mixture, combustion occurs.

The only way to truly understand combustion is to break the process down into its molecular components. In solids, the molecules are packed closely together by bonds that help give the material its shape and form. Some molecules form tighter bonds than others do. This difference is evident in how easy it is to break or pull the material apart.

> **TIP**
>
> The only way to truly understand combustion is to break the process down into its molecular components.

All atoms and molecules are in motion. When heated, this motion increases. As the rate of movement increases, so does the number of times that the molecules collide. The more heat present, the faster the molecular movement and the greater the impact of these collisions. Eventually, these collisions break some of the molecular bonds. As these bonds are broken apart, energy is released in the form of heat and light.

As the heat continues to rise and the collisions become more intense and more frequent, resulting in

FIGURE 4-8 Attempts to ignite a log with a match fail because the mass of the log is able to absorb the heat from the match; the heat is distributed but does not raise the log to its ignition temperature.

more bonds being broken, the heat from this process intensifies as well. As molecules break apart, some of their components combine with oxygen; this is the oxidation process. As a result of the oxygen bonding with these other molecules, heat is given off—an exothermic reaction. In time, this process becomes self-sustaining in the form known as combustion.

The heat applied to the solid matter must be sufficient to overcome the amount of heat that solid object is capable of absorbing. As an example, consider an 8-in.-diameter log placed in a fireplace by itself. You cannot simply place a match to the bottom of the log and expect it to ignite (**FIGURE 4-8**). In such a case, the heat from the flame of the match is absorbed by the mass of the log. Localized charring may occur, but the log will not ignite.

In contrast, if you place kindling in the form of wood shavings or balled-up paper under the log, a fire should readily ignite. When the flame of a match is applied, the kindling cannot absorb the energy because of its small mass. Thus, the heat overcomes this fuel, allowing for breaking of bonds, release of molecules that combine with oxygen, and eventual self-sustained ignition.

The surface-to-mass ratio explains what happens in the fireplace. As noted earlier, a large mass with little surface area, such as the log, readily absorbs the heat from the match, transferring it into its mass. But the paper has more surface than mass, and cannot absorb the energy of the match. Its molecules are rapidly heated to the point where their bonds are broken down; they bond with oxygen molecules in the oxidation process, and give off heat and light—that is, fire.

Smoldering fires involve a different scenario. Smoldering can be seen, for example, when upholstered furniture burns, as air can readily permeate the material and allow the material to char without a visible flame. The same can happen in sawdust piles and some combustible insulation products. Smoldering is even designed to happen in some products such as cigarettes and other tobacco products. Smoldering is also a phase of the combustion process, which is discussed later in this chapter.

It is important to explain what did or did not happen in the fire being investigated in terms of the fire tetrahedron. Was the heat source sufficient to ignite a particular item (match on a log)? Was the mixture of vapors with oxygen sufficient to create an ignitable vapor? Was the heat source sufficient to ignite the vapor? The investigator may very well have to account for all parts of this process in the development of a hypothesis and may eventually have to explain this integration of the various components to a judge and jury.

> **TIP**
>
> It is important to explain what did or did not happen in the fire being investigated.

Liquids. A liquid's ability to ignite and burn depends on its form. When diesel fuel in a large container is exposed to heat, to some extent that heat can be absorbed, preventing ignition. However, if you take that liquid and spray it into the air or onto a piece of paper (increasing its surface area), ignition would require perhaps only the energy from a match.

Liquids have two other important characteristics. The *density* of a liquid fuel dictates whether it will sit on top of water or sink and rest below the water's surface. Fuels such as gasoline and kerosene are lighter than water and will rest on the water's surface, allowing the fuel to continue its vaporization process and to burn if ignited. The *solubility* of a liquid fuel dictates whether it readily mixes with water. Polar solvents such as alcohol easily mix with water, which could be an important factor when suppression forces apply their water streams in and around the area of origin. Alcohols are soluble in water. If an alcohol was used as the first fuel ignited, the residue and unburned fuel may very well mix with the suppression water, essentially being washed away and leaving little trace of the fuel's presence.

Liquids will not ignite in the liquid state; instead, the fuel must be vaporized, with those vapors then burning. As with solids, investigators should consider the fuel in its molecular state. In basic terms, when heat is applied to the liquid, the molecules become agitated and move faster and faster. Atmospheric pressure and surface tension keep many of these molecules from flying from the liquid. However, as collisions between the molecules occur and the temperature rises, some molecules will have sufficient force to break free of the surface tension and enter the atmosphere, where they will be suspended in the air as vapor. This process—the collision of the molecules and the breaking of bonds—creates heat in the form of an exothermic reaction, the same as in solids.

> **TIP**
>
> Liquids will not ignite in the liquid state; instead, the fuel must be vaporized, with those vapors then burning. As with solids, investigators should consider the fuel in its molecular state.

These processes continue until the vapors given off are sufficient to support a flaming fire across the surface of the liquid, but not to allow combustion to continue. This is considered the **flash point** of the liquid.[2] As the temperature of the liquid rises, usually at just a few more degrees above the flash point the liquid generates sufficient vapors to allow the flame to continue to burn. This is considered the liquid's **fire point**.[2] In both circumstances, an external source of ignition is present. If the liquid is allowed to continue to heat without an external source of ignition and the liquid heats to the point that it ignites on its own (reaching its ignition temperature), it has reached its autoignition temperature.[2]

As with the vapors from a solid, the vapors from liquids must be present in the right proportions with oxygen to allow ignition. Likewise, the ignition source must be sufficient to ignite the mixture. If the ignition of the vapors is the cause of the fire or contributes to the extension of the fire, then the investigator must explain this relationship in detail in the investigation report.

Gases. Relatively few gases will ignite at room temperature. However, with increased temperature, other solids and liquids begin to break down and produce vapors. Vapors are indeed in a gaseous state, but a true gas is found in a gaseous state at normal temperatures and pressure. Vapors and gases can mix with air and ignite under the proper conditions. However, not all gases burn; in fact, some—such as carbon

> **TIP**
>
> Incidents involving gases require considerable scrutiny to ensure that all the facts match the hypothesis.

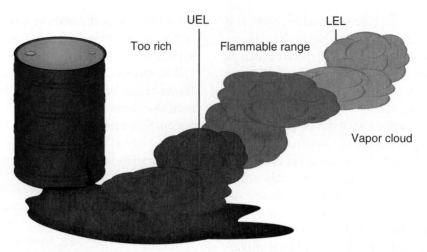

UEL   LEL

Too rich   Flammable range

Vapor cloud

FIGURE 4-9 Flammable/explosive range. The vapors closest to the spilled liquid are too rich to ignite. As the vapors mix with air, they may achieve the correct proportion of vapors to air to enable ignition to occur; this is the flammable range, also known as the explosive range. As the vapors spread farther from the spill, they dissipate (become too lean), becoming sufficiently diluted that the vapor-to-air ratio does not allow ignition. The threshold where the vapor concentration can be ignited but just before the vapor is too rich is known as the upper explosive limit (UEL). The threshold just before the fuel becomes too lean but can still ignite is known as the lower explosive limit (LEL).

dioxide—are used to extinguish fires. Incidents involving gases require considerable scrutiny to ensure that all the facts match the hypothesis. Many variables must be taken into account along with a lot more science, such as **vapor density**.

Vapor density is the weight of a gas when compared to air. This measurement is made at sea level, where air is considered to have a force of 14.7 pounds per square inch (psi) and is given a designation of 1. If a gas is heavier than air, it has a density greater than 1; if a gas is lighter than air, it has a density less than 1.

As investigators examine the scene, they may discover available fuel sources along with burn patterns or explosion patterns. In the process of developing the fire hypothesis, they must determine whether the fuel gas was capable of producing the damage seen at the scene. If they find evidence of an explosion at the upper levels of the structure, that may be an indication of a lighter-than-air gas, such as natural gas (vapor density = 0.55) from a kitchen range. If they see low-level explosive damage, that may indicate the presence of a heavier-than-air gas such as propane (vapor density = 1.6).[2] Investigators must exercise caution, however, because the type and method of building

construction may result in damage contrary to the expected properties of the gases or vapors. For more information on building construction, see Chapter 5, *Building Construction for the Fire Investigator*.

As shown in **FIGURE 4-9**, both gases and vapors can ignite only when mixed with an appropriate amount of oxygen. Each gas has an explosive (flammable) range. If the concentration of the gas or vapor is below that range, it is considered too *lean*. If the concentration is above the flammable range, the mixture is considered too *rich*. In either case, there is little likelihood of an explosive ignition. The flammable ranges for some gases are narrow. For example, propane has a **lower explosive limit (LEL)** of 2.15 percent and an **upper explosive limit (UEL)** of 9.6 percent. Other gases can ignite in almost any atmosphere. For example, acetylene has a LEL of 2.5 percent and an UEL of 81 percent. These figures can change dramatically as the temperature increases.[2]

Heat and Temperature

As everyone should know, molecules are always in motion. This motion, and the collision of molecules, is always producing energy in the form of heat. When an object is touched with the hand, it may feel cold to the touch, but that is only relative to the fact that the object has a lower surface temperature than the hands. However, that object still contains heat. When a release of energy—an exothermic reaction—occurs, it could simply make the object warmer. This could eventually lead to **thermal runaway**, a phenomenon in which an

TIP

As investigators examine the scene, they may discover available fuel sources along with burn patterns or explosion patterns.

increase in the temperature leads to an even greater increase in temperature. This greater increase in temperature then reacts again to increase the temperature even more. Along the way, energy is created that eventually results in the release of heat and light (fire).

Heat is kinetic energy, whereas temperature is the measurement of that energy that can be obtained by using an instrument such as a thermometer. Heat and temperature are related but are not the same: We measure the amount of heat using a device that gives us the temperature of the object.

TIP

Heat is kinetic energy, whereas temperature is the measurement of that energy.

Ignition Temperature and Ignition Energy

The ignition temperature is the minimum temperature that a substance must attain before ignition can occur. Many variables must be taken into account for a wide range of circumstances. With the exception of a few solid materials that can burn at the surface, such as coal or magnesium, all other solid fuels must be heated sufficiently to release vapors; these vapors then ignite and burn. Liquids are the same: Mostly, they need to be heated to release vapors, which in turn can be ignited at the proper temperature for that vapor. Some liquids produce sufficient vapors at their ignition temperature, which is less than the ambient air temperature. These fuels, such as gasoline, are capable of ignition if present in the right mixture with air and heat.

The ignition temperatures for materials listed in texts and publications have been ascertained through laboratory testing of the product. In the real world, however, the actual temperature may be higher than that experienced in the lab. Thus, it is imperative that an investigator research all aspects of any fuel that will be listed in the investigation report as the first material ignited. Unusual variables can affect the ignition temperature, making it higher or, in rare circumstances, lower. The investigator must also take into account

the size, shape, and form of the first material ignited, and all must be explained in the investigator's final hypothesis.

Just as important as the ignition temperature is the amount of energy necessary to make the ignition occur. For ignition to occur, only a tiny discharge is necessary. Take the example of someone walking across a carpeted floor. Under the right circumstances, that individual may pick up a sufficient static charge so that when he touches a grounded object, such as a door knob or another person, he experiences a discharge of energy. The individual being touched will feel the arc. This phenomenon indicates that the discharge has a high temperature; however, in this minute spark, no harm occurred.

If that same discharge occurs in an atmosphere of ignitable vapors, such as when fuel is handled at a gas service station, it can cause an explosive ignition. The fuel vapors may be within the explosive range (LEL and UEL) at the source of the energy, and enough energy may be present for a sufficient duration of time to ignite that vapor in and around the discharge area. The resulting fire or explosion could be devastating—yet it came from just the energy of a static discharge. The amount of energy necessary in such a situation is only 25 mJ.[2]

Sources of Heat

Heat is a factor in providing sufficient energy to release vapors from solids and liquids. It is also a factor in providing the energy needed to ignite a fire. Likewise, heat promotes flame spread and subsequent fire growth. Four sources of heat are distinguished: mechanical, chemical, electrical, and nuclear. Some are experienced every day and taken for granted, whereas others are more unusual and only infrequently the source of ignition.

An investigator must be familiar with all potential heat sources. Before identifying the actual heat source involved in a fire, all other heat sources must be eliminated. The investigator with a low level of knowledge about any of these heat sources is limited in the ability to form a viable hypothesis about the source of ignition.

TIP

An investigator must be familiar with all potential heat sources.

A multitude of resources and opportunities to learn more about sources of heat exist. College classes in chemistry and physics can help one understand

---

TIP

The ignition temperatures for materials listed in texts and publications have been ascertained through laboratory testing of the product. In the real world, the actual temperature may be higher than that experienced in the lab.

chemical and mechanical heat sources. Investigators can find that developing a professional relationship with college professors who specialize in these subjects can be invaluable. If an investigator needs to learn more about electricity, college physics classes can provide basic knowledge of electricity, electron movement, electrical potentials, and so on. An investigator can also learn about practical applications of electricity from a local electrical contractor. Spending a little time on a construction site can be very rewarding and can help the investigator understand various components and proper installation. Many seminars at the national level as well as local training sessions from various chapters of the International Association of Arson.

Mechanical

Mechanical heat is the heat that results from friction. Two or more objects rubbing together create friction. Friction, by itself, does not automatically mean a fire will occur. Factors such as the roughness of the surface, the types of materials rubbing together, the dryness of the materials, and the speed at which they come in contact are all variables that the investigator must take into account. For example, two smooth pieces of plastic coated with a silicone lubricant slowly rubbed together with little pressure will produce heat, but are unlikely to produce enough energy to start a fire.

Truck brakes are another example of two materials coming in contact that create mechanical heat. The surfaces of these materials and the fact that they are designed to slow down the rate of movement, combined with pressure of the brake cylinder, ensure that they will perform their intended function on a regular basis. However, with a mechanical failure of misalignment or if the cylinder creates pressure (gets stuck) when not activated during a high rate of speed, pushing the brake pads together, sufficient energy may be created to ignite nearby combustibles in the form of grease, plastic, and rubber components, or even debris caught in the workings. Brake fires by themselves may cause only local damage. However, in some circumstances the fire may extend into the vehicle, causing massive destruction of the vehicle and its contents.

Another form of mechanical heat is the heat of compression. This process is used in diesel engines, where the fuel is compressed to the point where ignition can be accomplished without an outside source such as a spark plug. Fire fighters can experience the phenomenon of compression heat when they fill self-contained breathing apparatus (SCBA) air bottles. The bottles heat up as they are being filled—a form of mechanical heat.

Chemical

A mixture of two or more chemicals can create heat (exothermic reaction) and, in some circumstances, sufficient heat to cause ignition. Although not an everyday circumstance, chemical heat can be found in industrial applications, and even some household chemicals are capable of creating such a reaction. Chemical heat can also be an arsonist's tool or even be used as a time delay. Regardless of the circumstances, investigators need to collect evidence in the form of debris for analysis along with possible examples of chemicals for further analysis.

Spontaneous Heating

One form of heating can occur without an external source of energy: Spontaneous heating can occur from biological actions as well as chemical reactions. Biological items such as damp hay, manure, or sawdust give off heat as they break down and decay. If these products are well insulated so that the heat is prevented from dissipating, the heat can increase until it reaches the ignition temperature of the material. When this happens in the presence of sufficient oxygen, spontaneous ignition can occur, such that the fire starts with no external source of heat or energy.[2] This phenomenon has been observed in objects such as hay, coal, charcoal, and linseed oil.

Hay. Freshly cut hay is usually allowed to set in the field before being baled, a delay that allows it to start the drying process. Once it is sufficiently dry, with less than 20 percent moisture content, the farmer will bale or roll the product and store it in barns and other areas safe from wet weather. The farmer places the bales and rolls tightly against each other and stacks them high to allow more storage in each structure.

If the hay contains too much moisture when baled or rolled, more than 25 percent, then a danger for spontaneous combustion exists. This wet hay creates an environment for microbiological growth. This activity will produce heat. In turn, this heat promotes more microbiological growth, which in turn creates even more heat. As the heat climbs to 150 degrees F, heat resistant bacteria (exothermic bacteria) will grow and start the process of chemical reactions which take over. If this heat is not allowed to dissipate, it can continue this chain reaction, creating thermal runaway. During this process there will be limited availability of oxygen within the area. Without the ability to sufficiently dissipate such heat and in combination with other contributing factors, the temperature within the bales may reach the hay's ignition temperature. As the interior burning reaches the outer edges of the stacked hay, available oxygen allows the fire to go into flaming combustion.

There are some outstanding research articles from universities and extension services that can provide a wealth of additional information on this process.

Coal. The average investigator may believe that coal mines are far from his or her jurisdiction. In reality, coal can be found in many industrial complexes at shipping ports along the coast. Even so, the investigation of a fire in a coal storage area is not a task that most investigators will undertake. For this reason, the investigator may want to seek an appropriate resource that can be available when needed. In addition, extensive studies and research on coal fires have led to multiple research articles from government entities and universities; these can be readily obtained through an Internet search.

Coal comes in specific grades, not all of which will spontaneously ignite. The lower grades of coal are more susceptible to spontaneous ignition. The storage of coal is a factor in fires: The larger the mass of the coal pile, the more potential there is for spontaneous ignition. Coal pile heights greater than 26 ft can be a contributing factor to the potential of spontaneous combustion. However, piles of low-grade coal only 10 ft high can spontaneously ignite. Other factors that contribute to this phenomenon include oxygen concentration within the coal, moisture content, and volatile matter content that adds to the oxidation process, resulting in ignition.

Oils. Linseed oil, also known as flaxseed or flax oil, is a drying oil. It has a multitude of uses, but is often encountered as a furniture finish that is applied with rags. When rags with linseed oil are bundled up and left in a pile, spontaneous ignition may potentially occur. This scenario leads to a large surface area where the process of the oxidation of the oil creates heat. With the rags bunched up, their proximity prevents the heat from dissipating. This oxidation of the linseed oil accelerates as the temperature increases, eventually reaching the ignition temperature of the rags and resulting in open flame. Were the rags hung out to dry without folds, the heat would not have an opportunity to build up, rendering the situation safe.

Spontaneous Human Combustion. It can be easy to be drawn into the notion of spontaneous human combustion. Some seemingly credible stories, articles, and docudramas have focused on this phenomenon. Some of these stories date back to the 1800s, with many citing the absence of another explanation as proof of spontaneous human combustion.

Fortunately, science provides plausible explanations for these cases. The key point of fire investigations is to be a seeker of truth and to use scientific principles. The International Association of Arson Investigators (IAAI) operates a website known as interFIRE (www.interfire.org) that uses science to explain what is actually happening in these incidents. The author of the *Spontaneous Ignition* report, Dr. John DeHaan, is one of the foremost experts in all forms of fire investigations. His studies have proven that there are logical, scientific explanations of the burns previously labeled as spontaneous human combustion.

Electrical

Electricity is all around us. Indeed, we just need to lose electricity to realize how much we depend upon it. Electricity and electrical devices produce heat either as a by-product of their workings or intentionally, as in electric baseboard heaters. Fires caused by electrical devices can result from malfunction because of poor design, improper construction, or poor installation. Likewise, and just as likely, a failure can occur from misuse of electricity and associated appliances. Investigators must be aware, however, that the mere presence of electricity does not mean it is responsible for a fire.

Electrical sources can be as small as a static electrical arc or as massive as a lightning bolt. Investigators most often come in contact with the other forms of electrical heating, which are generally human-made. **Resistance heating**, which occurs in two forms, is the result of an appliance or fixture design—for example, a lightbulb or a heater. Heaters can heat rooms or can cook food. Resistance heating also takes the form of an appliance or fixture failure when it does not act as designed, manufactured, or used. This unintended resistance heating is more than capable of igniting nearby combustibles. For more information on electricity and electrical fires, see Chapter 7, *Electricity and Fires*.

The electrical wiring designed to deliver energy throughout a structure can fail in the form of an overcurrent or an overload. Although this, too, can be a form of resistance, it can create arcs or sparks that are capable of being an ignition source.

TIP

Electricity as an ignition source is covered in more detail in Chapter 7, *Electricity and Fire*.

Nuclear

The first generation to deal with the term *nuclear* thought of it only in the context of an unbelievably devastating bomb. Today's populace is still aware of the military use of radioactive materials, but nuclear energy is used in electricity production as well as in industrial applications and medical treatments.

By their very nature, radioactive materials are unstable; they are constantly releasing atomic particles. Heat is a by-product of this process. When used in a reactor, the controlled release of these atomic particles can heat water, turning it into steam that powers the turbines and creates electricity. Various types of radioactive isotopes also have numerous industrial applications, such as in testing steel, quality of welding points, and the thickness of materials. Agricultural uses include improving the nutritional value of some crops, making pest-resistant plants, and conducting research. Medical uses include testing, treatments, and research.

Electricity?

Engine 21 responded to a fire in a small single-family residence. The two-bedroom house was vacant, with very few contents in the kitchen. The fire was limited to paint supplies in the corner of the kitchen, and a considerable amount of paper was spread around the floor, as if someone was preparing to paint.

In addition to the material in the corner burning, scorch marks appeared on each of the electrical outlets above the kitchen countertop, six outlets in all. The assigned investigator was nearby and responded to the scene. The investigator looked at the scorch marks, and then found an outlet in the wall exactly where the paint supplies were burning in the corner. That outlet showed heavy scorch marks as well, more prominent than what would have been caused by the extension of the fire.

The investigator told the engine company he would handle the classification of the fire cause on the fire incident report. Once back in the office, he listed the fire as accidental on the fire report and put down the source of the heat as electricity from a faulty electrical system. When the fire report was completed, it was sent to administration as policy dictated, allowing the public information officer access to it. The local newspaper had called the department and asked for particulars on the fire. The next morning a short article appeared in the local newspaper about the electrical fire.

By midmorning, the lieutenant from Engine 21, who was on the call the previous day, had seen the article in the paper. He called the assigned investigator's office and requested that the investigator meet him on the scene, which he did within the hour. While standing in the front yard, the lieutenant asked if the investigator noticed anything unusual. Much to the embarrassment of the assigned investigator, there on the power pole, approximately 15 ft in the air, was a coiled wire that had once gone to the house in question. This feeder line was neatly coiled and a bird's nest could be seen in the bottom loop. The power to the property had been disconnected for more than a month.

Walking around the house, they discovered there was no alternate power supply. Once inside the house, they took apart the charred outlets. The inside of the outlet was undamaged, not even showing sooting. The same was true for all the outlets, including the one at the area of origin.

The investigator had been trained to perform these steps, but did not take the time to do so. Instead, he chose to take the easy path based on limited knowledge. Had he followed the scientific methodology, he would have found this evidence prior to leaving the scene on the day of the incident.

A complete and thorough investigation was then done, and the laboratory analysis of the debris showed gasoline. After a full investigation, the homeowner confessed to burning the outlets with a plumber's torch and setting the fire. The motive was to collect insurance monies to pay for a kitchen remodel.

If an investigator is sent to look at a fire scene, he or she must be prepared to do a complete and thorough investigation and look at all the evidence. Using a systematic approach and scientific methodology can prevent similar events in the future. Furthermore, it is always a good policy for the assigned investigator to consult with the first responder officer, who may have insight as to the cause of the fire.

Just because something is capable of creating heat, that does not mean it is capable of creating enough heat to cause or sustain ignition. The amount of radioactive material one might find in a medical lab or in an industrial site most likely is not sufficient to create enough energy to be the heat source that starts a fire. Remember: It is just as important to be able to eliminate a heat source as to find one.

Heat Transfer

Heat transfer is a common topic in most fire suppression manuals. The intent is to teach fire fighter students how heat travels so that they can understand how to extinguish a fire and where to look for fire extension (**FIGURE 4-10**). Likewise, the investigator, whether for writing the fire report or for conducting a detailed investigation, needs to understand heat transfer to interpret the direction of fire travel. As the fire extends into one area or another, the investigator will need to explain how this happened or how it could not have happened through normal heat transfer.

Heat can be transferred through conduction, convection, radiation, and direct flame contact. At any particular fire scene, the investigation may find examples of each mode of heat transfer. While developing their final hypothesis, investigators may find themselves explaining how the heat from the heat source came in contact with the first material ignited.

Therefore, a thorough understanding of heat transfer is essential for any fire investigator.

Conduction

The transfer of heat through a solid object is known as conduction. Some materials are good conductors, whereas others are not. In the preliminary investigation, a key task is to decide whether a material at the fire scene was capable of conducting the heat from one point to another or whether that material acted as an insulator, preventing or retarding the flow of heat.

Metals are good conductors. When a flame is applied to one end of a metal rod, heat is conducted along that rod to the other end. Not many rods go through walls, but pipes do travel from one point to another, acting as potential sources of heat transfer. Metal trusses or I-beams are also good conductors. Because of their mass, it takes a while longer for the heat to build up before it is transferred across the beam.

Investigators must understand the properties of the materials they are considering as a potential source of heat transfer, so that they can explain how the fire traveled. A detailed investigation requires the investigator to understand all aspects of the thermal properties of materials; more importantly, the investigator needs to understand what is happening at the molecular level.

Convection

The transfer of heat through the movement of liquid or gases is known as convection. For the most part, the heated gases found in fires consist of smoke and fire gases. As these heated gases flow throughout the structure, they come in contact with solid objects and transfer heat to them, thereby eventually spreading the fire. Because hot gases tend to rise, the heat builds up in the higher levels of rooms or compartments. If they do not meet resistance, the heated gases continue to flow, moving into cooler areas and heating up the contents of the structure. As the fire increases, the gases become more buoyant and have a greater propensity to rise. This tendency is especially evident in vertical openings, where convection enables heated gases to rise into other areas of the structure.

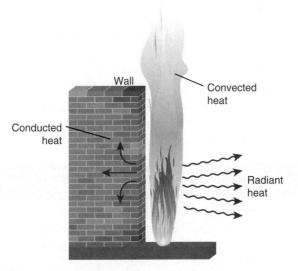

Wall
Convected heat
Conducted heat
Radiant heat

FIGURE 4-10 Examples of heat transfer in fire.

Investigators can look at the burn patterns to discern the direction of fire travel. Although the principle of heat travel through convection can help explain the extension of the fire, the investigator needs to be aware of the potential impact of ventilation. This can occur as a result of the fire breaking windows or the fire suppression crew implementing forced ventilation. The investigator must explain the resulting change in the movement of hot gases around the structure to test and complete the final hypothesis.

Radiation

Radiation is the transmission of energy through electromagnetic waves. Radiation travels at the speed of light, but only in straight lines. For example, radiated heat can be felt from the sun's rays or from the heat of a hot wood stove or fireplace. Radiant heat is capable of igniting nearby combustibles. If a fire is of sufficient volume, it may be capable of igniting combustible products on adjoining properties through radiant heat.

Fire fighters can readily see the damage from radiant heat when the vinyl siding of a home adjacent to the fire-involved structure deforms or melts. They will also feel the heat as they approach the fire. Indeed, radiant heat plays a role in every fire. In a compartment fire, this heat moves upward and outward. As this happens, it collects in higher elevations but cannot escape because of the ceiling of an enclosed space. Thus, heat, hot air, and gases collect in these areas, traveling by convection, but they also radiate downward from the heated area to the less heated area. This radiant heat can then play a role in extending the fire to combustibles below the heated layer.

As with all aspects of the transfer of heat, the investigator needs to first understand the underlying principle of radiation, be able to defend it scientifically, and be prepared to put that principle in simple language for a jury or judge to understand.

> **TIP**
>
> As with all aspects of the transfer of heat, the investigator needs to first understand the underlying principle of radiation, be able to defend it scientifically, and be prepared to put that principle in simple language for a jury or judge to understand.

Thermal Layering

The products of combustion, in the form of heated gases and smoke, rise from the fire because they are more buoyant than the surrounding air. If present in a compartment, they rise until they meet resistance, usually in the form of a ceiling. The rising gases form a plume, which hits the ceiling and then spreads horizontally. This horizontal spread continues until the plume comes in contact with a wall. As this buoyant mass of hot gases and smoke hits the wall, a slight downward motion occurs that is dictated by the amount of energy in the hot gas layer. Another phenomenon that occurs at this time is the creation of a ceiling jet, which consists of a thin layer of buoyant gases moving rapidly just under the ceiling in all directions away from the plume. Most often the ceiling jet is responsible for the activation of smoke detectors, sprinkler heads, and so forth.

> **TIP**
>
> The ceiling jet is responsible for the activation of smoke detectors, sprinkler heads, and so forth.

As these gases reach the upper levels in the compartment, they start to form thermal layers of varying temperatures. Sometimes these layers are visible until they are disturbed by ventilation, application of a suppression stream, or introduction of a larger volume of burning fuel. The temperatures of the layers can vary to the extreme between the temperatures at the ceiling and at the floor.

The upper thermal layer continues to bank down all the way to the floor unless it finds an escape route through a window or any other opening that leads out and away from the compartment. The damage created by this movement, including the marks on the structural components and contents, may be a valuable indicator to the fire investigator studying the burn patterns.[2]

Heat Release Rate

As fuels burn, they lose mass. This process results in the release of energy in the form of heat and light. The amount of energy produced over a period of time is known as the energy release rate (ERR), more commonly referred to as the heat release rate (HRR). The heat released can be measured in British thermal units (BTUs) or in kilowatts (kW).

> **TIP**
>
> The amount of energy produced over a period of time is known as the energy release rate (ERR), more commonly referred to as the heat release rate (HRR).

The rate at which heat is released is important to the fire investigator, but so is the total amount of energy released for a certain size of fuel package. The fuel package can consist of, for example, a sofa or a pile of wood shavings.

Knowing both the rate at which heat is released and the total energy that will be released can help the investigator in analyzing the fire scene and discerning whether the known contents could have caused as much damage as is actually seen at the site. These factors can also explain what should or should not have happened. A match at only 50 kW ERR cannot provide enough energy to ignite a solid 12-in.[3] block of wood—but it can readily ignite a piece of paper.

The investigator can consult a multitude of resources to learn about the HRR values of various objects. When searching the Internet for up-to-date information, use only reputable sites to ensure the accuracy of the information obtained. Most government sites, such as that of the National Institute of Standards and Technology, and some private websites, such as Underwriters Laboratories, have great reference materials. Nothing, however, can take the place of maintaining a good, up-to-date library. Publishers such as Jones & Bartlett Learning and the National Fire Protection Association offer many valuable texts that can serve as reference materials. The NFPA's *Fire Protection Handbook* is a good example of a text that is of immense value to everyone in the fire service.[2]

Compartment Fires

The configuration, construction, and contents of a compartment affect both the growth and the development of a fire. Every fire differs according to a multitude of variables, but some basic concepts and factors define the progression of all fires. Both NFPA 921, *Guide for Fire and Explosion Investigations*, and the NFPA's *Fire Protection Handbook*, 20th edition, describe in detail how fire progresses from the first flickering in a room to the destruction of the room and contents, and then decays and possibly self-extinguishes. Of course, there is always the possibility that the fire may be extinguished by sprinkler systems or the actions of the fire department.

As noted earlier, in the beginning stage of a fire, heat and smoke become buoyant and rise to the ceiling. This creates a plume, a column of smoke rising up to the ceiling. Upon hitting the ceiling, the heat spreads horizontally to create a thin layer of high heat known as the ceiling jet. It continues to travel until it hits resistance, such as a wall. If a natural vent is located at or near the ceiling, and if the fire does not increase the HRR or extend to other combustibles, this is the limit of the fire's damage and actions until the fuel has been consumed.

As shown in **FIGURE 4-11**, the fire often progresses, with smoke and heat escaping from openings in the room of origin. As the fire grows in size, more heat and smoke are produced than can escape, and both begin to bank downward (**FIGURE 4-12**). There, they

FIGURE 4-11 Compartment fire in the initial stage where the heat and smoke rise to the ceiling, and then travel horizontally, filling the area.
Courtesy of NIST.

FIGURE 4-12 Compartment fire in the stage when the fire progresses, releasing more heat and smoke, both of which bank down and affect more items in the room, possibly heating them up to their ignition point.
Courtesy of NIST.

TIP

The configuration, construction, and contents of a compartment affect both the growth and the development of a fire.

FIGURE 4-13 Through an open door, the hot thermal layer flows outward at the ceiling level. Below that cooler replacement, air flows inward that contains oxygen that aids in the combustion process.
Courtesy of NIST.

FIGURE 4-14 Compartment fire flashover where all exposed combustibles reach their ignition temperature.
Courtesy of NIST.

affect other items in the room and may heat them up to their potential ignition temperature.

As the fire continues to grow, the heat again rises. This process creates currents of hot air moving upward, which are then replaced by air at the lower level entering the room, filling the void. The process in which new air is introduced is called air entrainment (**FIGURE 4-13**). The fresh air aids in the combustion process, keeping the fire growing.

Occasionally, a flameover, sometimes called a rollover, occurs. In this event, the gases in the upper layer ignite, sending flames rolling across the buoyant layer. This may be—but is not always—a precursor to the phenomenon called flashover (**FIGURE 4-14**). In this transition, all the combustible materials in the room are at their ignition temperature and suddenly ignite. A flashover does not always occur. In fact, many fires start to decay because of a lack of fuels before a flashover takes place. The air (oxygen) supply also may be consumed before the room reaches a sufficient temperature to support flashover. In addition, if a fire in its early stage has natural ventilation, it never builds up sufficient heat to create the environment necessary for flashover to occur.

> **TIP**
>
> Occasionally, a flameover, sometimes called a rollover, occurs. In this event, the gases in the upper layer ignite, sending flames rolling across the buoyant layer.

If the compartment is in the post-flashover stage and ventilation occurs, either through a natural event or from suppression forces, a volume of smoke will escape from the compartment. This smoke, which consists of incomplete particles of combustion, is a fuel that is at its self-ignition temperature. Reaching the outside air provides the final side of the fire triangle—sufficient oxygen. In turn, the smoke may ignite. Witnesses sometimes report seeing flames within the smoke, or this phenomenon may manifest as a fireball above the structure.

In a situation where there is adequate ventilation, the fire will eventually decay as the fuel is consumed, and fewer and fewer flames will be noted. In situations with little or no ventilation, however, the compartment retains the heat, and oxygen content may fall to well below 16 percent, also leading to fewer flames. This can create a dangerous situation—a fire fighter's nightmare. If for any reason the compartment experiences an influx of air, the vapors in the compartment will be at the self-ignition point; they will mix with the newly introduced oxygen and ignite, sending a wall of fire out through the ventilation opening such as a door or window. Fire fighters call this event a backdraft or, depending on the source, a *smoke explosion* or *flashback*. This is a rare condition, but nonetheless does occur.

The likelihood that all variables involved in one compartment fire will ever be exactly like those in another incident is so remote as to be unimaginable. The description of the compartment fire in this text is only one scenario of many, but it suggests how the development and growth of fire proceed in a structure.[2,5]

Wrap-Up

SUMMARY

- Fire fighters and fire officers do not need to be scientists to investigate fires. However, those who neither understand nor have good scientific knowledge endanger themselves and the public they serve. Thus, all investigators should seek out additional training, which can be obtained from local fire organizations, community colleges, state fire training agencies, or national associations.

- Using scientific methodology is essential to a proper determination of fire origin and cause. Just as important, those conducting investigations should adopt a systematic approach when examining fire scenes. Doing so ensures that the entire scene is properly searched for all evidence necessary to formulate a proper hypothesis as to the fire's area of origin and cause.

- In the process of making the final determination of the fire cause, investigators must use good science—that is, science based on accepted principles and facts. They must constantly challenge what they have learned in the past and seek new knowledge based on current, up-to-date research and studies.

- This chapter has indicated how fire investigators can use basic firefighting science in their work. Items such as the fire tetrahedron model and the process of heat transfer show the fire fighter how to extinguish the fire; these same principles can be used to show others how the fire started, grew, and moved from one point to another.

KEY TERMS

Air entrainment An event in which, upon the ignition of a fire, the flames stretch upward and the developing heat causes the air to rise. The process of the air rising up creates a void at the lowest area of the fire; air then moves into this area, becoming part of the ignition process.

Autoignition temperature The minimum temperature at which a properly proportioned mixture of vapor and air will ignite with no external ignition source.

Backdraft An explosion resulting from the sudden introduction of air (oxygen) into a confined space containing oxygen-deficient superheated products of incomplete combustion.[3]

Ceiling jet A thin layer of buoyant gases that moves rapidly just under the ceiling in all directions away from the plume.

Conduction The transfer of heat through a solid object.

Convection The transfer of heat through air currents (or liquids).

Deductive reasoning Taking the facts of the case and thoroughly and meticulously challenging all facts known, using logic.

Empirical data Information collected that is based on observations, experiments, or experience.

Energy release rate (ERR) The amount of energy produced in a fire over a given period of time.

Exothermic reaction The release of heat from a chemical reaction when certain substances are combined.

Fire point The lowest temperature at which a fuel, in an open container, gives off sufficient vapors to support combustion once ignited from an external source.

Fire tetrahedron A solid figure with four triangular faces; each face represents one of the four items necessary for a fire (heat, fuel, oxygen, and uninhibited chemical reaction).

Flameover The point at which a flame propagates across the undersurface of a thermal layer; also called rollover.

Flash point The temperature at which a liquid gives off sufficient vapors that, when mixed with air in proper proportions, ignites from an exterior ignition source. Because of the limited supply of vapors, only a flash of fire across the surface occurs.

Flashover A transition phase of fire where the exposed surface of all combustibles within a compartment reach autoignition temperature and ignite nearly simultaneously.

Heat release rate (HRR) The rate at which heat is generated from a burning fuel.

Lower explosive limit (LEL) The lower limit of flammability of a gas or gas mixture at ambient temperature and pressure. It is the threshold just before the fuel becomes too lean, but still can ignite.

Oxidizer Any material that forms from a fuel and supports combustion.

Plume The column of smoke, hot gases, and flames that rises above a fire.

Radiation Transmission of energy through electromagnetic waves. It travels at the speed of light, but only in straight lines.

Resistance heating The heat produced when electrical current passes through a material of high resistance. This converts the electrical energy into heat.

Rollover The point at which a flame will propagate across the undersurface of a thermal layer.

Scientific method The systematic pursuit of knowledge involving the recognition and formulation of a problem, the collection of data through observation and experiment, and the formulation and testing of a hypothesis.[1]

Spontaneous heating A process in which a material increases in temperature without drawing heat from the surrounding area.

Spontaneous ignition Initiation of combustion from within a chemical or biological reaction that produces enough heat to ignite the material.

Systematic search A method or process used to search an area or a building that allows for complete coverage of the area, and that is done consistently at each fire scene or incident.

Thermal runaway A situation in which heat production increases during a fire; the exothermic reaction creates heat that increases the reaction rate, in turn creating more heat, which then increases the exothermic reaction.

Upper explosive limit (UEL) The threshold where the vapor concentration can be ignited, just before the concentration becomes too rich.

Vapor density Density of a gas or vapor in relation to air, with air having a designed vapor density of 1.

REVIEW QUESTIONS

1. Describe heating at the molecular level.

2. Describe the methods of heat transfer and give examples of each.

3. What are the differences between heat and temperature?

4. In detail, describe flaming combustion and smoldering combustion.

5. Define *spontaneous heating*.

6. Why is the heat release rate so important to the fire investigator?

7. What is flashover, and how can it affect the fire scene?

8. What is meant by a systematic approach for the fire investigator?

9. What are the four sources of heat? Define each.

10. What is thermal layering?

DISCUSSION QUESTIONS

1. Discuss scientific methodology. What is a hypothesis? Is it nothing more than an educated guess?

2. Discuss the necessity of following the systematic approach as defined in NFPA 921 or using a different approach.

3. How is the scientific methodology used in the everyday life of a fire fighter?

REFERENCES

1. National Fire Protection Association. (2008). *NFPA 921: Guide for fire and explosion investigations* (pp. 921–916). Quincy, MA: Author.
2. National Fire Protection Association. (2008). *Fire protection handbook* (20th ed., Sect. 1, Chap. 2). Quincy, MA: Author.
3. Nolan, D. P. (2006). *Encyclopedia of fire protection* (2nd ed.). Clifton Park, NY: Thomson Delmar Learning.
4. Scher, L. (2007). *The Texas city disaster* (Code Red Series). New York, NY: Bearport Publishing.
5. National Fire Protection Association, (2008). *NFPA 921: Guide for fire and explosion investigations* (Chap. 5). Quincy, MA: Author.

CHAPTER 5

Building Construction for the Fire Investigator

LEARNING OBJECTIVES

Upon completion of this chapter, you should be able to:

- Describe why knowledge of building construction is important to the fire investigator.
- Describe the five classifications of building construction.
- Describe the characteristics of walls and the types of walls found in structures.
- Describe characteristics of window and door assemblies and their relevance to the investigation process.
- Describe the characteristics of fire doors and fire windows.
- Explain the effects that interior and exterior finishes can have on a fire investigation.
- Explain the impact of occupancy and contents on the fire investigation process.
- Explain how experience in the past can give us insight on how we should work toward the construction of buildings in the future.

Case Study

As he had no cell phone service and was not familiar with the area, it took the young plumber some extra time to find the fire station and report that the building he had been working in was now on fire. He stopped at a couple of houses but, because it was a workday, no one was home. Finally flagging down a local deputy sheriff, the plumber told him about the fire, then quickly returned to the fire scene. The local volunteer fire department was on the scene in 5 minutes after their dispatch. Pulling up to the property, they found a two-story structure with flames coming from the first floor and the roof; they could also see fire through the opening to the crawl space below the house. Interestingly, heavy smoke was emerging from the second-floor windows, but no flames.

Given the rural setting, there were no exposures to consider other than overgrown fields on three sides of the house. The fire fighters decided to use their initial limited resources on the scene to prevent the fire from extending into the fields and the woods beyond. The chief called for a second alarm and made a special call for tankers (tenders). After a water shuttle from a local pond was established, the fire was finally extinguished, but not before the house was essentially flattened.

Once the fire was knocked down and while the crew was getting its equipment and apparatus back in service, the chief interviewed the young plumber who reported the fire. It was the plumber's first day on this particular job. The house was an old structure, built in 1900; all the structural wood was roughhewn. The plumber's assignment was to remove the old galvanized steel pipes and replace them with copper pipes. He decided to get the hardest part done first—that is, crawling under the structure to remove the pipes going up through the first floor and replace them with the new copper lines.

The plumber was under the house, heating one of the galvanized pipes to loosen it. The pipe was located next to the exterior wall. When the plumber turned off his plumbing torch, he noticed that the rough fibers of the floor joist had caught on fire. Almost immediately, the fire started to extend into the interior of the outside wall.

The plumber quickly worked his way out of the crawl space, ran to his truck and got a bottle of water, and then crawled back under the house. He tried to throw the water onto the fire, but had only minor success. He then went back out, grabbed his coffee thermos, and returned under the house again. Even pouring the coffee into the cup and throwing the coffee up into the interior of the exterior wall was unsuccessful. He made it back out, choking from the smoke, and drove off to find help.

The fire chief knew that this was an extremely old structure. Because of his familiarity with building construction from his fire service training, he quickly understood what had happened. When the chief first arrived, he was initially suspicious of the fire's origin because the fire was on the first floor and roof, but the second floor was producing only smoke. In fact, the extension of the fire was due to this structure's balloon construction, which meant there were no fire stops. The interior of the walls were open from the crawl space to the attic, allowing the fire to quickly extend from the bottom of the structure to the top. Eventually, it would extend into each floor, as it had in this case. The chief, acting as the first responder investigator, filled out the fire report listing the fire as accidental in nature; case closed.

However, the incident was not completely over. The house was in rough shape, and the owner had purchased the property by paying only $30,000 for the structure and $50,000 for the 10 acres of land. The owner planned on this being a "flip" project: He intended to purchase the property at a low price, fix it up, and sell it for a profit. The owner had taken out an insurance policy for $100,000 based on all the upgrades and improvements he intended to make. Residential insurance policies cover only the structures, not the land. With the structure now gone, to receive the insurance proceeds, the owner had to provide a proof of loss statement—a standard document in which the policy holder provides evidence that there was a fire and lists the items lost or destroyed.

In this case, the owner got a copy of the fire report from the local fire department showing that the fire was accidental in nature. The policy owner then provided the original bill of sale for the house and a list of the upgrades he had completed on the structure.

At this point, the situation changed. The owner provided a written list of the upgrades, including removal of all of the galvanized steel and cast-iron pipes and their replacement with copper supply and polyvinyl chloride (PVC) drain piping. He claimed that all of the plaster and lath walls and ceilings had been removed completely and new gypsum board (sheetrock) had been installed and painted, that the appliances had all been replaced with new units, and that new floors had been installed. He signed the proof of loss document seeking more than $100,000 in damages.

The insurance company adjuster went to the property to validate the proof of loss statement. Based on what he observed, he requested an independent fire investigator. When the investigator arrived, he witnessed that the two-story structure was on the ground as reported. The tallest portions of the structure were the stove and refrigerator, which were clearly made in the 1950s; they were not new modern appliances. All piping in the structure was older galvanized steel and cast-iron pipes (**FIGURE 5-1**), and the investigator found no copper pipe in the debris. A few burned pieces of wood were present, but most of it had been completely consumed. The vast majority of the debris (**FIGURE 5-2** and **FIGURE 5-3**) was residue plaster from the plaster and lath walls and ceilings; no gypsum board residue could be seen. After meeting with the local fire chief and speaking to the plumber, the fire investigator agreed that the fire was accidental in nature, but stated that the scene evidence gave a clear indication that the property owner had committed insurance fraud. The insurance company refused to pay the claim.

FIGURE 5-2 This is all the debris left from a two-story structure. The vast amount of the remaining debris is plaster from the plaster and lath interior finish that was reported to have been removed and replaced with gypsum board (sheetrock).
Courtesy of Russ Chandler.

FIGURE 5-1 Galvanized steel water pipe that was reported to have been removed and replaced with all copper pipe.
Courtesy of Russ Chandler.

FIGURE 5-3 Close-up of one of the pieces of the debris, consisting of plaster. Notice how it was applied wet (in the early 1900s): It was pushed onto the walls, allowing some to weep between the wooden laths and adhere to the walls.
Courtesy of Russ Chandler.

 JONES & BARTLETT LEARNING NAVIGATE **2** *Access Navigate for more resources.*

Introduction

It is critical that all first responder investigators and assigned investigators be well versed in building construction. Such knowledge is necessary to assess the safety of the structure before entering to conduct an investigation. It is just as necessary to understand building materials. The type of building construction, the interior configuration, wall coverings, contents, use of the building, suppression systems, compartmentation, and more can all have an impact on the fire growth, direction, and suppression, as well as the investigation of the incident. A solid knowledge base in building construction will aid the investigator in finding the area of origin.

This chapter is only an introduction to the topic—enough to allow for a basic understanding of building construction. More comprehensive information on building construction is available in Jones & Bartlett Learning's *Fundamentals of Fire Fighter Skills* and

Hazardous Materials Response texts. Another outstanding text for training and reference is *Brannigan's Building Construction for the Fire Service* by Francis Brannigan and Glenn Corbett.

We are all the sum of our experiences; as such, we may be strong in one area and not as knowledgeable in others. For this reason, investigators need to be prepared to leverage resources that can fill the gaps in their own knowledge when the need arises. If fire investigators have not had the opportunity to develop a strong background in building construction or in areas such as electrical systems, then they should seek out someone, such as the local building official or electrical inspector, who may be able provide more information if necessary. If fire investigators do not come from a fire service background, then they may not have had intensive training in building construction. If this is the case, then the local building official may be a great resource.

When looking at a structure, the type of construction is just one item that must be considered by the fire investigator. Building components, compartmentation, wall coverings, and contents are all critical factors in the overall investigation. Likewise, the fire investigator must take the occupancy and the building's use into account. The fire investigator may also have to investigate the reason for any fire fatalities and ways that they could have been prevented.

Types of Construction

One of the great outcomes from the creation of the National Fire Protection Association (NFPA) was the creation of the various NFPA codes and standards. Notably, this organization provides model codes such as building and fire codes that can be adopted by localities. Even though some other code creation groups compete with the NFPA model codes, NFPA standards are used in all of the model codes employed today. One such standard, NFPA 220, Standard on Types of Building Construction outlines five types of construction based on how much the building either resists or contributes to the initiation or extension of a fire.

Type I: Fire Resistive

With at least 2 hours of fire resistance, Type I is the most fire-resistant type of construction. The floors, walls, partitions, components, and roofs of such a structure are noncombustible, usually made with concrete and protected steel-frame construction. The contents will burn, but the structure itself should not. That said, concrete and protected steel-frame

FIGURE 5-4 Type I fire-resistive construction is commonly found in schools, hospitals, and high-rise buildings.
© John Foxx/Alamy Stock Photo.

construction can sustain superficial damage as the result of excessive heating. This type of construction can be found in schools, hospitals, and large and/or tall office complexes (**FIGURE 5-4**).

With Type I construction, the contents will limit the fire itself. With a large fuel load, the structure could suffer damage, so investigators should exercise caution at the site. Because most of the building components are fire-resistive, burn patterns on walls, floors, and ceilings can help point the investigator toward the area of a fire's origin.

Type II: Noncombustible

Similar to, but less stringent than Type I construction, the Type II classification requires noncombustible structural components such as steel. This type of construction is usually reserved for larger structures (factories and warehouses) that may contain fire walls to provide additional protection from fire spread. Fire walls are separations that have a certain fire rating as defined in building codes. These separations must maintain structural integrity for an amount of time as defined in the code, resisting the spread of the fire (**FIGURE 5-5**).

Although Type II construction leads to a noncombustible structure, heat can impact the stability of the structure. Expanding steel from the fire could cause enough damage to put the stability of the structure at risk, so investigators must use caution when working at the site. Structural damage from heat could point to the area of origin. However, care should be taken because heavy damage could appear to be an area of origin, but actually may have just been an area with a large fuel load. More investigation may be necessary.

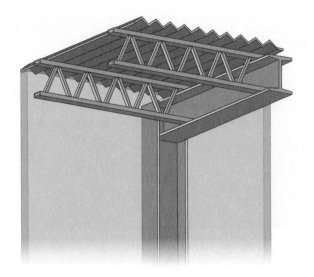

FIGURE 5-5 Type II construction is referred to as noncombustible construction.

FIGURE 5-7 Type IV construction was once used for buildings suitable for manufacturing and storage occupancies. Today, new buildings of Type IV construction are rare.
© Helen Filatova/Shutterstock.

FIGURE 5-6 Type III construction is referred to as ordinary construction because it is used in a wide variety of buildings.
© Ken Hammon/USDA.

Type III: Ordinary

Usually limited to buildings not exceeding four stories in height, Type III construction has masonry exterior walls, which are an integral part of the structure, supporting the floors and roof. Depending on the age and the building code, the interior structural components are usually wood with gypsum board finish, which usually provides at least 2 hours of fire resistance. The fuel load for Type III construction includes both the contents and the combustible building materials (**FIGURE 5-6**). Of interest to the fire investigator, void spaces in such a structure may allow a fire to spread both vertically and horizontally.

Because the interior components will most likely be made of wood, they are not protected and can experience heavy fire damage. As the investigator examines the burn patterns within the Type III structure,

he or she must be sure to compare patterns on similar surfaces. For example, surfaces protected with gypsum board should be compared with other walls covered by the same component. In addition, if the investigator finds heavy char on wood components, he or she needs to check whether they were protected or not protected at the time of the fire.

Type IV: Heavy Timber

Type IV construction could be as tall as eight stories. Most of these structures were built more than a hundred years ago; relatively few are built today thanks to alternative, safer building materials. Most Type IV structures were constructed as mills and warehouses. They typically have noncombustible exterior walls, usually masonry, that will have a 2-hour fire resistance.

The interior beams, girders, and columns of a Type IV structure will be made of, at a minimum, 8-in. by 8-in. heavy timber. Floors and roofs will consist of, at a minimum, 3-in. by 6-in. planks. This creates an increased fuel load, although the mass of the structural components makes them resistant to ignition. However, once ignited, a fire in this structure will be hard to extinguish. Fortunately, original Type IV structures rarely have any voids to allow hidden extension of a fire. If occupancy subsequently changes to smaller compartment spaces, though, there will most likely be voids allowing an extension of fire (**FIGURE 5-7**).

Just as with Types I, II, and III, the exterior of a Type IV structure will be masonry. Close examination of the patterns on the interior columns and beams may point back to the area of origin. However, investigators should make sure heavy char patterns were not created by excessive fuel load, rather than being near or at the area of origin.

Type V: Wood Frame

Type V is the most common type of construction and is used widely for one- and two-family dwellings and small commercial establishments. Most components of these structures, including the floors, walls, and roofs, are made of wood or other combustible materials (**FIGURE 5-8**).

Such structures usually contain void spaces that can allow fires to extend beyond their area of origin. The building's exterior could be made of wood, brick, or other masonry, but most likely it will just be a veneer. Common construction features may include engineered items such wood I-joists and trusses.

Wood-frame structures built before the mid-1900s are most likely based on balloon construction. In this building approach, the wood studs are continuous from basement to the roof, which creates a void from bottom to top. Should a fire occur at lower levels, it can easily extend to the roof and other parts of the structure (**FIGURE 5-9**).

Newer wood-frame construction methods are called *platform frame*. This approach entails just what its name implies: A platform is built, and on top of that platform is a framework of walls. These walls will consist of a bottom sill plate, with studs going up to a top plate completing the framework, which are the walls. The next floor (platform) is then constructed on top of these wall components (**FIGURE 5-10**).

All five types of constructions have inherent hazards and danger of collapse. Investigators must take special care to assure structural integrity before entering any structure to conduct the interior examination. The process of examining the scene and related safety factors are covered elsewhere in this text.

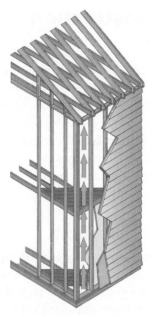

FIGURE 5-9 In a balloon-frame building, the exterior walls create channels that enable a fire to spread from the basement to the attic.

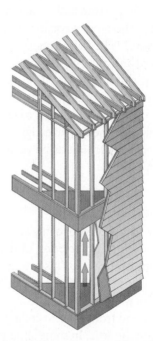

FIGURE 5-10 In a platform-frame building, the floor platform blocks the path of fire, preventing it from spreading from one floor to another.

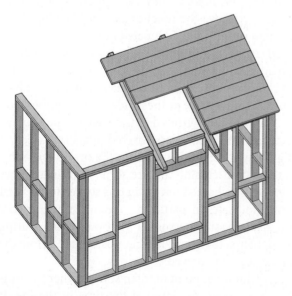

FIGURE 5-8 Type V wood frame construction is the most common type of construction used today.

Hybrid Construction

Some buildings may contain more than one type of construction. In such a case, investigators will need to pay close attention to the type of construction when working their way through the structure. Different construction types may impact the interpretation of the burn patterns.

Other Construction Features

The investigator should also examine other components, features, and construction assemblies of building construction when conducting the fire investigation. Their presence, or absence, can impact the fire growth and direction. The investigator must consider each component in making the determination of the fire origin and cause as well as their potential contribution to any loss of life.

Compartmentation

Fire compartmentation prevents the rapid spread of fire, providing sufficient time for occupants to escape. By preventing the growth of the fire, it also reduces the danger to the first responders who enter the building. Of course, compartmentation reduces the amount of fire damage as well.

The type of occupancy, fire load, and presence of fire suppression systems will impact the degree to which the compartmentation actually slows the growth of a fire. The type of construction of the compartment is critical. The fire rating of a covering such as gypsum board and any openings will impact the degree to which it slows down the fire. In the best-case scenario, any doors, windows, and openings are fire rated and have automatic closures to contain the fire as much as possible. The fire investigator must take all of these factors into account as he or she determines the fire spread.

Walls

Walls are a key component of compartmentation. The type of materials and composition of the wall will establish the rating of the wall. This rating indicates how long it will delay the spread of the fire. To be considered a fire wall, a wall must meet the minimum ratings and needs to extend from floor to ceiling and continue beyond, through the attic. In some situations, it should extend through the roof, creating a parapet. Some fire separations are also rated as walls; they protect vertical openings. This group includes utility chases, elevator shafts, and exit stairwells. Such barriers prevent the extension of both fire and smoke, ideally allowing time for occupants to escape.

Walls can be load bearing, which means they bear the weight of the structure above by conducting that weight to the foundation. The failure of a **load-bearing wall** could risk the collapse and destruction of the structure itself. Walls that are not load-bearing are essentially partitions. **Non-load-bearing walls** support only their own weight, and can be removed or breached without compromising the building's structural integrity. Some walls may separate two occupancies; the shared wall is usually load bearing. In many occupancies, it will also be a fire wall.

Doors and Windows

Most doors are constructed of wood or metal. Interior doors are usually wood and can be solid, but are more likely to be hollow. Even a hollow door is capable of acting as a fire barrier for a short time. Solid doors are more likely to be installed for security or as a fire barrier.

Rated doors will be labeled as such, and usually feature a metal plate on the side of the door (protected by the frame). These doors are specifically designed to prevent the passage of flame, heat, and smoke. NFPA 80, *Standard for Fire Doors and Other Opening Protectives*, provides minimum standards for construction, testing, maintenance, and inspection of **fire doors**. It references the full **door assembly**, including hardware, closures, and activation. These doors are designed to automatically close should a fire occur; once closed, they must remain closed through the use of latching hardware. Should there be any indication of door failure that resulted in the extension of fire or smoke, then the investigator should examine the reason for that failure.

The NFPA 80 standard also addresses other openings such as windows. If there is a desire to have an exterior line of sight or sunlight enter into an area through a fire-rated wall, the windows themselves must have the same rating as the doors in that area. Windows may appear as wired glass or glass blocks; with new technology, they can appear as any other window yet have the functionality of blocking flame heat and smoke for a designated duration.

Of course, most structures have nonrated windows. When these windows fail, investigators need to examine them closely. In case of broken glass, was it broken from the outside or the inside? If glass is found on the floor or on the ground outside, what type of break occurred, and is the glass covered in soot? If clean, it might have been broken prior to the fire. Later in this text, when discussing the fire scene examination, we will discuss what to look for in this scenario and what it may indicate. In particular, the investigator needs to ask suppression personnel what was broken during fire fighter activity versus what was broken before the fire department arrived.

An additional point of interest is whether the windows were open or closed at the time of the fire. If the ambient outside temperature was below freezing, one would not expect to find the windows open. The investigator must be extremely careful not to come to conclusions until all the facts are uncovered. As a seeker of truth, the investigator would need to eliminate any reason why these windows might have been left open. This information could be revealed during investigative interviews and confirmed during the follow-up investigation.

Interior Finishes

Interior finishes are any building material on the surface of the walls, floors, or ceilings. It includes paint, hardwood floors, carpet, tile, and in some cases even cabinets and countertops. Some of these items can dramatically add to the fuel load.

Many of these items are designed to have minimal impact on the extension of a fire, providing they are used appropriately. Carpeting is required to meet federal regulations such as the Consumer Product Safety Commission's (CPSC) tests for carpeting. For years, the only evaluation method was a methenamine pill test, in which a burning pill was dropped on the carpet; the rate at which the fire extended then dictated whether it passed this test. Other evaluations now include the nut test and a radiant panel test, which provides a more realistic assessment of carpet to determine if someone could escape in a carpeted corridor with radiant heat from overhead.

Carpet installed on the floor that has passed a burn test may not be acceptable if installed on walls, where rising flame impingement on fibers above will rapidly spread the fire. However, some carpet products can safely be mounted on walls. Be cautious about making assumptions and thoroughly research these issues before reaching a final conclusion about the fire cause or fire spread. All interior finishes must also be examined to determine their contribution to the spread of fire and to the creation of smoke or other contaminants.

Exterior Finishes

The exterior surfaces of a structure, including walls and roofs, can be covered in a wide range of products. If they comply with a national building code, most surfaces will be fire resistant or at least will reduce the potential fire spread. However, not all materials will resist flame, such as wood siding. Other materials, such as vinyl siding, may melt, thereby exposing other combustible building materials.

Some siding may be stucco, which consists of two or three layers of concrete (Portland cement) creating about a 1-in.-thick finish over metal-reinforcing mesh. This provides approximately 1 hour of fire rating to that wall. In years past, stucco might cover cement block or a plywood finish. Because it is concrete, stucco will not burn, but the outer layer of paint may burn or scorch. The material behind the stucco may include foam sheets or block.

Stone and brick veneer are not structural materials. When properly applied, they are tied to the wall with metal wall ties, which are then nailed to the sheathing (in case of a concrete wall, the ties are mortared in), keeping it anchored to the structural wall. This is a concern for operations crews as well as the fire investigators: Unanchored walls can fail, putting all at risk.

Sheet metal and aluminum siding are other products that may be used as exterior finishes. Early on, they were advertised as protection from fire brands generated by brush fires or burn barrels. They were durable as protection from the elements. Aluminum was advertised as low maintenance. Today, metal finishes can come in all styles and shapes; they can replicate wood siding or can be artistic and modern. The key concern is that they can conduct electricity, so fire fighters must take care to assure they are not energized.

Roof Assembly

In addition to protecting the interior of the building, roofs can improve building stability. The major issue in a fire is the potential failure of the roof, which could dramatically change the intensity and direction of fire travel, complicating the investigation.

Many roof configurations and types of construction materials are possible. Steel is the most commonly used choice for industrial and commercial occupancies, whereas residential buildings usually have wood construction. Common rafters are heavy lumber running at an angle from the top of the wall to a ridge board that becomes the peak of the roof. When properly built with the right size of timber, this construction is quite strong and creates additional living space or storage. A less expensive means of supporting a roof relies on roof trusses, which consist of rafters, posts, and struts, usually held together with gusset plates. Their strength stems from the multiple penetrations from the plate into the wood. However, these penetrations are shallow. When heated, they loosen, creating the potential for roof failure.

Occupancy

The occupancy is not part of the building construction per se, but it certainly plays a role in the suppression and the investigation of the fire. Occupancy and

Building Code Versus Art

A new retail area was built to resemble European architecture. One of the two-story buildings had stucco walls with stone on the corners along with clay tile roofs. A small fountain was placed on the front just to the right of the front door; the small fountain pump was plugged into an exterior outlet next to the fountain.

One day, smoke was seen coming from near the outlet and an employee called 911. While reporting the smoke, the caller advised that the outside wall had quickly caught fire and flames were now all the way to the roof. The fire department arrived to find the roof on fire, as well as about one-third of the front outside wall.

Once the fire was extinguished, it was discovered that the corner granite and roof tiles were made of fiberglass. The exterior stucco was actually a plastic laminate designed to resemble stucco. The same finishes were used on the other three buildings—even the wood beams on the Swiss Alps store were fiberglass. Fortunately, the buildings themselves were concrete, which protected the interior of the structure. Fire damage was limited to just the exterior finishing and no injuries occurred.

When the structure was built, both the fire and building officials had visited and approved all the permits. However, the construction plans and blueprints that had been submitted failed to mention the composition of the materials to be used. At a glance, the structure looked as if it had real stucco walls and a ceramic tile roof. The day after the fire, the building officials advised the owner that the exterior wall and roof finish had to be replaced to meet the code. The building owner offered to install an exterior fire suppression system instead of changing the finish. Officials accepted that alternative.

contents are discussed in several other areas of this text in more detail.

The building and fire codes apply differently to the wide range of occupancies. When the use of a building changes, there may be a change in how the code applies to the existing construction, and certainly to the fire codes. In some cases, a change in occupancy may not be allowed due to the construction features of the existing building, or the changes necessary may be cost prohibitive.

Knowing the type of occupancy is critical before starting the scene investigation. The original structure could have been built as a warehouse for masonry supplies, which are essentially noncombustible materials. Now, however, the structure might be used for fiberglass fabrication, which involves combustible materials and flammable liquids. If the fire suppression system has not been installed or upgraded, then it could change the way a fire progresses. This consideration, in turn, would change what the investigator would expect to find during the scene investigation. Knowing that methyl ethyl ketone peroxide (MEKP) is necessary for fiberglass work, the investigator would not be surprised to find indicators suggesting the presence of an ignitable liquid. Knowing what belongs or might not belong in a structure is important in coming up with a viable hypothesis.

Contents

The contents of the structure should be consistent with the type of occupancy. Their volume will vary. For example, some warehouses may be full to capacity, whereas others are only marginally filled. The type of contents can make a difference in the investigation, especially given that some contents can be more volatile than others and can create a more pronounced burn pattern that may not necessarily be the area of origin.

Sometimes the type of contents may be unexpected. In one case, a local noncombustible structure had long been used for metal fabrication, so no suppression system was necessary. The engine company responded to an emergency medical call at that facility and discovered that it now was a storehouse of military surplus formaldehyde, which is flammable and poisonous. Several hundred barrels were now occupying almost the entire structure. The barrels were strapped to wooden pallets, and the pallets were stacked on top of each other almost to the ceiling. The owner had purchased the surplus with hopes of bottling and selling it to local funeral homes.

The engine company reported their findings to the local fire marshal. Working with the owner, the structure was eventually upgraded to have a fire suppression system, the pallets were removed, and the barrels were stored on metal racks. In the meantime, a 24-hour fire watch was established.

Although no fire occurred in this case, the observations of the engine company officer led to the correction of a very serious hazard. This event, as well as many other events, added to the sum of the fire fighters' experiences—a development that can aid in responding to future events, including investigating fire scenes (**FIGURE 5-11**).

FIGURE 5-11 Improper storage of military surplus formaldehyde.

Courtesy of Russ Chandler.

Future of Construction

There will always be those who seek to improve on the construction materials on the market today. Likewise, we can anticipate further efforts to decrease building costs and to use new materials that reduce our carbon footprint. In addition, we can expect great strides in developing construction materials that are lightweight, prevent fire spread, and offer cost savings.

The capability of using wood-fiber products instead of solid wood has been around for decades. Over the years, steel has had to make way for new aluminum trusses and supports. The building construction industry will continue to make improvements in both. New developments now focus on combining wood and metal for trusses and joists. Society as a whole is seeking lower building costs along with making lightweight materials.

Concerns for fire safety will require diligence in testing and research to keep up with the evolution of the building construction materials industry. In the past, we have experienced adverse events such as fire fighters falling through composite roof sheathing, fire extending up the exterior of high-rise buildings, and structural failure due to some of these products. To prepare for these future innovations and developments, industry, research facilities, and the fire service must work collaboratively to assure a fire-safe world for the future.

Wrap-Up

SUMMARY

- The National Fire Protection Association has established classifications for building construction that range from fire-resistive to wood-frame types. These standards provide guidance on construction as well as information necessary for the fire investigator to consider in the investigative process.

- Properly built and maintained compartmentation can slow down the progress of a fire. For the compartmentation to be effective, properly constructed walls or partitions are needed, and any openings must have fire-rated doors and windows.

- Interior and exterior finishes may create further extension of the fire and must be considered by the fire investigator before making a final determination of the fire origin and cause.

- The fire investigator must also examine the occupancy and contents of the structure. In particular, the investigator must determine if the occupancy is appropriate to the type of structure, and if it meets all building and fire codes.

- Examination of the contents will determine whether they were appropriate for the type of occupancy and if they contributed to the spread and growth of the fire.

- The future of fire safe building construction depends on a partnership between industry, researchers, and the fire service.

KEY TERMS

Balloon construction A type of building construction in which wood studs are continuous from basement to the roof, creating a void from bottom to top. Should fire occur at lower levels, it can easily extend to the roof and other parts of the structure.

Door assembly All of the component parts of the door from the framing which include the sill (bottom of the frame), the jamb (sides of the door frame) and the head. This will also include the door itself and its components from the hardware including the door knobs, hinges and strike plates. This assembly may also include additional components such as windows beside or above the door.

Exterior finishes The outer finish of a building that provides protection against the elements.

Fire door Any door with a fire resistance rating; it is used to limit the spread of fire and smoke to other areas of the building.

Fire wall A wall or partition designed to limit the spread of fire. The specific construction will dictate how long it will prevent the spread of fire.

Gypsum board (sheetrock, dry wall, plaster board) A wall covering, usually with a gypsum core, bonded between two layers of paper that serves as the first and basic finish on walls and ceilings.

Interior finishes The exposed surfaces of walls and ceilings; can be wood, paint, or wallpaper applied over gypsum or sheetrock.

Load-bearing wall Any wall that bears structural weight (load) resting on it, transferring that weight to the foundation below.

Methenamine pill test Also known as the "Standard Method for Ignition Characteristics of Finished Textile Floor Covering Materials." A standard test of carpets to determine whether they meet product safety standards, in which a pill is ignited and dropped on the tested material to see if it will ignite the material and cause a spread of the fire beyond a limited measurable mark.

Methyl ethyl ketone peroxide (MEKP) A catalyst added to resins (usually in fiberglass construction) which through chemical reaction creates heat which hardens (cures) the resin.

Non-load-bearing wall A wall that supports only itself.

Plaster and lath A wall and ceiling covering created by thin strips of wood closely mounted on studs where plaster is spread on the boards, allowing it to adhere, leaving a smooth plaster finish.

REVIEW QUESTIONS

1. Describe the five classifications of building construction, and provide characteristics of each, including what the investigator may have to take into consideration during the investigation.

2. Describe the types of walls that may be found in a structure.

3. Explain the impact that various interior finishes may have on the spread of fire.

4. Explain why the investigator needs to know the type of occupancy.

5. Explain why the investigator needs to know the expected contents of a structure.

DISCUSSION QUESTIONS

1. The local judge needs additional space to store his massive collection of law books. His library and office are full. There is an exit hallway that is never used, labeled "For Emergency Only." The building is fire resistive; the finish in the hall is painted gypsum board and the hall has a tile floor. The judge places his books on metal shelves, and even with the books lined on both sides, the width of the isle is within code. You respond to a very small fire at the courthouse with no damage. The cause was an electrical short on a lamp cord. However, when leaving, you enter this hall and see the new bookshelves. Are you concerned, and if so, what do you do?

2. A local vacant commercial building is now occupied by scores of homeless families, including children. Your jurisdiction has no more room in its existing shelters. It is winter, and the weather is freezing. Your engine company responds to a smoke alarm activation. You find that it was set off by one of the families, who were lighting some paper bags to warm their hands. This is an intentionally set fire, not an accident. There is no malice so the fire was accidental. What action do you take next?

CHAPTER 6

Fire Protection Systems

LEARNING OBJECTIVES

Upon completion of this chapter, you should be able to:

- Describe the basic components and functions of a fire alarm system and its examination during a fire origin and cause investigation.
- Describe the basic components of a commercial sprinkler system and its examination during a fire origin and cause investigation.
- Describe the basic components of a residential sprinkler system.
- Describe special hazard suppression systems.
- Describe smoke control systems and their impact on life safety.

Case Study

The fire department had preplanned this high-rise structure extensively. This building had always been a concern for the fire department because of the difficulty of reaching the upper floors of the 20-story building. In addition, this structure was a concern because it also included five levels below grade. A fire at the lowest level would require fire fighters to work their way down through access stairs that could be acting as chimneys for the upward movement of fire, gases, and smoke. The first three below-grade levels were allocated for tenant parking, and the fourth and fifth levels consisted of individual storage cages for the tenants.

Each storage area had four rows of cages: a row built against the entire west wall, one row built against the east wall, and another two rows of cages built back to back down the middle. This arrangement created two aisles running the length of the room. All four rows were 10 ft deep; each of the dividers was also 10 ft. This created caged cubes that were 10 ft by 10 ft. The middle rows had 70-ft spaces for walking between the two aisles.

The cages were formed from 2 × 4's at floor level and at the ceiling, with an additional horizontal board at the 4 ft level. Each cage had an access door framed by 2 × 4's. All walls, dividers, and doors were covered by cage wire. Each door had a hasp for a padlock. This entire area was protected by a wet sprinkler system and ionization detectors connected to a central alarm.

On Saturday morning, a tenant went to the fourth lower level to retrieve some items from her storage cage. When she got off the elevator, thanks to an automated system, the lights came on. She could detect a slight smoke smell, but it was weak, so she assumed someone was smoking and shrugged it off. She had noticed cigarette butts on the floor in the past. She had also caught teenagers smoking in this area in the past.

As the tenant walked down the aisle, she noticed that the smoke smell was getting stronger. When she was approximately 50 ft from the elevator, she could see smoke coming from floor level in the middle row. The cage was locked, so she ran back to the elevator area; it had a phone that went directly to the security booth on the first floor. The tenant reported that she could see and smell smoke. Security personnel advised her to immediately go up the stairs to the main level, making sure the stairway door closed behind her.

As security was calling 911, the fire alarm system activated for the lower fourth floor, although no water flow alarm for the fire sprinkler system had occurred as of yet. Engine 22 arrived and security reported that there had not yet been a water flow alarm. The lieutenant and his crew started walking down to the fourth lower level with high-rise packs. A second engine arrived and started setting up to supply the standpipe and sprinkler system.

Arriving at the fire floor, the fire fighters found heavy smoke and flames hitting the ceiling in one of the storage cages approximately 50 ft from where they entered. Black smoke was rolling across the ceiling. Hooking up to the standpipe system, the crew stretched the attack line and completed a quick knockdown on the fire. The fire had extended into the cages surrounding the cage of origin and just a little beyond that. The biggest surprise was that although the sprinkler heads had activated, there was no water flow. The lieutenant sent one of the fire fighters to the outside stem and yoke (OS&Y) valve on this floor; he found it unlocked and shut off.

Although the cage was still intact, the fire fighters easily pulled it apart to access the area. The scene itself showed that the fire had started from inside the locked storage cage. There was a perfect up and out pattern, a V pattern from inside the cage at floor level. Directly on the ceiling was an area where the soot had burned away, giving a clear indication of the area of origin (**FIGURE 6-1**). The cage was locked with a padlock. In the debris, the fire fighters saw cigarette butts on the floor in the cage, near the wall against the aisle. Looking further, they found cigarette butts just inside several other cages, of different brands. The contents were common household items. Inside the cage where the fire started, there were no chemicals, no electricity, absolutely nothing that could be a heat source—except the numerous cigarette butts on the floor. The debris appeared to consist of a lot of empty boxes, Christmas wrappings, and other miscellany.

FIGURE 6-1 Burn pattern pointing back down to area of origin

The lieutenant was looking further into the possible cause when the battalion chief (OIC) and a security supervisor for the building arrived. Security solved the cigarette butt mystery. The teenagers in the building used this space to hang out and smoke. There had not been any vandalism, just the cigarettes put out on the floor. Staff swept up the common area, but did not do any other cleanup. Cigarettes that dropped or were thrown into the cages stayed in the cages unless the tenant assigned to that cage cleaned it up.

The tenants assigned to the cage where the fire started were out of state on vacation. Security called an emergency contact for the couple; when the couple called back, they identified the padlock, which was unique, and advised that they had the only keys. They also confirmed that there was nothing of value in the cage and no one would have a reason to enter that cage while they were away. This suggested that the fire was accidental in nature, but the mystery of the closed OS&Y valve remained to be solved.

The officer in charge (OIC) let the lieutenant continue looking into the origin and cause because an assigned investigator was not yet available. The lieutenant asked security personnel about the closed OS&Y. In the main office, they found a log book for service, maintenance, and tracking of all suppression systems in the building. To everyone's surprise, the log reported that the sprinkler inspection company had unlocked and closed the OS& valve for system maintenance four months earlier. There was no indication in the log book that the OS&Y had been turned back on—so it appeared to have been off for four months.

After replacing the activated heads and replacing some fire damaged pipes, the system was turned back on. Thanks to this initial investigation, the facility now has electronic monitoring (supervisory switches) on all valves.

Access Navigate for more resources.

Introduction

In every residential and commercial structure, one item that must be examined by the fire investigator after an incident is the fire protection system. This consideration becomes more vital when the incident involves a loss of life. Was the system properly designed and installed? Did the system operate properly? If so, why did the fire grow? Why didn't the victim escape?

In larger commercial occupancies, the various fire protection systems can help identify the location where each component became activated. In more modern complex systems, they can even identify the time when each device activated, which can provide essential information in the investigation of the fire incident, essentially enabling the investigator to see the direction of fire and smoke travel.

Even the smallest business or residential occupancies should at least have a smoke alarm. Part of any fire investigation is examining the device(s) and determining whether they were activated—and if not, why they did not work. Fortunately, this information is requested as part of every incident report filed in the National Fire Incident Reporting System (NFIRS).

All certified fire fighters will have received comprehensive training on fire protection systems and will have gained even further knowledge working in the field. If the investigator's jurisdiction does not have many systems, then the investigator will need to review this knowledge base periodically to stay current. One resource would be to go back to a fire fighter text, such as Jones & Bartlett Learning's *Fundamentals of Fire Fighter Skills* and *Hazardous Materials Response*. However, every investigator at every level always needs more knowledge. An outstanding text that should be in every investigator's and fire officer's library is Jones & Bartlett Learning's *Fire Protection Systems* by A. Maurice Jones, Jr.

Alarm Systems

Alarm systems range from a single-station alarm in a residential unit to a complex system of detectors (ionization, photocell, or rate-of-rise detectors) in a commercial establishment. They can be combined with extinguishing agents such as Halon or suppression systems such as fire sprinklers.

Residential Fire Alarms

In a typical residential occupancy, the alarm system will be a single-station alarm—usually an **ionization detector**, a photoelectric device, or a combination of both. These alarms may be linked to other smoke alarms in the home, allowing them to go off simultaneously, and thereby alerting occupants in other sections of the home such as the basement or second floor. The future holds some promise of improvements in these devices, given the new Underwriters Laboratory standards was implemented in 2020. These new testing criteria will require a detector to be able to distinguish between a nuisance alarm such as burned food and an actual fire. Researchers are also looking for other ways to improve the function of smoke detectors and alarms.

Detector Versus Alarm

Smoke detectors do just what the name implies. Their function is to detect the presence of smoke based on an ionization device, rate of rise, a photo-electric cell, or some other means. The smoke alarm is the sounding device that activates upon receiving a signal from the detector. In residential smoke alarms, both detector and alarm are contained within one device. In commercial systems, the detectors are independent devices connected into a central system. Upon activation of any one of the detectors, a signal is sent to the alarm system, setting off an audible alarm and potentially a visual alarm as well.

Fire alarm devices may be destroyed as a result of the fire, so it may be impossible to determine if and how one activated. The investigator should always look at the power source to assure that if properly working, it would have activated. If the alarm uses batteries, then check whether if the battery was present (**FIGURE 6-2**). Even if the fire directly affected the alarm, there will often be residue of the battery consisting of the metal casing and the hard carbon. If the system was 110 volts, check the panel to assure that the device had current.

Many modern systems use a combination of both batteries and 110-V current, which means the investigator should check both power sources. If a battery is in place with no or minimum damage, it can be tested with a multi-tester to see if it has any energy. The presence of energy means only that the device had the power to activate. If the battery is dead, that state may have been caused by damage from the heat, so this test will be inconclusive. It does not take much heat to render the device so damaged that no testing is possible.

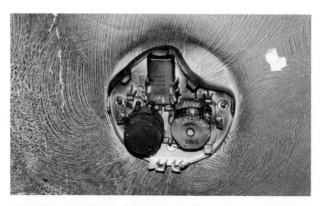

FIGURE 6-2 Residential ionization alarm still in position but damaged; battery is in place.

Courtesy of Russ Chandler.

Any survivors from the fire may be able to remember whether they heard the alarm. Also, check with neighbors or witnesses to see if they remember hearing an alarm. If the alarm activated and a fire fatality occurred, then the investigators need to expand the investigation to determine why the victim did not escape.

If the level of destruction prevents the investigator from determining if there was a residential alarm, then look in the debris below where you would expect the alarm to be located. If the wall and ceiling finish is still intact, tell-tale signs may indicate where the alarm could have been located prior to the fire.

Residential alarms can also be complex, with a control panel, zones, and combination detectors that can include smoke, heat, and carbon monoxide detectors. They could also be combined with security (burglar) and medical alarm devices. These systems are usually 110 V powered with a battery backup.

Tragic Loss of Life

All too often we find that the batteries were removed from fire protection systems, rendering them inoperative. One tragic event involved the loss of four young children, ages 2½ to 9 years, in an older motel where a single mother was renting a room by the week. The mother had no other family, and had just left her abusive husband, who had kept her isolated; as a result, she had no real friends. After leaving him, because she had no support and little money, she would lock (keyed deadbolt) the children in the motel room and go to work to support her family. There was one hot plate in the room, and the room was equipped with a smoke alarm.

By the time the fire was discovered, flames were already coming out of the windows. The investigation revealed that the motel had neither an extinguishing system nor a central alarm system. Closer investigation showed that there was no battery in the smoke alarm. The area of origin for the fire was a hot plate on the dresser, but the cause of the fire was never determined.

Nothing could change the results of the fire; no comfort could be given. However, a thorough investigation could provide information to help others and make sure a similar incident did not occur, in either that jurisdiction or neighboring jurisdictions. By entering the results of the investigation in the NFIRS, data are compiled with other similar incidents to "fight fire with facts" and help others across the nation.

Commercial Alarm Systems

Alarm systems for commercial establishments are designed for the type of occupancy and the hazards they may be protecting. When a building's occupancy changes, the alarm system many need to change as well. The investigator should take this factor into account in case of a system failure.

For these larger systems, staff at that facility usually are familiar and assist with its operation. They can identify what should happen as well as what has been happening or not happening for a time before this incident. The assigned investigator may also want to contact the alarm company installer who maintained that system to set up a meeting for the installer to review the system and share his or her findings.

If commercial alarm systems are properly designed, manufactured, installed, and maintained, ideally any incident in the facility will remain a small one, thanks to early warnings. However, if a large event occurs, then the system must be examined to determine if any faults caused a delay in the alarm. This examination may unearth information that can help identify the area of origin for the fire.

Initiating Devices

There are two ways to activate an alarm. Some alarms activate automatically based on their design and purpose. Other devices activate manually—that is, they require human interaction, such as with a manual fire alarm box.

Automatic Initiating Devices

Automatic fire detection devices will operate based on what they detect: heat, flame, fire, smoke, or gas. Their use depends on the hazard they may be protecting against. The first responder investigator will initially be concerned about what activated the device and why. For the most part, the answer will be a simple one—for example, the ionization detector activated in the motel continental dining area when a patron overcooked a piece of toast. Not every case is so straightforward, however; thus, the activation of a device may require the first responder investigator to call for the assigned investigator.

Two different types of heat detectors are available on the market today: fixed-temperature heat detectors and rate-of-rise detectors. Some devices combine the two technologies. Fixed-temperature devices are available that function at a wide range of temperatures. Similarly, rate-of-rise detectors can have different settings to match needs specific to the environment or product being protected. Other specialized devices are also available, such as the rate-compensation heat detector. The investigator will need to explore the type of system at the fire scene and—more importantly—should seek additional knowledge of these systems.

Smoke detectors are automatic initiating devices. Because they become activated when they detect smoke particles instead of heat, they are considered life-safety devices. Smoke detectors have two primary types of operation. Ionization models detect smoke through the use of a radioactive element. The element creates a constant electrical charge in the chamber between two plates. When smoke enters the chamber, it reduces the current flow, allowing the device to activate. The photoelectric detector works on the principle of having a light source and a receiver. In the *light scattering* unit (**FIGURE 6-3**), a light beam is aimed across the chamber; when smoke enters the device, it scatters that light beam, which is then picked up by a photocell. In the *light obscuration* unit, a light beam hitting a photocell is the normal state; when smoke enters the chamber, it reduces the amount of light hitting the photocell, thereby setting off the device.

Several other types of initiating devices have been developed for specialized applications, such as the duct smoke detector. Air moving throughout a commercial facility could potentially spread smoke from one part of the building to many other locations. To prevent this distribution of smoke, smoke detectors are mounted on the interior of the duct work. When such a detector becomes activated, it closes a damper, shutting off the flow of air and thereby stopping the distribution of smoke to other sections of the building. The duct smoke detector also sets off a notification device (alarm).

Air sampling smoke detectors (ASSD) are designed to pull air samples from the duct work using a small fan. This air is sampled for any indication of the presence of smoke. If smoke is detected, the air ducts are then automatically shut down, preventing the smoke and

FIGURE 6-3 The operation of a photoelectric smoke detector.

eventual fire from spreading throughout the structure. Air sampling systems are usually set up as smoke detectors. Other specialized systems are designed to pick up a specific gas associated with the structure, such as leaking ammonia from processing plants. Air sampling systems in ductwork can also be designed to measure specific concentrations of a vapor when over open vats.

Other types of detectors include flame detection (**FIGURE 6-4**), which senses infrared or ultraviolet light waves. These devices come in handy when the facility requires explosion detection and subsequent activation of explosion suppression systems.

Gas detectors are designed and calibrated to detect a specific gas or combination of gases. This type of system usually requires more frequent calibration. Like the other detection devices, gas detectors can be designed to set off a general alarm throughout the building or to trigger a specialized alarm that goes to a specific location. Gas detection is also used in residential occupancies in the form of carbon monoxide (CO) alarms for combustion furnaces or gas hot water heaters. These devices are designed to activate and set off an alarm when they reach a certain critical concentration. The devices that the engine company uses to detect CO are based on the same technology.

Manual Activation Devices

Manual activation devices (**FIGURE 6-5**) can be local, perhaps with the alarm going to an on-site station

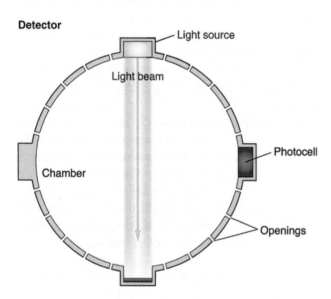

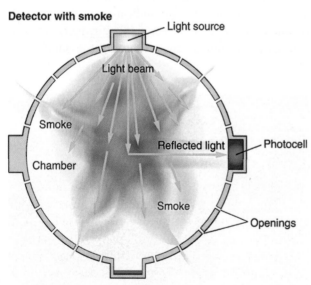

FIGURE 6-4 Flame detectors detect specific light waves as a basis for detecting and alarming. In some systems, these devices will activate the suppression system.

FIGURE 6-5 Manual pull stations.
A. © rob casey/Alamy Stock Photo, B. Courtesy of Honeywell/Fire-Lite Alarms

such as the front desk of a small hotel. Another type of manual device is monitored by an outside entity, and even in some cases monitored by the fire department. Most of these devices take the form of a manual fire alarm box. Pulling on the handle sends an alarm indicating that there is a fire, which also usually sets off an evacuation audible alarm.

Most localities have adopted one of the model codes; most of these codes, in turn, reference or directly adopt the NFPA 72, *National Fire Alarm and Signaling Code*. NFPA 72 specifies the recommended location of the manual fire alarm box, such as within 5 ft of an exit, and even specifies the mounting height so the device can easily be found during an emergency.

Fire Alarm Control Unit

Large structures—mostly commercial or industrial facilities—may have multiple initiating devices, along with multiple zones. These devices will be tied to a **fire alarm control unit (FACU)** that monitors these devices (**FIGURE 6-6**). The FACU's capabilities depend on its age. Older units provide basic services, activating a notification device such as a horn, buzzer, or flashing lights. Newer FACUs provide for alarm activation and also send trouble signals, indicating when any of the devices needs adjustment or calibration, or when a device fails, such as going offline. An advanced FACU may even interact with other building devices such as elevators, door locks, smoke control systems, and fire protection systems. When one of these systems goes

down or is not functioning, the FACU may indicate this failure by activating a supervisory signal.

The FACU can either consist of a local panel for staff to monitor or be designed to notify a monitoring service that will notify authorities. Some systems may even be monitored by the fire department or hooked up to automatically call the fire department.

From a suppression standpoint, these modern FACUs can narrow down the location of the fire based on the devices that have activated. Remote FACU monitors may be located at the entrance to the building, allowing suppression personnel to access that information upon arrival at the scene. Suppression crews may benefit from having done engine company inspections or preplanning, allowing them to become familiar with the system in a particular structure. This consideration is important because each manufacturer's panel will be different as may be its functionality.

The data from the FACU are also of importance to the investigation. If the system has time stamp functionality, it can show the fire investigator where the fire started and how the fire progressed through the structure. If the system has been silenced or reset frequently, it may indicate that the system has a problem or that something in the facility has been tripping the activating devices. If the system is not functioning or was turned off, the investigator should check the power source, including the battery backup. Also check whether the circuit breaker is dedicated to just this system and has not been turned off. Check the condition of initiating devices in areas not impacted by the fire to see if they are maintained, dirty, or blocked, which could indicate why activation may have been delayed in the fire area.

Alarm Notification Appliances

The ambient noise in the facility may dictate which appliance should be used to notify occupants of a potential fire. The appliance could just be a simple bell. However, if bells are already used for other applications in the facility, then the alarm should use a different sound, such as a horn or voice alert. Systems may also include a visible alert in the form of flashing lights (**FIGURE 6-7**) in a particular pattern, even using brighter strobe lights to get the attention of the occupants.

More recently, some hotels have installed light strips along the floor at the wall. These lights are similar to the running floor lights in the passenger area of an airplane. Both show the occupant the direction to the nearest escape. Some facilities are also putting exit signs near floor level in case the ones over the doors are obscured by smoke.

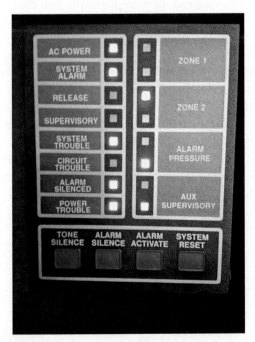

FIGURE 6-6 Modern fire alarm control unit (FACU) indicating the zone where initiated.

Courtesy of A. Maurice Jones, Jr.

FIGURE 6-7 An alarm notification appliance combining a speaker and strobe light.

False Alarms

False alarms from automated systems can stem from a faulty device or poor design—but they may also indicate that the system is operating properly. The latter possibility is why suppression units should continue their response to false alarms even if the management or occupants of that location want to cancel the call. In some incidents, although occupants could not see or smell smoke, the fire was actually present in its early stages. Thus, the system was acting as designed and the event was not truly a false alarm. Alternatively, upon arrival, the engine company may find that the system did send out a false report. In both cases, the engine company's findings must be entered into the fire incident reporting system. These statistics can be of vital importance in the future. In situations where a multitude of false alarms occurs at a particular facility, actions may be taken to reduce those incidents. Some jurisdictions enact local ordinances to charge a fee after a certain number of these false notifications to encourage facilities to repair or maintain their systems.

Manual pull stations can be a target for vandals or pranksters. In some jurisdictions, these kinds of events happen more frequently in college dorms. Deactivating or silencing the pull station is not an option due to the life-safety nature of the occupancy. Investigation of this type of incident can be conducted by the local police or fire marshal (assigned investigator). Chemicals (dyes) can be put on the interior edge of the pull station that will stain the perpetrator's hand, making it easier to catch the culprit.

Disgruntled Student

For three nights in a row, the local college female dorm had the pull station box activated in the late evening hours. The fire marshal covertly applied marker dye to the interior of the handle on the double-action pull station. The dye itself was clear when applied, but turned a bright purple after coming in contact with skin. By putting it on the backside of the interior handle, the fire marshal ensured that only the person pulling the alarm would come in contact with the dye.

The pull station was located in the lobby on the first floor, which had glass all along the south wall. After waiting until little activity was occurring in the area, the fire marshal installed a small hidden camera outside, facing the glass-enclosed lobby.

The false alarms causing evacuation of the building were just as frustrating to the students as they were for the fire department. There was no secret that the box in the first-floor lobby was the device being pulled. Dorm residents and guests seemed to keep a vigil in the lobby for a few nights. On the fourth night at 0100 hours, the alarm came in for the same dorm. Upon arrival, the engine company found that the pull station had been activated. The engine captain scanned the crowd; it was a very warm evening, and he checked for individuals who had their hands in their pockets.

The fire marshal arrived and pulled the hidden camera from the bushes where it had been swapped out each day. Sure enough, it caught what appeared to be a college-age male, who pulled the alarm and walked out of the lobby. When the fire marshal showed the video to the dorm's resident advisor, she identified the young man and pointed at him, where he was standing in the crowd outside the lobby. The dye had done its job, spreading all across the middle three fingers on his right hand.

It did not take long for the young man to confess that he had set off all of the alarms. His girlfriend, a resident of that dorm, had broken up with him the previous week. He just wanted to make sure she did not have any visitors.

The dean of students recommended charging the student as a deterrent to others who would be so foolish. The judge fined the student, and he was given 50 hours of community service, which he completed by cleaning up the fire station and the apparatus.

Additional training on alarm systems is available from local alarm companies or from associations such as the Automatic Fire Alarm Association or the Monitoring Association (formerly Central Station Alarm Association).

Sprinkler Systems

Most sprinkler systems today are considered commercial sprinkler systems. They can be found in apartments, offices, stores, and industrial complexes. Residential sprinkler systems are installed in private homes. These systems require a smaller water supply, lower-yield water flow, and different piping. Such residential systems will go a long way toward reducing fire damage and fire fatalities.

Hollywood has misrepresented sprinkler systems: Movies often depict all the heads going off at one time in an occupied facility. Although such deluge systems do exist, in which all heads go off at the same time when activated, these systems are limited to water curtains in live show theaters and specialized industrial hazards. This inaccurate portrayal of all the heads going off at once has made many individuals and companies hesitant to install sprinkler systems in both residential and commercial facilities. More than 100 years of history and science has shown that sprinkler systems save lives, however, and any water damage is far less of a concern than the smoke and fire damage.

A properly designed, installed, and maintained commercial sprinkler system will put out the fire, or at least keep it from spreading. If the system fails to do so, an investigation is warranted. The first responder investigator and the assigned investigator may not have the skill set to determine proper design or maintenance expectations for a sprinkler system, but they must understand its basic concept and components. Beyond that, the investigator should cultivate resources that can provide appropriate assistance when needed.

A typical sprinkler head has its own (activator) release mechanism that is usually a fusible link—typically two pieces of metal held together with solder (**FIGURE 6-8**). The solder is designed to melt at a specific temperature, allowing the head to activate. The next head will not activate until it, too, is exposed to that specific temperature. Then and only then will that head activate, allowing water to flow. Should the fire continue to spread, usually due to excessive fire load, then additional heads will activate as they also reach the predesignated temperature.

Not all sprinkler heads are equal. Ordinary occupancy heads will activate at temperatures ranging from approximately 135°F to 170°F. Some industrial

FIGURE 6-8 Upright automatic sprinkler heads consisting of frame, heat-sensitive element, orifice, orifice cap, and deflector.
Courtesy of A. Maurice Jones, Jr.

processes may require heads to activate at lower temperatures to assure quick activation. More often, because industrial processes create heat, the detectors may need to be set at a higher temperature to prevent accidental activation. Sometimes, the heads with fusible links have their preset activation temperatures printed on the head. They will also be color coded (**TABLE 6-1**). This allows the inspector or maintenance personnel to see from a distance what temperature head is at that specific location.

Ordinary sprinkler heads in an office environment will have no color. In contrast, the heads over an open-heated vat in an industrial complex will need to work in a condition of high ceiling temperatures, so these systems may need to have a head that will not activate until a higher temperature, such as 350°F (325°F to 275°F head), is reached. Those heads should be colored red.

The right side of Table 6-1 indicates the colors used for frangible bulb sprinklers. These heads rely on a liquid-filled glass bulb (**FIGURE 6-9**) that breaks when heated, allowing the sprinkler to activate. Each bulb color is set to activate at a specific temperature. Should there be a false alarm from a sprinkler head activation with no fire, the heads need to be examined. For example, if the false activation occurs over an industrial-heated vat area and an investigation reveals that the head was white instead of red, the fault may be an improperly installed head.

For locations with a wet sprinkler system that may have heads in areas that may be exposed to freezing temperatures, specialized dry-pendent sprinkler drops are available. They consist of a standard pendent head

TABLE 6-1 Temperature Ratings, Classifications, and Color Codings

Maximum Ceiling Temperature		Temperature Rating		Temperature Classification	Color Code	Glass Bulb Colors
°F	°C	°F	°C			
100	38	135–170	57–77	Ordinary	Uncolored or black	Orange or red
150	66	175–225	79–107	Intermediate	White	Yellow or green
225	107	250–300	121–149	High	Blue	Blue
300	149	325–375	163–191	Extra high	Red	Purple
375	191	400–475	204–246	Very extra high	Green	Black
475	246	500–575	260–302	Ultra high	Orange	Black
625	329	650	343	Ultra high	Orange	Black

Reproduced with permission of NFPA from NFPA 13, Standard for the Installation of Sprinkler Systems, 2016 edition. Copyright© 2015, National Fire Protection Association. For a full copy of NFPA 13, please go to www.nfpa.org.

FIGURE 6-9 Frangible-bulb sprinkler heads activated when the liquid in the bulb expands and breaks the glass.

attached to a pipe that is filled with air under pressure. When the head opens, the air escapes, which in turn allows the seal on the top of the drop to open, letting water flow (**FIGURE 6-10**).

Sprinkler System Grid

The sprinkler heads, which usually have a ½-in. thread, are attached to a pipe of the commercial sprinkler system. This system consists of an array of overhead pipes, sometimes above the ceiling, but always below the roof. Pipe size is calculated based on the amount of water needed at the farthest point. The sprinkler heads may be mounted directly to these pipes or may extend down from a smaller pipe. These heads are considered upright if the deflector is at the top and the threads are at the bottom. If the head is installed with the deflector at the bottom, pointing downward, then

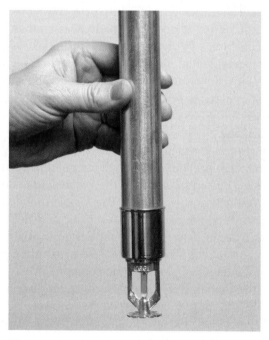

FIGURE 6-10 A dry-pendent sprinkler head, a dry drop, used to protect areas that are exposed to a freezing environment.

it is considered a pendent. The upright and pendent types of heads have different deflectors and cannot be interchanged—a distinction that may be important to the investigator. If the sprinkler system did not perform as required, the investigator must determine if the cause was a flaw in the system design, installation, or maintenance.

FIGURE 6-11 shows a complete commercial sprinkler system. Water supply is controlled either from

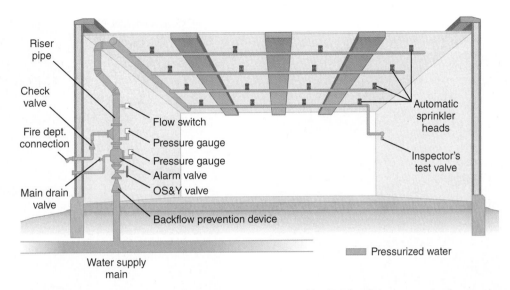

FIGURE 6-11 Automatic sprinkler system including sprinkler heads, piping, control valves, and water supply.

FIGURE 6-12 Post indicator valve (PIV) used to open and close an underground valve.

FIGURE 6-13 Outside stem and yoke (OS&Y) valve to control the flow of water into a sprinkler system.
© Jones & Bartlett Learning. Photographed by Glen E. Ellman.

a **post indicator valve (PIV)** (**FIGURE 6-12**), which is located outside the building, or an **outside stem and yoke (OS&Y)** (**FIGURE 6-13**), which is usually found inside the building. Other components include a backflow preventer valve to assure that contaminated water inside the sprinkler system does not flow back into the water supply. The alarm valve is the heart of the system. Its large clapper valve allows water to flow into the system. Pressure gauges above and below the alarm valve indicate both the water pressure of the water supply and the pressure of the water inside the sprinkler system.

Should the system fail, the first item to check is the water supply. If public water is flowing, then the investigator should check the indicator valves to assure they are in the "on" position. Although a window in the PIV indicates whether the valve is open or shut, it is always a good idea to unlock and rotate the valve to assure it is truly open. The investigator should also check the OS&Y to confirm that the stem is extended, showing that it, too, is open. In addition, the fire investigator should inspect the fire department connection (**FIGURE 6-14**) to assure that nothing had been put in this connection to block the flow of water.

FIGURE 6-14 Fire department connections enable the fire department to hook up an engine to increase the pressure and volume of water supplying the sprinkler system.
Courtesy of A. Maurice Jones, Jr.

FIGURE 6-15 A water-motor gong sounds when water is flowing in a sprinkler system.

When a sprinkler head opens, it allows water to flow. This, in turn, opens the alarm valve. With the clapper of the valve open, water flows to a water-motor gong outside the building. This gong will activate, sounding an alarm-like bell (**FIGURE 6-15**). An electronic alarm may also be attached to the riser. It relies on a butterfly valve inside the pipe that moves when water flows, sending an electronic signal to a FACU.

Dry-Pipe Sprinkler Systems

Dry-pipe sprinkler systems are installed in structures that are unheated and located in areas with the potential of freezing weather, such as large warehouses. Instead of water, this system is filled with pressurized

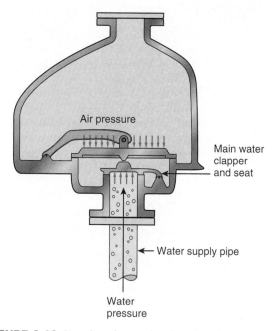

FIGURE 6-16 Dry-pipe alarm valve showing air pressure holding the clapper down to prevent flow of water until a sprinkler head activates.

air. The air comes from a compressor with a switching mechanism that starts the compressor when the air pressure in the system registers a preset level. The dry-pipe alarm valve (**FIGURE 6-16**) is designed so that air pressure holds down the clapper, keeping water from entering the system until an activation occurs. When a head is activated, from either a fire or an accidental discharge, air flows out, releasing the pressure on the clapper and allowing it to open. The system then fills with water, allowing water to flow out the sprinkler head.

Accelerators and Exhausters

Because a dry-pipe sprinkler system is filled with air and has only one head opening, there can be a delay in water flow, especially if the open head is located at a distance from the valve. To overcome this disadvantage, the system mayincorporate devices to accelerate the water flow or exhaust the air, thereby allowing water to flow more rapidly. The accelerator is placed at the alarm valve, where it can sense the drop in air pressure from a head activation. When this happens, the accelerator allows air pressure to flow to the supply side, enabling the system to react more quickly—that is, to accelerate.

An exhauster is another type of accelerator that is usually installed at some distance from the alarm valve. When it senses a drop in air pressure from an opening sprinkler head, it becomes activated. The exhauster's larger-diameter opening allows the air to escape more quickly. Once it senses water, the device will close.

Preaction Sprinkler System

A preaction sprinkler system is similar to a dry system in that there is no water in the pipes and the air is not pressurized. When an initiating device such as a smoke detector sends a signal to the FACU, this unit sends a signal to the preaction valve, allowing it to open the clapper so water can flow into the system. This type of system is a good choice in areas where damage might occur from hitting heads or pipes.

Deluge Sprinkler System

Deluge sprinkler systems are installed in areas where a large volume of water is needed, such as an airplane hangar or large industrial processes. In these situations, this system may also have the capability of flowing foam. All the heads for this system are open. The system can be activated by an initiating device through an FACU or from a pilot-line detector head. When activated, it opens a deluge valve, flowing water through all the heads.

Residential Sprinkler Systems

Residential sprinklers are designed to protect egress pathways; they use quick-response residential sprinkler heads (**FIGURE 6-17**) to control incipient-stage fires. These systems use plastic pipe, and they flow

FIGURE 6-17 Residential sprinkler heads react more quickly than commercial heads and use less water.
Courtesy of A. Maurice Jones, Jr.

water from the domestic water system. For the fire fighter, knowing how to shut off the residential sprinkler system is a priority to reduce water damage once the fire is extinguished. The fire investigator must take these sprinkler heads into account when looking at burn patterns or the direction of fire travel.

Dry and Wet Chemical Extinguishing Systems

Seen in many commercial kitchens, dry and wet chemical extinguishing systems are installed primarily to protect areas where flammable liquids are used. Newer facilities tend to rely on wet chemical systems (Class K extinguishing agent) that are more effective with cooking fats. Class K extinguishing agents have quick extinguishment capability and are easier to clean up after the fire. Many of these systems also protect the ductwork and have manual activations.

Dry systems also protect against cooking-fat fires and shield ductwork. In addition, they shut off the fuel/electricity to contain the situation. Some of these systems provide for automatic activation of alarm systems locally and sometimes send alarms to monitor companies. Dry systems can be used at fueling terminals and over specialized processes.

Special Hazard Suppression Systems

Not all occupancies are best protected by a standard sprinkler or chemical suppression systems. Specialized systems have been developed to protect everything from a copy of the U.S. Constitution (Halon system) to commercial airplane hangars (specialized foam systems).

Halon-based extinguishing agents were an innovation when they were initially introduced as fire suppression systems. Today, Halon and its future replacement agents are used in computer server rooms and data centers, as well as in vaults protecting historic documents, books, and other valuables. However, concerns about Halon's side effects have prompted the industry to research and identify safer extinguishing products. Notably, this toxin can damage eyes and skin. Individuals exposed to a Halon release can experience cardiac and central nervous system effects. In high concentrations, Halon can create an oxygen-deficient environment. New products such as FM200, Inergen, and FE13 are now replacing Halon agents.

Some newer specialized systems are designed to react and apply an agent more quickly than existing fire suppression systems do. This speedy response is important with processes such as electrical components manufacturing. Some commercial kitchens continue to use outdated dry chemical systems in kitchen hoods. Although the newer powder systems are somewhat easier to clean up, the introduction of wet chemical systems represents a major improvement. The liquid in such systems immediately cools the cooking oils and, like the powder systems, creates a foam layer over the top of the cooking oil, thereby preventing re-ignition. Thus, wet systems are more efficient and easier to clean up after activation.

Smoke Control Systems

Mechanical systems control the movement of smoke. Some of the earliest of these systems consisted of controlled openings in the roof directly over the stages in theaters—and some are still in use today. Because early 20th-century theaters used candles for illumination, along with an abundance of cloth and paper props, it was common for fires to start on the stage. In such an event, these early smoke control devices would open, allowing the smoke to escape rather than flow in the patron area and endanger audience members' lives.

Today, smoke control systems come in all shapes, sizes, and designs. Many of these systems are part of an HVAC duct system. When fire is detected, the system shuts down the ductwork using dampers, which prevents smoke from traveling throughout the structure. Smoke control systems are often found in places like atriums or as part of a stair-pressurization system. The primary purpose of these systems is protection of life, especially when occupants are expected to shelter in place.

From the investigation aspect, each alarm and detection system present in a fire-involved structure needs to be checked to ensure that it activated and performed properly. This examination is especially critical if the incident led to injuries or fatalities. Even if there were no victims, the investigator should check the system thoroughly. Making sure that the system worked at the time of the fire could also assure that it will properly activate in the future.

Fire Investigative Perspective

If a fire occurs in a facility that has a fire protection system, the investigator will need to closely examine that system and determine whether it functioned as expected. Each system has its own unique features that are critical for operational efficiency. The investigator must be familiar with the various types of systems to be able to ascertain whether they operated properly, malfunctioned, or were tampered with. Of course, investigators cannot be proficient in all aspects of an investigation, but should have resources available to assist when necessary. As they progress through their career, investigators will continually add to the sum of their experiences and become more proficient themselves. In the meantime, they should not hesitate to reach out to local fire sprinkler and suppression installation companies. Many offer training opportunities to fire service personnel. Topics can range from explaining the components of the systems, how they operate, and how to spot malfunctions.

Wrap-Up

SUMMARY

- Fire alarm systems are a key to fire life safety. They can vary from the single-station alarm found in residential homes to the complex systems installed in larger manufacturing facilities. Alarm systems in structures with large occupancies, such as theaters and hotels, can be activated manually from a pull station or automatically from sensors.

- In larger facilities with comprehensive systems, a fire alarm control unit may assist in the suppression operations. This unit can also prove valuable in the investigative process, potentially showing where the first alarm activated and, therefore, the area of origin as well as the direction of fire travel.

- False alarms need to be investigated to determine if a failure of the system occurred. Such information is important to provide data for future improvements. Of course, the primary reason is to assure the facility has its faulty system repaired.

- False alarms can also stem from the action of individuals. This, too, needs to be investigated to correct that behavior. When malice is involved, it may be advisable to let the courts handle the incident, with the perpetrator's punishment intended to act as a deterrent to other vandals.

- Commercial sprinkler systems can be found in apartments, hotels, stores, manufacturing, and storage facilities. Properly designed, installed, and maintained sprinkler systems may extinguish the fire, or at least keep it from spreading. The investigator needs to examine any such system that fails to control a fire to determine whether a failure occurred, especially if the incident led to any injuries or fatalities. In a large complex facility, the sprinkler system can be daunting for investigators who do not have extensive training in these systems. The sprinkler industry can provide insight into most of these systems and potentially training for the fire service.

- Residential sprinkler systems are not new to the industry, but their rate of adoption has not grown as quickly as the fire safety world would want. These systems are designed to use less water, which enables them to operate from the domestic water supply in the home. They use deferent sprinkler heads and piping specifically designed for residential occupancies.

- Special hazards require special protection, and the fire protection industry has provided products to meet the needs of many of these hazards, ranging from foam systems to protect airplane hangars to chemicals that can react almost instantly and save a computer server room. Historic and valuable documents and books could be damaged just as badly by water as by fire; thus, the vaults where these valuables are stored and protected by special chemical systems. As with other fire protection systems, the investigator will need to identify the cause and impacts of any failure of specialized fire protection system.

- Smoke control systems either block smoke to keep it from traveling to other parts of a structure or remove the smoke to enable occupants to safely evacuate a building.

- Systems designed to prevent or reduce the damage from fire are intended to allow occupants to evacuate or be protected in place should a fire occur. The fire investigator's role is to examine these systems to determine whether they operated properly and if not, how they failed. A failure could be the result of a malfunction, improper maintenance, or even poor design. This determination will be even more important if any injuries or deaths occur as a result of the fire.

KEY TERMS

Accelerator A component of a dry sprinkler system that is designed to induce air (from the dry pipe system) into a chamber of the dry pipe valve to reduce the amount of time it takes the system to activate allowing the flow of water into the system.

Class K Extinguishing agents for fires in cooking appliances involving cooking oils.

Deluge sprinkler system Fire suppression method using open sprinkler heads; when a valve opens, it supplies to all heads simultaneously, immediately flowing water.

Duct smoke detector Device that samples the air in the air distribution system for products of combustion. When activated, it closes dampers, thereby preventing smoke from traveling through the system.

Exhauster An early action device installed on a dry sprinkler system. When a sprinkler head activates, this device will remove the air in the system,

allowing a more rapid delivery of water to the sprinkler head.

Fire alarm control unit (FACU) A device that serves as the "brain" of the fire alarm system, monitoring and managing its operation. It serves as a link between initiating devices and notification devices. Most systems have primary and secondary power.

Fixed-temperature heat detector Device that activates at a predetermined temperature.

Heat detectors A device designed to sense and respond to thermal energy (heat) at a predesignated rate and or temperature. There are two classifications, *rate-of-rise* which will respond to a given rate of temperature increase and a *fixed temperature* device that will react at a specific temperature.

Ionization detector Initiator that uses a small amount of radioactive material to ionize air between two metal plates, creating a constant current between two plates.

When smoke enters the chamber, it reduces the current, setting off the alarm.

Outside stem and yoke (OS&Y) Sprinkler control valve with a visible stem that allows an investigator to identify whether the valve is open or closed by observing the length of the stem (screw) exposed. Occasionally referred to as outside screw and yoke.

Photoelectric detector Detector using a light beam and a receiver. Visible particles of smoke entering the chamber either scatter the light or block the light, causing the alarm to activate.

Post indicator valve (PIV) Sprinkler control valve with an indicator that will read open or shut.

Rate-of-rise detector A device that responds when heat exceeds a predetermined rise in temperature.

Smoke detectors Automatic initiating devices that detect visible particles of smoke and send a signal to initiate an alarm.

REVIEW QUESTIONS

1. What are the three types of smoke alarms?
2. How does a smoke detector differ from a smoke alarm?
3. What are initiating devices? Provide examples.
4. What is a manual activation device?
5. What is the purpose of the duct smoke detector?
6. What is a FACU?
7. What are indicating valves? Give two examples.
8. Describe a deluge system.
9. What is a dry drop and why would it be used?
10. What fuels are involved in Class K fire?

DISCUSSION QUESTIONS

1. What can you do in your community to encourage residential sprinkler systems?
2. You responded to a medium-sized manufacturing facility for an alarm; it turned out to be an accidental activation. While there, you notice that this new business has an extremely clean and well-maintained facility, but you see that its flammable liquid storage cabinets have had holes drilled in them and sprinkler heads are mounted on the inside of the cabinets. As per the manufacturer specifications, the safety feature of these cabinets has been violated by the holes. This is also a violation of the fire prevention code. However, the sprinklers seem to go above and beyond to provide safety. What do you say or do?

Electricity and Fire

LEARNING OBJECTIVES

After completing this chapter, you should be able to:

- Describe basic electrical concepts.
- Describe a typical residential electrical system and its components.
- Describe some common electrical failures.
- Describe electrical wire size, sheathing, and usage.
- Describe how the use of improper electrical components can create sufficient heat for ignition.
- Describe circumstances in which conductors can melt.
- Describe the role of static electricity in an ignition sequence.
- Describe generators and solar electricity, as well as the associated potential for a fire.

Case Study

The house was brand new, sitting in an affluent new subdivision. It was a two-story, wood-frame, vinyl-sided house. The electricians had finished their final work only a few days prior to the fire. The homeowner was moving in, and had turned on the air conditioning for the first time as boxes were being carried into the house.

The fire was discovered by neighbors who were out walking their dog. When the fire department arrived, flames were visible in the left rear corner of the home. By the time the fire was knocked down and under control, it had consumed a good portion of the back entrance, laundry room, and utility room. The remainder of the home had both smoke and heat damage.

The fire officer in charge identified the area of origin based on burn patterns. He also started getting statements from witnesses before they disappeared. The couple who reported the fire stated that they first saw the flames in the front right corner of the house—the exact opposite side of the house from where the most damage was found. Furthermore, the front right corner suffered little to no damage. With witnesses contradicting the burn patterns, the officer called for an assigned investigator.

The investigator discovered a clear, definitive area of most damage in the back left corner of the house. Exactly in the center of this area, on the underside of where the flooring would have been, was a charred electrical junction box under the debris. The floor joist where this box would have been secured was completely consumed. However, this site did not match witnesses' reports of the fire.

Even though it was daylight, the fire fighters were using large floodlights inside the structure. The investigator moved these lights to the area with heavy damage, and aimed them down at the hole in the floor, into the crawlspace. The fire scene evidence indicated that this area was the origin of the fire. The investigator then went to the sidewalk and stood in the exact place where the neighbors said they were standing when they saw the fire. The mystery was solved: The neighbors were looking at the house from across the street in a direct line of sight from the right front corner to the left rear corner. The house was on a slight rise that was approximately 5 ft above the level of the sidewalk. From the witnesses' perspective, they were looking into the open vent for the crawlspace. Directly back from this angle of view, the investigator could see the glow from the fire department lights located in the left rear of the structure.

With this confirmation, the investigator returned to his designated area of origin. The crawlspace burn patterns clearly indicated that the fire went upward from this point, in the area of the junction box. The only available heat source was the number 6 aluminum electrical wire and a junction box where the wire fed from the panel down to the box. The wire then attached to the two outside heat pump units located on the right side of the house.

The aluminum wiring in the area of origin had melted, and the junction box was sitting on the ground in the carbon debris from the burning structure. It was at the bottom of this debris, indicating that it was involved in the early stage of the fire. The steel box was heavily damaged but still had the metal cover plate attached. After taking the screws off the cover plate, the investigator discovered that the only items remaining inside were the metal spring residue from six wire nuts (wire connectors), along with some carbon and melted aluminum.

Tracing the wire back to the undamaged areas showed that it was an aluminum number 6 conductor (wire) designated as two 6s and 8. This indicated that the plastic vinyl sheath contained two insulated number 6 wires with a number 8 bare ground wire, all aluminum.

The inside of the cover plate showed clear signs of electrical arcing. After taking the box and the wire nuts as evidence, the investigator compared them to the code requirements. It was discovered that the box was too shallow for the number of wires and connections being made. A comparison of the wire nuts also indicated that they were too small for the size wire they were trying to connect. **FIGURE 7-1** shows the box

FIGURE 7-1 Electrical box from the area of origin showing the size of the box.

Courtesy of Russ Chandler.

FIGURE 7-2 New electrical box showing the size required by the National Electrical Code for the size of the wire and number of connections to be made in the box.
Courtesy of Russ Chandler.

from the area of origin, which was a little more than 1½ in. deep. In contrast, **FIGURE 7-2** shows the correct size box for the wire size and number of connectors; it is more than 2 in. deep.

As noted earlier, the electricians had finished the job of wiring only a short time prior to the fire. Taking the electrical box, the burned wire nuts, and photographs, the investigative team met the electricians at the scene the day following the fire scene examination. The master electrician was the owner of the electrical company. He was accompanied by his journeyman electrician, who was not quite 20 years old.

The fire investigator knew that the interview must include even the most obvious questions. By conducting the fire scene investigation first, before the interview, he obtained many of the answers to those questions. The response to those questions from the person being interviewed can indicate if he or she is being honest or deceptive. The master electrician forthrightly stated that he had not personally wired or checked every connection done on the job. He turned to his apprentice and made it quite clear that he expected the assistant to answer every question honestly and clearly so that the fire investigation team could find out what really happened.

The fire investigators showed the electricians the area of origin, explained their process of coming to such a conclusion, and showed them the evidence collected at the area of origin. The apprentice hung his head and said that he was worried that that connection might have been a problem. His truck still had the supplies used on the last day he worked at the house. The apprentice pulled out new wire nuts and stated that he knew he should have gotten the proper size but carried none on the truck and wanted to get the job done. He also stated that he found a nick in the wire leading to the heat pumps and discovered there

was enough play in the wire to allow him to cut the wire and create a splice.

The apprentice knew he needed a larger box, but none was available on the truck. Rather than drive the 20-plus miles back to the shop, he made do with the box he had. He confirmed that the box was smaller than the one needed and that he had secured the box to the floor joist in the crawlspace in the area where the fire started.

Not once did the owner of the electrical company, the master electrician, stop the younger employee during his story. In fact, he encouraged the young man to continue. Such ethics were certainly commendable.

When the wire nut did not fit on the two wires, the apprentice shaved the aluminum wire down to get the two wires to fit in the connector, but he did not get as good a connection as he would have liked. After making the connections, he found that no matter how he twisted or turned the wires, he could not get the cover plate in position over the connections. Laying on his back in the crawlspace, he found that he could put his knee on the cover, pushing it up in place enough to get the screws started, connecting the cover to the box. He then tightened it down and crawled back out.

At that point, the apprentice turned on the HVAC unit and it powered up; since it worked, he shut it down and left the job. After telling his story, it hit home to him that he was responsible for the fire. The master electrician apologized on behalf of his company and stated he would contact his own insurance company. Before leaving, he let the investigator keep the wire nuts from the truck, which were placed with the burned wire nuts from the scene (**FIGURE 7-3** and **FIGURE 7-4**).

FIGURE 7-3 The remains of the wire nuts from within the electrical box in the area of origin, alongside wire nuts taken from the same box used by the electrician when the box was wired.
Courtesy of Russ Chandler.

FIGURE 7-4 An attempt to get two wires of the same size as were used on the scene to connect, using the wire nuts used by the electrician.
Courtesy of Russ Chandler.

Given the lack of malice on the part of the electrician, the county did not take any legal action. With an accidental fire, there was no further action on the investigators' part other than completing the fire incident report (NIFRS). The homeowner's insurance company covered the entire loss. It was later learned that the insurance company for the electrician did reimburse the homeowner's insurance company in a process called **subrogation**.

Introduction

Electrical fires frequently represent one of the top four leading causes of fires in the United States. This chapter provides the first responder with the basic knowledge of electricity and its relationship in fire cause.

Electricity plays into many of the other chapters of this text, from spoliation to sources of ignition. A thorough understanding of electricity is critical for every investigator, from the fire officer doing the initial cause and origin investigation for the fire incident report to the assigned investigator looking at an incendiary fire. The investigator cannot eliminate electricity as a fire cause unless he or she understands how electrical systems operate properly versus fail.

> **TIP**
>
> An incendiary fire cannot be affirmed until all other potential heat sources are eliminated.

The first step toward understanding electricity is to understand the basic electrical system components and their installation. That includes not just the components installed in new construction today, but also those that met the code in the past and remain in use. This chapter covers basic electricity concepts and points to electrical failures that can lead to a fire. Just

as important, it points out some findings that are not necessarily proof that the fire started from an electrical source.

Neither this chapter nor an entire book could provide the information necessary to supply all of the knowledge for an investigator to be competent in this area. Assigned investigators must understand **Ohm's law** and know how to apply it to understand the relationship between current and resistance. They must be able to work the formulas to obtain **amperage**, **voltage**, and **watts** for any given situation. It is recommended that every person conducting an investigation should seek additional training in this area.

> **TIP**
>
> Assigned investigators must understand Ohm's law and know how to apply it to understand the relationship between current and resistance.

A further recommendation is to establish a professional relationship with a local electrical contractor, or an electrical inspector, who can be very helpful in explaining different issues with electricity as well as give the investigator an opportunity to get some hands-on training on how components are supposed to be installed and how they are supposed to operate.

The investigator should look within his or her own department—it is surprising how many off-duty fire fighters do electrical work on their days off as a second career.

Basic Electricity

Our homes and industry use alternating current (AC) for their power needs. Direct current (DC) is restricted to batteries in vehicles and various portable devices. At the most basic level, both sources of electricity provide the energy necessary to run common electrical devices. Both also provide electricity as the result of the flow of electrons, and have hazards that can cause injuries and start fires.

> **TIP**
>
> Today, direct current is restricted to batteries in vehicles and various portable devices. In contrast, homes and industry use alternating current for their power needs.

Although electrical energy is capable of being a source of ignition, it is really the appliance or wiring that creates the hazardous situation. When appliances, electrical components, or wiring have been improperly designed, manufactured, or used, electricity has the potential to be the energy that starts the unwanted fire.

> **TIP**
>
> Although electrical energy is capable of being a source of ignition, it is really the appliance or wiring that creates the hazardous situation.

Flow of Electrons

Typically, comparing water to electricity is a good way to understand how electrical current flows though wires. To keep this analogy as simple as possible, consider a faucet at the kitchen sink. The water flows through the pipes, up to the faucet, and into the sink, where the water is used; it then flows out through the drain. The pipes are like electrical wires, and the water is under pressure, which we measure in pounds per square inch. In wires, we measure this pressure in volts. The larger the pipe, the larger the amount of water that can be carried, which is measured in gallons per minute. The larger the wire, the more electricity that can be carried, which is measured in amperage.

When it reaches the sink, water is used to wash dishes or clean vegetables; essentially, it is used to do work. Likewise, the electricity powers a light or runs a fan; thus, it does work, just like the water in the sink. From this point, both the water and the electricity flow back to the ground, neither under pressure.

Electrical failures can also be compared to water problems. If the pipe has a leak, we have less pressure and not as much water as we need to do the washing. A failure of the wiring insulation that allows the electrical energy to leak to ground will do the same: It will provide less power to do work. In turn, lights may dim and electrical motors may slow down. Conversely, if the pipe is over-pressurized, it may burst, or maybe an appliance will fail and start leaking. An overload of electrical current on a wire could also cause it to fail, melting the insulation. The overload could then go into an electrical appliance, causing that appliance to fail as well.

The electrical energy flow can also be taken down to the molecular level. Consider a pipe with an interior diameter of 1 in., which is filled with marbles having a 1-in. diameter. (The only reason for using such large marbles is to keep them from falling out in this thought experiment.) With the pipe full, push just one more marble into one end of the pipe. When this happens, one marble will be pushed out of the pipe on the other end. The energy to make this happen resulted from the marble that was pushed in, which affected the first marble. That marble pushed on the second marble, and the second marble pushed or impacted the third marble, on and on down the line, pushing out the last marble.

Although the process is not exactly the same as marbles in a pipe, the effect is similar when considering electrons. Electrons do not go in a straight line like the marbles in the pipe, but as one more electron is introduced into the wire, the impact is felt all the way down the line, pushing out electrons on the other end, back into the ground. The electrons flowing through the wire are the energy potential that can make things such as lightbulbs and appliances work. We want to keep the electrons under control, they love ground: They would like nothing better than to escape to ground, and will do so through any possible avenue.

> **TIP**
>
> Conductors are items that allow the flow of electrons. Insulators are items that block the flow of electrons.

Keep in mind that this analogy is an overly simplistic way to look at electron flow and other aspects of electricity. Electrons do not feel any real emotion of *like* or *dislike*, nor do they demonstrate any thought whatsoever; instead, science and the laws of physics govern the effects of electricity and the actions that make things happen. It is imperative for investigators to seek additional training in this area.

Today's Electricity

In North America, the primary delivery of electrical power for distribution and for larger occupancies, such as malls and factories, is referred to as three-phase power. Three-phase power is ideal for large electrical motors because the delivery system provides for consistent power, allowing a smoother operation and reducing vibration and motor movement.

Should the fire involve three-phase power, it will be a major event requiring a full-time investigator's efforts. The first responder investigator is more likely to investigate fires with single-phase power, which is the focus of this chapter. This type of power supply is more frequently found in residences and small businesses.

Single-phase power is delivered with two wires and a ground either to the weather head, sometimes called a service head, for above-ground delivery or up through the ground to the meter for underground service. The wire coming from overhead is called a **service drop**, and the wire coming from underground is referred to as a **service lateral**. The meter—the next part of the delivery system—allows the utility company to

measure the amount of electricity used. Removal of the meter from the **meter base** is also a way of cutting off the electricity to the structure.

> **TIP**
>
> Logic would suggest that the absence of the meter means that the electricity to the structure has been cut—but this may not always be the case. Either intentionally or owing to a failure in the system, some current can leak through, energizing the circuits of the structure. Never rely on the power company meter being removed as the basis of deciding that the electricity is completely cut off. Always use a multimeter to test the circuitry at the panel and in other locations where working around electrical lines.

> **TIP**
>
> Never rely on the power company meter being removed as a basis of deciding that the electricity is cut off.

Service Panel

From the meter base, the electricity enters the structure to the main electrical **service panel**. The power comes in on two wires, with a bare neutral wire also being present. Each power wire provides 120 V, so together the pair provides 240 V. Once inside the panel, the current is broken down into branch circuits (**FIGURE 7-5**).

FIGURE 7-5 Diagram of an electrical circuit breaker service and sample branch circuits.

For larger appliances, 240 V is supplied. For lights, outlets, and smaller appliances, 120 V is supplied.

Inside the service panel, a ground bar allows the connection of all neutral and ground wires. The sides of the panel box have partial precut knockouts that, with application of a slight force, can be turned into openings for the insertion of bushings that protect the wire as it enters the panel. An assortment of these connectors are available on the market, from metal screw types to plastic units that snap in place. Blank covers for these holes are available in case a hole is no longer necessary or was knocked out by mistake.

Grounding

It is essential that the entire electrical system is grounded. Ground with the earth is part of the system's safety function and helps avoid fires, especially during a surge such as a lightning strike. The grounding block in the service panel should be connected via an adequate-size wire to a grounding source. The ground wire should be attached to metal plumbing pipes if they extend at least 8 ft into the ground or to a ground rod, which must be driven into the ground to the depth where the ground stays moist. Usually, this requires a ground rod that is at least 8 ft of copper or anodized steel. Every section of the electrical service must be grounded. The continuous use of the bare conductor will lead to the system maintaining ground. If metal conduit is used, it must also be grounded to ensure a safe environment.

> **TIP**
>
> It is essential that the entire electrical system is grounded.

Floating Neutral

The service connection coming into the structure consists of two hot wires and one neutral. With a properly working system, the neutral helps balance the system, by providing a fixed point of zero, and each hot wire provides 120 V. When the neutral wire separates, it is called a floating neutral. Without the neutral, the voltage can vary between the two hot lines. It could be 180 V on one line and only 60 V on the other, although these amounts will vary. Thus, when testing with a voltmeter, the investigator may notice that the voltage is only 80 V on one branch, yet when the investigator tests that same branch again in a few minutes, it could be 140 V. The varying voltage will become obvious to the occupants because incandescent lights will get bright or may go dim. The problem could be too much current, which could cause fires, or too little current, which could cause damage to motors. A floating neutral is not a grounding issue, but one strictly related to the parting or breaking of the neutral wire that comes in from the power company.

"Floating neutral" has another meaning associated with generators. In those applications, the floating neutral is an intentional design in the system to allow for proper operation. This should not be confused with the floating neutral discussed in this chapter, which is a potential source of the failure in an electrical distribution system.

> **TIP**
>
> When the neutral wire separates, it is called a floating neutral.

> **TIP**
>
> A floating neutral is not a grounding issue, but one strictly related to the parting or breaking of the neutral wire that comes in from the power company.

Overcurrent Protection Devices

For a common residential structure, overcurrent protection devices come in the form of either fuses or circuit breakers. These devices are designed to protect the system from a short circuit or dead short, which could cause overheating of wires, equipment, or devices and result in damage or fires. However, a 20-ampere (amp or A) fuse does not automatically blow, nor does a 20-amp circuit breaker trip, when current reaches 21 amps or even 30 amps. These protection systems must be more complex to handle the various components in a residence. Motors, when starting up, will take considerably more energy than when they are operating. The system must be designed to handle the surge of a motor starting, yet protect as rapidly as possible against an arcing wire to prevent a fire.

> **TIP**
>
> Fuses and circuit breakers are devices designed to protect the electrical system from a short circuit or dead short, which could cause overheating of wires, equipment, or devices and result in damage or fires.

Overcurrent protection devices may be activated by an extreme draw of electricity or by heating. Thus, if a fire occurred near an electrical panel, the external heat could trip the breakers. This possibility must be kept in mind when conducting a fire investigation and examining the electrical panel.

Fuses

The fuse service panel, depending on the age of the panel, may be anywhere from 30 amps to 400 amps. Older fuse panels could have as few as two screw-type plug fuses. More frequently, older panels include two primary cartridge fuses that act as a disconnect, and another set of cartridge fuses that feed one appliance. Branch circuits enable one fuse to serve each branch. The National Electrical Code or the local electrical code will dictate how many devices (light fixtures or outlets) can be fed off one branch, and this will be based on the size of the wire, among other things.

The two primary types of fuses are **plug fuses** and **cartridge fuses**. The plug fuse has a brass screw base; a safety version has a smaller-diameter ceramic screw base and screws into a brass holder base that is the same size as the standard plug fuse base. The top of the plug gives information as to the amperage and type of fuse, such as a delay fuse. The clear-view center window allows one to look inside and check the status of the fusible link (i.e., intact or blown). If the fuse blew as the result of a dead short, the lead from the link will splatter on the inside surface of the fuse window (**FIGURE 7-6**). If the fuse blew because of other reasons, the link will simply melt and separate, stopping the flow of electricity.

The window of the fuse will be either round or hexagonal (six-sided). A hexagonal shape indicates a 15-amp fuse, whereas a round window indicates a 20-amp or higher rating. Many manufacturers also place the amperage on the button, which is found on the bottom of the fuse (**FIGURE 7-7**).

FIGURE 7-6 Plug fuse blown as the result of a dead short. A splatter of ink can be seen on the underside of the fuse window.
Courtesy of Russ Chandler.

FIGURE 7-7 Samples of both plug and cartridge fuses.
Courtesy of Russ Chandler.

Cartridge fuses also have a fusible link on the inside connecting the two ends of the fuse. If too much electricity flows through the link, it will melt, breaking the connection. Some cartridge fuses have replaceable links, but most are disposable. The ratings of the fuses are printed on the outside cylinder. Some also have the amperage stamped on one of the copper ends. To check whether a cartridge fuse has blown, one would have to use a multimeter using the ohms settings to see if the circuit can be completed. If it cannot, then the fuse is blown.

Circuit Breakers

Circuit breakers perform the same function as fuses, but more efficiently. When a fuse blows, a new one must be purchased and installed. By contrast, breakers trip when a failure occurs and can be reset by just flipping the switch back on. The rationale for use of breakers is not based so much on cost as on convenience.

Many manufacturers of breakers provide a way for the breaker to indicate its status. Once tripped, the breaker either goes to the center position (**FIGURE 7-8**) or has a small window in which green will show when the breaker is operating normally and red will show if the breaker has tripped. To reset the breaker, the homeowner simply turns off the breaker, and then turns it back on.

The downside of breakers for the investigator is obvious: Once a breaker has been flipped back to the on position, there is no way to know what happened to that breaker or which branch failed. This constitutes spoliation, as discussed in Chapter 8, *Spoliation* (**FIGURE 7-9**). However, if left untouched, the breakers can be secured and sent to a laboratory for X-rays and examination.

> **TIP**
>
> The downside of breakers for the investigator is obvious: Once a breaker has been flipped back to the on position, there is no way to know what happened to that breaker or which branch failed.

Circuit breakers for 120-V branches have a single pole—essentially, one place to attach a wire. Breakers for 240 V tap into both lines coming into the panel and are considered double pole; that is, they have places for two power wires to be attached. The double-pole breaker sometimes has one switch, but may also have two switches that are attached with a bridge at the factory. Specialized breakers such as **ground fault circuit interrupter (GFCI)** breakers are available for use in outside or interior locations subject to water use, such as kitchens, bathrooms, and laundry areas, to provide greater protection from electrocution. Likewise, delay breakers are used when motors start up and draw a large amount of electricity at that time.

Breakers come in smaller sizes as well. A half size allows for two breakers to fit in the space of one full-size breaker, and a quad size allows four 120-V breakers to fit in the space of two breakers. When two half-size breakers are joined together, they are sometimes referred to as **tandem breakers**.

FIGURE 7-8 Breakers in the center position have tripped. The investigator can examine those branches to determine whether they had any involvement in the fire's cause.
Courtesy of Russ Chandler.

FIGURE 7-9 Circuit breakers removed from the panel and then flipped on and off by an enthusiastic fire fighter. This constitutes spoliation and could hamper the investigator's ability to make an accurate determination of fire origin, cause, and responsibility.
Courtesy of Russ Chandler.

A great advancement in fire safety was the 2002 National Electric Code's requirement that arc fault circuit interrupters (AFCIs) be installed on all bedroom circuits in new construction. These breakers not only sense a short and shut down, but also sense an electrical **arc** from problems such as wiring with damaged insulation, frayed cords, or overheated cords. This change was implemented as part of an effort to prevent fires in residential occupancies during the night that would endanger the occupants during sleep.

TIP

A great advancement in fire safety was the 2002 National Electric Code's requirement that arc fault circuit interrupters (AFCIs) be installed on all bedroom circuits in new construction.

Disconnect Box

Additional electrical distribution boxes are fed by the main electrical panel but are designed to service a specific machine or appliance. In homes, these boxes may be used for furnaces or hot water heaters. They are seen even more frequently in industrial occupancies, where they are used to provide electrical current to specific large machines.

The additional boxes can use plug fuses, cartridge fuses, or circuit breakers. Those with breakers rely on the breaker itself to act as the mechanism that shuts off the appliance. Those with fuses usually have a handle assembly of some type to allow it to be shut off in one motion.

Fuses Doing Their Jobs

Sometimes what looks like a bargain can prove very costly in the long run. A small printing company needed to replace its commercial printing machine. To handle more orders, the company needed a larger machine, but because it had not been paid for those new orders, there was a cash flow problem. The solution was to purchase a used printing machine, which is exactly what the company did.

When installed, the larger machine needed more electrical current to operate. The existing subpanel for the older printer could not be upgraded. Replacing it with new breakers would cost more than the capital the company had on hand. However, the electrician advised the owners that they could use an older cartridge fuse panel disconnect that he had in his shop.

Installation went well, but the machine constantly blew the cartridge fuses. The operator would pull out the large blown fuses and install new fuses. The blown fuses were just dropped on the floor. After several weeks of continuously replacing fuses, and having to wait for someone to run to the store and get new fuses, the operator made a discovery that he thought was the solution to his problems.

When the installation was done on the machine, it was an industrial application, so the National Electrical Code required that all wires be run in conduit. Small pieces of the metal conduit that had been trimmed off when doing the installation were found on the floor near the burned-out fuses. The operator noticed that the two items, fuses and conduit, were almost the same diameter. He took the conduit to his shop and cut off pieces of the conduit to

be the same length as the fuses. He then inserted the pipe in the disconnect box in place of the fuses, closed the box, and pulled the handle to the on position.

Things went great for about an hour, but then employees noticed a slight smell of smoke. In minutes, smoke was coming from the printing press. The workers called 911, but before the fire department could arrive, the entire printing press had gone up in flames. When the engine company interviewed the operators, nothing was said about the fuses. However, the lieutenant noticed the old fuses on the floor near the disconnect and called for an assigned investigator. During the investigator's fire scene examination, the disconnect shown in **FIGURE 7-10** was found.

FIGURE 7-10 Electrical disconnect fuse box with electrical conduit inserted instead of the required fuses.
Courtesy of Russ Chandler.

Common Failures Associated with Breakers and Fuses

Sometimes there may be an obvious indicator of an electrical problem, such as a stack of blown fuses sitting near the electrical panel. This can prompt an entire string of questions to be asked in the interview with the structure occupant or owner. Attempts have been made to bypass the protection offered by fuses by placing a penny behind the fuse, allowing it to come in contact with both the center button and the outside surface of the brass screw base. Other techniques to bypass the plug or cartridge fuse may also render any protection useless, all because someone either did not want to buy more fuses or take the time to constantly change them out or correct the electrical problem causing the overload.

Circuit breaker design has improved over the years to prevent tampering and bypassing. Older breakers could be prevented from tripping if the switch were held in the closed position by tape or even jamming materials into the breaker to keep the switch from moving. Today's breakers will trip regardless of the physical position of the switch, preventing this type of tampering.

> **TIP**
>
> Circuit breaker design has improved over the years to prevent tampering and bypassing.

The investigator should know the basics of the national and local electrical codes and know there is a reason for each section of these codes. For example, one requirement states that the covers on fuse or breaker panels must be in place with doors closed. Besides the obvious potential for electrocution, hot melted metal fragments might fly from the box, igniting nearby combustibles, should a massive system failure occur without a cover in place.

Another reason for covers being placed on panel boxes, disconnects, and even over electrical boxes is to reduce the amount of debris (potential fuel) that enters these devices. Contaminants such as dust could also serve to impair the proper operation of switches and devices.

The first part of an electrical system examination should be to ensure that each branch circuit has the proper level of protection. A common branch with electrical outlets may be protected with a 20-amp breaker or fuse. If the branch is providing energy to just light fixtures, it may have a 15-amp breaker. However, if either is protected by a 30-amp breaker, that finding should prompt further examination—not just on that branch, but on other branches, disconnect boxes, and fixtures.

Always check for attempts to bypass the overcurrent protection devices, such as nails through cartridge fuses. Just as homeowners, or even arsonists, can be unlimited in their imagination to overcome protection devices, the fire investigator must be open minded and unlimited in creative thinking when looking at a system where the overcurrent protection devices have failed.

> **TIP**
>
> Always check for attempts to bypass the overcurrent protection devices, such as nails through cartridge fuses.

Wire

When thinking about conductors, metals are the first thing that come to mind. It is not the strength of the metal but its ability to share electrons that allows current to pass freely. This ability is measured as relative conductance,[1] where silver represents 100 percent conduction. Obviously, silver is too expensive for common structural wiring, so copper (98 percent relative conductance) has been the metal of choice for wiring homes and businesses.

> **TIP**
>
> It is not the strength of the metal but its ability to share electrons that allows current to pass freely.

Many stores sell audio and video cords with gold connectors, advertising that they are the best choices. This claim is not made because gold is a better conductor; in fact, gold has only 78 percent relative conductance. Instead, the gold connectors are used because gold is the least chemically reactive of the metals. In other words, it will not corrode or oxidize as readily as will other metals, so it makes for better audio and video connections.

Fiscal concerns explain the failure to use copper for large transmission lines. That is, aluminum, at

68 percent relative conductance, was chosen for the task because of its much lower cost. In the 1960s, the cost of copper began to rise. To lower the costs of new construction, the industry started providing aluminum wiring for residential and commercial structures. To get the same amount of electricity through the lines as copper, the electrical codes called for aluminum wiring to be used that was one size larger when replacing copper or used instead of copper.

Wire size numbering may not seem to make sense. For example, small bell wire is size 22 gauge, whereas the wire size to power an electrical outlet is size 12 gauge. There is a logical explanation, however. The American Wire Gauge (AWG) is used to standardize the size of electrical conductors. Note that an entirely different scale is used to measure steel wire, the W&M Wire Gauge.

To make wire, it is drawn (pulled) through draw plates with a preset hole. Usually, the wire is drawn from one size down to the next, then down to the next, and so forth. The size of the wire comes from the number of times it takes to draw the wire down to a particular diameter. It takes more draws to create size 22 wire than it does to create size 4 wire. Wire sizes from gauge 5 through 14 are also subject to another measurement in this process: The wire is equal to the number of bare solid wires it takes to span 1 in. For example, suppose you take the same size solid conductors and lay them side by side, and it takes 10 conductors to span 1 in. In this case, the size of each conductor is 1/10th of an inch, or size 10.

> **TIP**
>
> The size of the wire comes from the number of times it takes to draw the wire down to a particular diameter.

Solid wire is used for permanent electrical systems. This is true up to the larger size wires, such as 8 AWG or larger. At this size, the larger solid wire would be hard to handle, especially when making bends to get to panels or electrical equipment. Manufacturing the wire in strands enables better flexibility for better handling. Further, around this size, use of aluminum wire leads to a considerable cost savings.

Smaller wires used for plugging in appliances and lamps are also stranded, largely for flexibility and durability. Any wire bent time after time will stress and eventually break. The smaller strands are more flexible and resilient to a certain point, but even they will

fail after a period of time. In most motor vehicles, the electrical wires are stranded to account for the constant movement and vibrations.

Wire Insulation

Electrical conductors today are coated with a plastic vinyl to insulate the wire and thereby prevent the escape of electrons, which could cause injuries or heating that could lead to ignition of nearby combustibles. Insulation also prevents the two conductors from coming in contact with each other and causing a dead short, which has a high probability of starting a fire. These plastic-coated conductors are designated as NM, which stands for nonmetallic. The pliability of the plastic enables the wire to be bent at 90-degree angles without damage or failure. Of course, constant bending will cause the insulation to crack or tear.

For structural wiring, each conductor carrying and returning the electricity is individually coated. A ground wire is present as well, but is usually not coated. All of these wires are then coated with another layer of plastic vinyl for protection. Some manufacturers place paper between the coated wires and the outer vinyl sheath, with additional paper surrounding the ground wire as well.

A different type of sheathing is used for underground wire. To protect it from the damp conditions of being in the ground, the sheathing is solid plastic vinyl, rather than not just a coating. It encompasses each insulated wire as well as the bare ground wire. This type of wire carries the designation of UF, which stands for underground feeder.

The sheathing (**FIGURE 7-11**) also provides information about the wire, which is printed on the

FIGURE 7-11 Sheathing of wire with plastic vinyl protects the insulator, ensuring no leakage of current occurs. The outside sheathing indicates the size of the wire, number of wires, and usage of the wire.
Courtesy of Russ Chandler.

surface or imprinted in the sheathing itself. This label indicates whether the wire is NM or UF, and it identifies the size of the wire along with the number of insulated conductors. The bare wire is not included in this number but rather is designated as "G" or "Ground." A wire designated as 12/2 G contains two insulated number 12 wires with a bare ground; 10/3 G contains three insulated number 10 wires with a bare ground wire.

Some wires do not need the outer plastic sheathing. For example, individual insulated wires are available to be used in plastic or metal conduit (**FIGURE 7-12**)—that is, tubing designed to carry wiring from one point to another. Conduit is typically used in areas where the wiring may be exposed and susceptible to damage. A flexible metal sheath called Greenfield is also available, though it is more aptly described as flexible armored cable.

The color of the insulation on the individual wire usually indicates its use or function. The bare wire or wire with green insulation is the ground wire. Black and red wires carry current. White wire is neutral and carries current at zero voltage, completing the circuit back to the panel. Be aware that in some wiring, the white wire will carry a full load. When this is done, the electrician should tape or paint the end of the wire black for this designation; however, this is not always accomplished on the job. Unfortunately, not all wiring is done to code, so any color wire may actually be carrying the load.

Older Wiring

Older wiring is not necessarily dangerous. Some complications arise in residential homes that have older, ungrounded wiring and outlets (the two-prong-only type). Using adapters can sometimes provide grounding, but this is not always the case. Even in some older homes that are upgraded to newer wire, electricians may not replace all circuits, instead leaving the older disconnected wire in the walls.

The left side of the photograph in **FIGURE 7-13** depicts the turn-of-the-century wiring known as knob and tube. Both the knob and the tube were insulators made of ceramic that held the wire in place as it was run throughout the structure. The wiring on the right side of the photograph is post–World War II wiring that was insulated with a rubber and cloth covering, eliminating the need for ceramic insulators. It includes only two wires. In the center is a piece of conduit, a metal tube containing the wires; the conduit itself acts as an additional ground if properly installed.

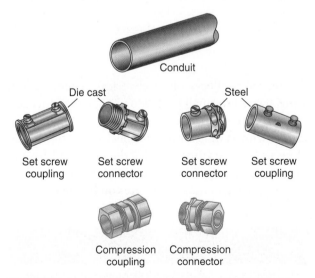

FIGURE 7-12 Conduit is required in some industrial and business applications. The conduit and connectors also serve as a grounding path back to the electrical panel. Greenfield, also sometimes required by code, provides flexibility during the installation process and in areas where the conduit may experience flexing or slight movement.

FIGURE 7-13 Examples of older knob and tube wiring, post–World War II wiring, and newer conduit wiring.
Courtesy of Russ Chandler.

National Electrical Code

The National Electrical Code (NEC) is actually a product of the National Fire Protection Association (NFPA) code development process. This panel of experts requires items in the code based on the fact that omitting them would be a hazard. Many of the items in the NEC represent lessons learned from known fires; incorporating corrections into the code is an attempt to prevent future similar occurrences.

> **TIP**
>
> The National Electrical Code is a product of the National Fire Protection Association code development process.

For example, when the NEC requires a certain size box for a certain size wire and number of connections, that specification exists because it is the safest way to do that particular installation. The connectors, known as wire nuts, are also specifically designed for a certain number of wires of a certain size. To exceed that number of wires results in an unstable joining of the wires—one that will fail at a later date when the wires creep out of the connection, resulting in a loose connection and arcing that could result in a fire.

Any poor connection could result in heating. When a wire is attached to an outlet or light switch, there should be sufficient contact to ensure that all current passes through it without excess heating. When looking at an outlet or other similar connection, observe the condition of the wire and screw for pitting or damage. Loose connections will result in arcing, which in turn can cause damage.

A tight connection is critical to the proper operation of any device and for the safety of the system. Sometimes a loose connection can occur as a result of the type of wiring being used. At other times, other factors in its use and properties of different metals come into play in creating a hazard.

> **TIP**
>
> A tight connection is critical to the proper operation of any device and for the safety of the system.

Resistance Heating

Resistance is present in all aspects of electrical energy delivery. Less resistance occurs with copper than with steel, which explains why we use copper for the distribution of electricity. In many appliances, we want the resistance to carry out the task for which the appliance is designed. For example, a portable electric heater is designed to have resistance across the heating element, which in turn causes the element to heat up, turn red, and give off heat, accomplishing the intent of the device.

At other times, resistance is not wanted. For example, both a loose connection and a corroded wire limit the amount of surface area for the flow of the electricity. With the same amount of energy going through a smaller area, there will be resistance. The increased resistance, however, creates heat—sometimes to the point of creating a glowing connection. Should combustible materials be present in the vicinity, an unwanted fire may occur. The rest of this chapter looks at other resistance issues that can create problems.

Common Failures Associated with Wire and Insulation

It is impossible to cover all types of failures, but some of the more common possibilities warrant discussion here. Most important, the investigator must remember that the fuses and circuit breakers cannot prevent all electrical failures. A failure of an electrical extension cord will be insufficient to trip a 15-amp breaker. As such, the cord could continue to heat, starting a fire, and the breaker will not trip.

> **TIP**
>
> A failure of an electrical extension cord may be insufficient to trip a 15-amp breaker.

Common failures of an electrical system include using a wire too small for the load to be carried, causing the wire to overheat. However, a considerable safety factor is built into most electrical systems. Just using a 14-gauge wire instead of the required 12-gauge wire may cause the wire to warm up, but is unlikely to produce enough energy to ignite combustibles.

> **TIP**
>
> A considerable safety factor is built into most electrical systems. Just using 14-gauge wire instead of the required 12-gauge wire may cause the wire to warm up, but is unlikely to produce enough energy to ignite combustibles.

Damaged insulation, which can happen during installation, may allow continued degradation and may eventually lead to a short circuit. Wires that are nicked or stretched cannot carry the normal load; they, too, could heat up to the point of ignition for nearby combustibles. Stretching can be caused by electricians pulling the wire during installation. Other activities can also stretch wires, such as when plumbers or carpenters find wires in their way and pull and stretch them to get them out of the way of their project. Nicking of wires or the insulation can be caused by pinching or improper use of staples to hold the wire in place—a properly installed staple will just hold the wire snug.

Wires are run in the walls and through the studs through drilled holes. Without careful planning, the nails used for attaching the sheetrock or other materials could penetrate the wall and be driven into the wires, causing harm or short circuits. Finding this problem after the fact is not always easy, but it is possible with a thorough investigation. Sometimes finding such damage in other areas of the structure, such as that shown in **FIGURE 7-14**, can support the findings that this could have happened. However, by no means does finding such a hazard elsewhere support this reason as the cause of the fire in another location. Instead, the investigator must look for direct evidence.

Other items that may indicate the condition of the electrical system or use of electricity in the structure include excessive use of extension cords and use of frayed or patched electrical cords (**FIGURE 7-15**). Degradation of structural wiring from vibrations causes chafing of the insulation. This can result in some slight arcing from the hot wire to ground, but

FIGURE 7-15 This wiring was not involved in the fire but gives an indication of the potential condition of wiring elsewhere in the structure.
Courtesy of Russ Chandler.

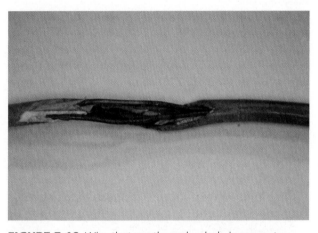

FIGURE 7-16 Wire that ran through a hole in concrete chafed to the point of removing the insulation. Occasional arcing occurred on the wire, which caused the beading on the side of the wire.
Courtesy of Russ Chandler.

may not actually cause the breaker to trip. The wire in **FIGURE 7-16** ran through a hole in concrete pipe and was attached to an electrical water pump. The constant vibration of the pump also led to vibrations of the wire, which chafed it on the concrete. Notice the small beading on the side of the wire: This is where occasional arcing took place. Fortunately, this problem was discovered before it caused more damage.

Other causes of damage to the insulation of power cords can come from an exterior source of heat, such as a portable heater. There can also be heat buildup from cords being insulated by carpets or other items in a living space such as newspapers or magazines. Power cords for most small appliances and lamps and small extension cords are designed to dissipate the heat. When this does not happen, the heat buildup could lead to failure of the insulation, potentially exposing wires, which could lead to an electrical short.

Sleeving of the insulation is an indicator of internal heating, which itself is a sign of overcurrent in that

FIGURE 7-14 Nails put in wall penetrate wiring within the wall. The fire started in another room, but appeared to be electrical in nature. Thus, the investigator undertook a search to see if there were other electrical problems in the structure.
Courtesy of Russ Chandler.

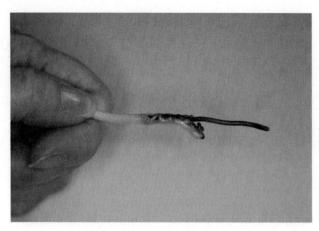

FIGURE 7-17 Sleeving on a single conductor, the result of the wire heating up the interior, allowing the insulation to loosen. Wire slides freely within the insulation.
Courtesy of Russ Chandler.

branch. Sleeving becomes evident when the insulation melts enough to allow it to slide easily back and forth over the wire (**FIGURE 7-17**).

> **TIP**
>
> Sleeving of the insulation is an indicator of internal heating, which itself is a sign of overcurrent in that branch. Sleeving becomes evident when the insulation melts enough to allow it to slide easily back and forth over the wire.

Aluminum Wire Problems

As mentioned previously, the increasing cost of copper raised the cost of wiring in the mid-1960s. Aluminum was found to be a good substitute for copper. It was considerably less expensive, was an abundant resource, and was flexible enough to allow for ease of handling in running the wire and making connections. The only drawback of aluminum was that it does not have the same ability to carry current as copper does. An easy fix was to use a slightly larger aluminum wire that would allow it to carry the same load as a smaller copper wire. A number 12 aluminum wire can carry the same current as a number 14 copper wire.

> **TIP**
>
> Aluminum does not have the same ability to carry current as copper does. An easy fix is to use a slightly larger aluminum wire that allows it to carry the same load as a smaller copper wire.

Aluminum wire does have a characteristic that proved detrimental in the early years of its installation. As the new wire became more widely adopted, electricians just substituted it for copper wire, but gave no thought to the potential problem of using copper connectors with aluminum wiring. Where the aluminum wire wrapped around the screw post on an outlet or other fixture, it was found that unlike copper, the aluminum would expand slightly when heated. The heating was caused by the normal flow of electricity to the device. When the aluminum wire expanded, it put pressure on the copper or copper-clad screw post. Over time, the screw would ever so slightly back off, causing it to loosen on the wire. When this happened, less surface area came in contact between the wire and the device. This caused the same amount of electricity to flow through a smaller area, creating even more heat. Eventually, the problem created a *glowing arc*, producing excess heat and producing light. These conditions have been known to cause fires.

Another issue with aluminum wire is that the surface of the wire, when exposed to the atmosphere, will corrode and form aluminum oxide, a white powder. This powder is nonconductive, so it creates more resistance to the flow of electricity. Copper wire also oxidizes on the surface when exposed to the atmosphere, but the compound created on the surface of copper is conductive.

In 1972, Underwriters Laboratory and the wiring device manufacturers came to an agreement on the testing to be used on devices designed to be used with aluminum wire. If a device could pass this test, it would be designated as **CO/ALR**. Prior to this time, manufacturers were allowed to label their devices for use with aluminum with no testing or documentation, with the devices being designated as **AL/CU**.

Thanks to the latest advancements, there have been fewer fire-related events with aluminum. Recent testing done by the Wright-Malta Corporation[2] included 1000 devices with 4000 wire terminations; half of the devices were the older technology AL/CU and the others were the newer CO/ALR (**FIGURE 7-18**). The samples used were also from the various manufacturers. All devices were wired properly, according to the National Electrical Code. There were many failures and burnouts with the AL/CU devices, and only one failure with the newer CO/ALR device. Note, however, that the use of aluminum wire, installed properly, is not a fire hazard in itself.

Other devices and connectors now allow copper and aluminum to be spliced together. These devices, according to the National Electrical Code, must be marked explicitly to indicate that they can be used in such a situation. The investigator should check the

FIGURE 7-18 An AL/CU connector. The CO/ALR connectors use newer, safer designs.

Courtesy of Russ Chandler.

marking on any connectors found; if such a connector has been damaged, the investigator should attempt to get an example of what was used to assure it was a proper connector.

Arcs

When electrical current jumps across a gap between differing electrical potentials, the resulting discharge of energy, called an arc, produces a high-temperature luminous discharge. The distance between the gaps can be dictated by the amount of energy potential. NFPA 921, *Guide for Fire and Explosion Investigations*, describes arcs as a high-temperature luminous electric discharge across a gap or through a medium such as charred insulation.[3]

> **TIP**
>
> When electrical current jumps across a gap between differing electrical potentials, the resulting discharge of energy, called an arc, produces a high-temperature luminous discharge.

Usually the arc is of such short duration that it will not be a competent ignition source for most nearby combustibles. However, under the right conditions and with the proper fuel, a fire could occur. In an atmosphere of ignitable dust or vapors in the correct proportion, arcing could lead to ignition.

A parting arc occurs when the path of the electricity is disrupted, as when an electrical wire separates from a connection. For the brief moment when the wire starts to leave, the flow of electricity will attempt to continue to flow across the gap, producing an electrical discharge. This can be seen when an electrical plug is pulled from the outlet, especially if there is a load on the cord attached to the plug.

> **TIP**
>
> A parting arc occurs when the path of the electricity is disrupted, as when an electrical wire separates from a connection.

Arc welding utilizes a parting arc. The circuit is completed when the rod is touched to the grounded metal. Pulling the rod away from the metal creates a parting arc; this energy is enough to melt metal. Of course, this arc is not sustained with common electrical current. A special arc welder must produce the current specific to make this a continuous process.

A spark from an arc is a molten piece of metal that is capable of flying some distance and still having sufficient heat to act as a competent ignition source for ignitable materials. The distance traveled depends on many factors, including the energy being carried by the wire and the size of the molten globule.

Arcing Through Char

The insulation material surrounding the wire is combustible. In its designed form, it prevents the escape of electricity. However, when wiring is exposed to fire, the insulation will burn and char. As the insulation turns to carbon, it will allow the flow of electricity, thus arcing through char. This condition usually occurs as a result of the fire; that is, it is not necessarily the cause of the fire.

Because the amount of current flowing through the char may be insufficient to trip breakers or fuses, the arcing may continue and create a path of damage toward the panel. As the arcing takes place, the splatter of metal will be noted. Careful examination may point back to where the arcing originally occurred, farthest from the panel, which can help pinpoint the area of origin.

> **TIP**
>
> Careful examination may point back to where the arcing originally occurred, farthest from the panel, which can help pinpoint the area of origin.

Static Electricity

Static electricity is just as it implies: It is electricity that is *static*, not in motion. In the process of movement, different materials may pick up extra electrons. This can occur when a person walks across a carpet on a dry day, but a conveyer belt or even the flow of liquids may also pick up electrons. The accumulation of electrons creates a negative charge; then, as electrons are given off, a positive charge develops.

The electrical potential exists on the surface of a nonconductive body; this body is said to be charged. The electrons are essentially trapped and have no potential of movement until they meet a conductive body. At that time, the excess electrons flow to seek balance, and a small discharge occurs. This discharge is sufficient to create light and sometimes it will even make a snapping or popping noise—for example, when the finger touches a door knob. Under most circumstances there is no damage, although some people may experience a momentary pain if they are part of the completion of the circuit. The static arc discharged from the tip of a finger will not burn the skin or cause any damage because it is extremely small in amperage and very short in duration. In contrast, if this discharge takes place in an atmosphere of ignitable dusts or vapors in the right proportion, the discharge may potentially contain enough energy to be a competent source of energy to cause the ignition.

When investigating whether static electricity was involved in the ignition sequence, the investigator must determine if a potential for such a discharge existed. Several safety regulations cover the flow of ignitable liquids that require bonding and grounding. For example, a 55-gal drum of an ignitable liquid sitting on its side in a cradle is required to be bonded to the ground. If liquid is to be flowed from the spigot of the drum to a container, a ground strap attached to the drum must also be attached to the container so that both are on the same potential and do not collect additional electrons. Without this bonding and grounding, a discharge could build up and cause a static discharge. Under the right conditions, it could be volatile.

TIP

When investigating whether static electricity was involved in the ignition sequence, the investigator must determine if a potential for such a discharge existed.

The investigator should be familiar with all the regulations associated with any process being used in and around the fire scene. With this knowledge, a line of questioning for the witnesses and victims will be more insightful and beneficial for the ultimate discovery of the fire origin and cause.

Melting of Conductors

Not directly associated with the ignition sequence, but something that will aid the investigator in the fire scene examination, are issues dealing with the melting of conductors. Copper melts at 1981°F,[4] and aluminum melts at 1220°F.[4] Because of its low melting temperature, aluminum is likely to melt in any fire. Any patterns created by the melted aluminum wire will be of little use.

TIP

Copper melts at 1981°F, and aluminum melts at 1220°F. These temperatures can vary slightly based on the amount of contaminates within the copper or aluminum wires when they were manufactured.

Copper can provide information based on its physical condition during the fire scene examination. If there is no impact on the copper and insulation is intact, the investigator knows that limited heat was present in that area. When bare copper is found, that is a clear indication that the heat was more intense in this area. Because the insulation burned off, the investigator should see some discoloration of the copper, such as dark red or black coloring, as a result of the oxidation.[5] The copper wire itself can also start to show damage as heat increases. Blisters will appear as the copper starts to break down. Then, the wire will start to melt, showing dripping on its underside. As this process continues, more melting will start to thin out the wire, reducing it in size. There will be no sharp line of demarcation, and the end of the exposed wire may appear pointed.

Tracing the wire damaged from external heat from the undamaged area to the damaged area will aid in determining the direction of fire travel—namely, from areas with more damage to areas with less damage. If the wire is missing, it will be difficult to find evidence of the melted copper. Nevertheless, the investigator can likely find evidence of wire installation, such as drilled holes and staples, that may help in tracing the wire path.

When an electrical short occurs in an energized wire, the wire will have a different appearance than when it melts. The energy of the short may cause small globules of wire to fly off and fall back on the wire, resolidifying. The size of the globules may be easier seen with a magnifying glass. The end of the wire will have a **bead** showing a clear line of demarcation.

> **TIP**
>
> The energy of a short may cause small globules of wire to fly off and fall back on the wire, resolidifying.

Some electrical arcing can be less severe, causing a notch in the wire instead of a separation of the wire. This notch may also have small globules attached to the area near the notch, and the notch will appear melted. In contrast, a mechanical notch may have scratch marks or misshaped wire.

A laboratory can examine the wire to make a more definitive evaluation as to whether the wire was energized, and whether the damage resulted from an electrical short or melting from flame or hot embers. The most important question in regard to an electrical short and the fire scene is determining which came first: Did the fire cause the electrical short, or did the electrical short cause the fire?

> **TIP**
>
> A laboratory can examine the wire to make a more definitive evaluation as to whether the wire was energized, and whether the damage resulted from an electrical short or melting from flame or hot embers.

Generators, Solar Electricity, and Future Enhancements

Generators

Electrical power backup, a resource once reserved for the high-tech industry, is now available for residential occupancies thanks to its lower cost. Along with this new widespread use of generators come new hazards and issues. Many systems are automated, coming on when the main power fails, provided they are properly installed. Unprofessional installations, usually in homes, have resulted in surges when the main power came back on, leading to fires or injuries.

Fuel systems have increased hazards in and around homes. Portable generators require larger stores of gasoline. Larger whole-house generators could require propane or natural gas, which translates into lines to city systems or larger storage tanks. The operation and refueling of these systems have created some problems. Notably, pouring fuel into the generator tank while the unit is hot or while it is still running has resulted in the ignition of the vapors.

Grounding will usually be mentioned in the manufacturer's literature. It is always a sound practice, but smaller units and those with a frame usually do not need to be grounded because the frame itself creates its own ground. However, if the unit is being used in a stationary location, it never hurts to create a ground. Something as simple as stretching vehicle jumper cables from the frame to a ground point could always serve as an extra layer of protection.

The investigator working a fire involving small equipment such as generators would be wise to not assume their operational features. By searching the Internet, a manual of operation can be found from most manufacturers and for most models. This could be a valuable tool to determine if the equipment was properly set up and maintained.

Solar Power

Photovoltaic (PV) energy systems are becoming more prevalent in both commercial and residential occupancies. Residential home systems are capable of producing 1000 V or more. If incorrectly installed or maintained, such solar power systems could easily be a source of energy to start fires or cause electrocution. This is an area of constantly changing and improving technology. With each major incident, electrical codes are updated to provide additional protection.

Solar systems use battery banks for energy storage, which could provide a source of energy for a future fire. Science and industry are working to develop more and larger units for both commercial and residential use.

Even someone with considerable knowledge about residential electricity may find it challenging to delve into a failure of a solar system that resulted in a fire. As with all general areas, it is always helpful to have a resource to assist in such an investigation, in the form of the electrical inspector or solar system installers. Fire investigators should also consider checking with their colleagues in neighboring jurisdictions to see if they have experience with these systems.

Electric cars are discussed in Chapter 14, *Vehicle Fires*. Some of the technology discussed in that chapter will be relevant to some battery-operated systems in homes and industry. Again, the investigator will need to obtain more training on these units and work to develop a knowledge base.

Wrap-Up

- The old adage that "a little bit of knowledge could be dangerous" is never more true than when talking about someone making a determination of origin and cause in a fire that involved electricity. Every investigator, but especially the full-time assigned investigator, must become very knowledgeable about electricity, the electrical code, the potential for electrical failure, and electrical issues to ensure that he or she can conduct an accurate and credible investigation of the fire origin and cause.

- Electrical theory is important to understand. The investigator must be able to explain these underlying principles to others to sell them on a theory of a fire's area of origin and electrical cause. A fundamental understanding of the components of an electrical system, the code mandates, and common practices are also a great tool for the investigator.

- Knowing what is not a potential heat source can be just as valuable and important as knowing what can be a competent source of heat for a fire. The investigator must also recognize the limitations on overcurrent protection devices—what these devices cannot protect as well as what situation they should be able to prevent.

- Wiring is designed to carry a certain electrical load. Knowing the limitations of all wiring can assist the investigator in recognizing how much of a load could create sufficient heating to cause a failure. Connections such as wire nuts, outlets, and switches associated with the wiring also have a designed load limitation. Examining all of these components in and around the area of origin is crucial to either identify a failure as the competent heat source or eliminate it as being involved in the fire cause.

- Static electricity can be a competent heat source, but only when the available fuel—such as ignitable vapors or dust—is present in the right proportion where the energy can cause ignition. The investigator must determine if there was the potential for the discharge of electricity.

- The fire investigator needs to examine the condition of the electrical wire conductors and their sheathing, as the extent of their damage can indicate the intensity of the fire. The appearance of the wire will differ depending on whether it is damaged from exterior heat versus heat from an electrical short. It will be up to the investigator to assess all of the evidence and determine whether the fire caused the electrical short or whether the electrical short was the heat source for the cause of the fire.

- Alternative energy sources such as generators and solar electricity have their own characteristics that the investigator needs to consider. The fuel source for the operation of the generator will need to be examined as well as the associated electrical connections.

- Solar electricity systems are constantly improving and changing as new technology is introduced. A typical residential system is capable of producing 1000 V or more, providing sufficient energy to be a competent heat source. Although investigators will need to become better educated about these systems to enable them to recognize the expectations of these components, the overall operation still involves the basic electricity system.

- The most important tool that the investigator has at his or her disposal in examining a fire scene for electrical evidence is the same that has been mentioned in every chapter: The fire investigator must approach every scene with a clear and open mind and use a systematic approach to ensure that there is an accurate and plausible finding in the final hypothesis developed.

KEY TERMS

AL/CU (CU/AL) A designation for electrical connectors indicating that the device is safe for use with aluminum and copper wires. This designation is made by the manufacturer, and no definitive test by a third party can validate that the devices were designed or safe for aluminum wire.

American Wire Gauge (AWG) The standard used by the electrical industry to determine the size and designation for electrical wiring.

Amperage The strength of an electrical current.

Arc A discharge of electricity from one conductive surface to another, resulting in heat and light.

Bead The melted end of a metal conductor that shows a globule of resolidified material with a sharp line of demarcation to the remainder of the wire.

Cartridge fuse An electrical device designed to interrupt electrical current when an overload occurs. It consists of a tube encasing the fusible link.

Circuit breaker A protection device within an electrical system that will automatically cut off the electricity should an excessive overload occur.

CO/ALR A designation given to electrical devices that can safely use aluminum wiring. It is used only with those devices that meet a predetermined test accepted by both Underwriters Laboratory and the industry.

Conductor Anything that is capable of allowing the flow of electricity; a term commonly used to describe the wires that provide electricity within a structure.

Conduit A tube or trough made from plastic or metal that is designed to provide protection for electrical conductors.

Electrons Negatively charged particles that are part of all atoms.

Floating neutral A neutral that is not grounded. In a single-phase circuit, it can cause an unbalanced flow between each of the two current-carrying wires. This results in an unbalanced load, which can cause an overload and failure in appliances attached to the circuit.

Fuses Devices used as screw-in plugs or a cartridges that contain a wire or thin strip of metal designed to melt at a specific temperature associated with certain overcurrent situations. The melting of the wire or metal strip stops the flow of electricity as designed.

Greenfield The manufacturer's name for a flexible conduit. This name has become a common term to describe any flexible conduit, either metal or plastic.

Ground fault circuit interrupter (GFCI) A circuit breaker that is more sensitive than the standard breaker, tripping at a slight ground fault with the intent of preventing electrocution.

Grounding block A mass of metal within an electrical panel, affixed with holes and screws to allow for the insertion of wires; the tip of the screw binds and holds the wire in place.

Insulator A nonconductor of electricity; commonly thought of as a material such as glass or porcelain, but today includes plastic and rubber (as in the coverings of electrical wire).

Knob and tube An early method of running electrical wiring through a structure. Wires were supported by porcelain knobs, which isolated the wire from the structure. Holes were drilled into beams and studs, where a porcelain tube was inserted and acted as an insulator, allowing the wire to safely pass through. The porcelain knobs and tubes acted as insulators that kept the electricity from going to ground.

Meter base The receptacle for the electrical meter installed by the power company.

NM Stands for *nonmetallic*; refers to a covering over wire conductors intended for use within a structure to deliver electricity to various points within the structure.

Ohm's law Mathematical equation that describes the relationship between the voltage (V), current (I), and resistance (R); it can be expressed in three ways, depending on what needs to be solved:

$$I = V \div R$$
$$V = I \times R$$
$$R = V \div I$$

Plug fuse A threaded fuse that screws into a socket in an electrical panel; also known as a fuse plug.

Service drop Overhead wiring from the electrical company that is attached to the weather head to deliver electricity to the structure.

Service lateral Underground wiring that comes from the electrical company and enters the meter base.

Service panel The destination for electrical wire that travels from the meter base; it provides a means for overcurrent protection and the ability to distribute electricity throughout the structure through branch circuits.

Sleeving The slight separation of the insulation from around the electrical conductor(s), producing an effect that allows the insulation to loosely slide back and forth on the conductors.

Spark A glowing bit of molten metal debris.

Static electricity The buildup of a charge, negative or positive, as the result of items coming in contact and then breaking that contact, either taking away or leaving electrons, and resulting in the change of the static charge of the items involved.

Subrogation A legal process usually associated with insurance companies in which the insurance pays the insured for his or her loss, then seeks reimbursement from the responsible party. This process makes obtaining a settlement under an insurance policy go smoothly.

Tandem breaker Two single 15 (or 20 amp) breakers manufactured together, allowing both breakers to fit in a single slot in the electrical panel. Used to save space in the panel. They cannot be used for 208/240 volts because they are both on the same bus bar in the panel. These units are sometimes referred to as piggybacks, cheaters, or slimline breakers.

UF The type of insulation on the electrical wiring that is used as an underground feeder. The insulation tends to be thicker and solid up to the insulated conductor.

Voltage Electromotive force or pressure, difference in electrical potential; measured as a volt (V), which is ability to make 1 ampere flow through a resistance of 1 ohm.

Watt A unit of power equal to 1 joule per second.

Weather head The weatherproof device where the electrical service from the overhead (or underground) power enters the building; sometimes referred to as a weather cap.

REVIEW QUESTIONS

1. What is the difference between a conductor and an insulator?

2. What is a floating neutral? Is it found in the structural wire or on the service wire from the power company?

3. How can a static electrical arc that can hardly be felt be a source of ignition of an ignitable vapor or dust?

4. What can be seen on the inside of the glass on a 20-amp plug fuse in case of a dead short?

5. Circuit breakers will trip from an overload. What other event or condition will cause a breaker to trip?

6. In which room of a house are arc fault circuit interrupters usually installed, and why?

7. Why does the code require covers to be placed on electrical panels and boxes?

8. If aluminum wire is to be used instead of copper, what size aluminum wire should be used if the code requires size 14 copper wire?

9. Describe a copper wire that has melted as the result of exposure to a flame.

10. Describe a copper wire that is damaged as the result of an electrical short.

DISCUSSION QUESTIONS

1. Why is it so important for the investigator to have a strong knowledge of electrical components, their operation, and their failures?

2. What actions can or could be taken by the fire department (government) against an individual who improperly installs wiring that does not meet the National Electrical code and ends up being the source of energy that starts a fire in a residence?

REFERENCES

1. Research and Education Association. (2002). *Basic electricity* (p. 126, Table 7.2, "Relative Conductance"). Piscataway, NJ: Author.

2. InspectAPedia/ElectricAPedia. *Using COALR or CU-AL electrical outlets and switches with aluminum wire.* Retrieved August 25, 2008, from http://www.inspect-ny .com/aluminum/COALR.htm

3. National Fire Protection Association. (2008). *NFPA 921: Guide for fire and explosion investigations* (pp. 921–11, Chap. 3, Sect. 3.3.7, "Definitions"). Quincy, MA: Author.

4. Custer, R. L. R. *NFPA field guide for fire investigators* (pp. 4–80, Table 4.38). Quincy, MA: National Fire Protection Association.

5. National Fire Protection Association. (2008). *NFPA 921: Guide for fire and explosion investigations* (pp. 924–77, Chap. 8, "Electricity and Fire"). Quincy, MA: Author.

CHAPTER 8

Spoliation

LEARNING OBJECTIVES

Upon completion of this chapter, you should be able to:

- Describe and understand spoliation issues.
- Describe how spoliation issues are affected by first responders as well as by the assigned investigator.
- Describe and understand the remedies that courts have ascribed to those who have created spoliation issues.
- Describe what actions may not be considered spoliation.
- Describe the role of private-sector investigators.
- Describe the process of and reason for creating specific training programs and policies on spoliation.

Case Study

When working a fire scene, it is not uncommon for investigators to share feelings of loss with the family who has just experienced an absolute disaster. It is even more disturbing when one family member blames another for the incident, creating more pain on top of the fire damage.

On one particular Saturday evening, Engine 15 responded to a reported kitchen fire. The crew did a quick knockdown of the flames, but the kitchen was a total loss, and smoke spread throughout the entire house. Thankfully, everyone was safe.

The fire scene clearly indicated that the fire had started on top of the stove. In the process of documenting the scene, the fire officer noticed a frying pan with an electric burner imprint on the bottom of the pan, indicating that the burner had been on for some time. It had broken down the frying pan metal, creating the imprint. The frying pan had black carbon residue, which suggested the pan had some contents prior to the fire. Burn patterns on the wall pointed right back down to the frying pan. All of the range's control knobs had been consumed in the fire. The question: Was the burner left on, or did the stove fail, activating the burner?

During a brief interview, the officer discovered that the oldest child, a 15-year-old girl, was babysitting her two younger brothers while her parents were out for the evening. She indicated that she had fried some potatoes in a frying pan on the front right burner earlier in the evening but had not cleaned the kitchen yet. The leftover potatoes were still in the pan.

It would have been simple to mark this down as an accidental fire in which the teenager had left the burner on under the fried potatoes and then exited the kitchen. But this was not the case: The girl insisted that she had turned the burner off. She even said that she had gone into the kitchen to get a snack, and because the potatoes were cold, she got something else instead. She stated that the burner on the stove must have gotten warm on its own without anyone turning

the knob. She said her youngest brother had been blamed for leaving the burner on some weeks ago, but she knew that he did not turn the burner on, and the stove must have been malfunctioning.

The parents arrived home while the fire department was still on the scene, and the devastation was quite a shock for both the mom and dad. They were clearly relieved that everyone was safe. But when the fire officer advised the father of his findings, the mood changed. The father began chastising the daughter, screaming at and threatening her. The fire officer had no easy task getting the father calmed down.

Meanwhile, the assigned full-time fire investigator arrived, and the fire officer explained the scene, his findings, and the situation. The investigator explained to the father that they could test the switches on the stove to see if the switch was in the on position when the fire occurred. That response seemed to mollify him.

The investigator took another set of photos of the range, each control, each burner, and the overall area. In this process, the investigator noticed that the soot around the knobs appeared disturbed. When the investigator checked with the engine company officer and his crew, one of the younger fire fighters admitted that he turned the knob posts to make sure the range was off. Of course, the fire fighter could not remember the position of the knobs before touching them. After some silence, the investigator explained to the crew that any chance of discovering the original position of the knobs was now nonexistent. Worse yet, he had to explain this situation to the parents and their children.

Granted, this was a minor incident. Nevertheless, that family will never have closure; the father will have trouble trusting his daughter again in the future, and the daughter will always feel she was unjustly blamed for something that was not her fault. Further, the fire service lost an opportunity to work toward preventing future events. On the upside, there is no doubt that the engine crew member developed a new understanding of spoliation that day.

JONES & BARTLETT LEARNING
NAVIGATE 2   *Access Navigate for more resources.*

Introduction

This chapter covers important concepts related to the protection of **potential evidence** and the safety of those handling and collecting the evidence. **Spoliation** (spoh-lee-ay-shun[1]), or the spoiling of evidence, is not

a new term, but one that has seldom been heard in the fire suppression field in the past. Regardless of the cause, spoliation is the destruction of evidence that can constitute an obstruction of justice. In the fire investigation setting, *destruction* may simply mean the alteration of any evidence or potential evidence that prevents other experts from seeing the value of that evidence.[2] The insurance industry and the civil courts have been dealing with spoliation issues for years. The fire service is not immune to these issues, nor is it immune to the repercussions should spoliation occur out of careless disregard or ignorance.

Every first responder, fire officer, and fire investigator must be well aware of spoliation issues and take steps to prevent them. Each and every department should establish a written agency policy and a training program on the steps to take to prevent spoiling evidence at any fire or explosion scene. Part of that training should include other agencies, such as dispatchers and other emergency responders, because they all may handle evidence of one type or another.

> **TIP**
>
> Every first responder, fire officer, and fire investigator must be well aware of spoliation issues and take steps to prevent them.

Failure to maintain and protect evidence can result in court-ordered fines, sanctions, or dismissal of the court case. Intentional or negligent acts that result in spoliation could result in criminal charges against the first responder or investigator for obstruction of justice. The primary reason to protect evidence, however, is because it is the right thing to do. Fire investigators protect and serve the people—and there is no better way to do this than to ensure we protect the evidence that may lead us to the discovery of how a fire started so that we can prevent future incidents.

> **TIP**
>
> Failure to maintain and protect evidence can result in court-ordered fines, sanctions, or dismissal of the court case.

What Is Spoliation?

The National Fire Protection Association's (NFPA) *Guide for Fire and Explosion Investigations* defines spoliation as "the loss, destruction, or material alteration of an object or document that is evidence or potential evidence in a legal proceeding by one who has the responsibility for its preservation."[3] The most important term in this definition is "potential evidence."

Almost any item within and around the structure could potentially become evidence in the determination of the area of origin, fire cause (cause of the explosion), or identification of who or what may be responsible for the event. Thus, fire fighters must make every effort to minimize the alteration of the scene before the investigator or investigative team has an opportunity to examine the scene.

Oftentimes, the placement or position of an object is just as important as the object itself. So many things can be considered spoliation that it is impossible to list them all here.

> **TIP**
>
> Any object on the scene could have evidentiary value. Oftentimes, the placement or position of that object is just as important as the object itself.

To show how easy it is to commit spoliation, consider two examples. The simple act of flipping circuit breakers eliminates their evidentiary value in the investigation. Once the breaker switch is moved, the investigator cannot immediately see its position at the time of the fire. However, if the breaker is taken to the laboratory and X-rayed, its position may be determined. The next option could be to drill out the pins holding the plastic housing together (**FIGURE 8-1**). If due care is taken in the process, the interior may potentially provide evidence of the position of the interior workings of the breaker (on, off, tripped) at the time of or during the incident. The same can be said of the position of control knobs on appliances (**FIGURE 8-2**). Once moved by suppression forces,

FIGURE 8-1 A circuit breaker interior as seen after drilling out the pins holding the two pieces of the plastic housing. If done with due care and according to spoliation recommendations of NFPA 921, Guide for Fire and Explosion Investigations, it can provide valuable evidence.

Courtesy of Russ Chandler.

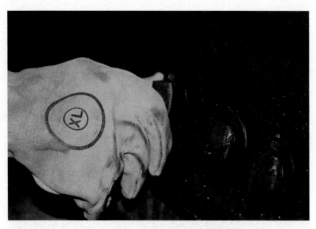

FIGURE 8-2 The simple act of turning a knob on an appliance can destroy the knob's use as potential evidence by creating spoliation.
Courtesy of Russ Chandler.

they lose their evidentiary value, possibly preventing the accurate and final determination of the fire cause and responsibility.

Spoliation can have a dramatic impact on everyone associated with a fire or explosion scene. Destruction of evidence can affect the ability of both the public and private investigators to determine accurately the area or origin and cause. Spoliation can prevent the owner from seeking remedy should the fire have been caused by a third party such as an appliance manufacturer. The insurance companies representing all parties involved are affected financially because they might not be able to recover funds they paid out for the claim, when they could have otherwise taken civil action against the responsible party. The spoliation of evidence could also lead to key evidence not being allowed to be presented in court, potentially allowing an arsonist to go free to commit other crimes.

What Spoliation Is Not

As a general rule, suppression personnel, in the normal course of their duty while searching for victims, extinguishing the fire, and searching for fire extension, are not committing spoliation. During the overhaul stage, suppression personnel must be cognizant of spoliation and limit the moving or alteration of debris to only that necessary to accomplish their task of searching for pockets of hidden fire, thereby ensuring that the fire is completely extinguished and no rekindling will occur. Carrying out these duties does not constitute spoliation. Most importantly, spoliation concerns should not get in the way of safety.

Once the fire is placed under control and the incident switches from suppression to investigation mode, a different set of rules applies about what can or cannot be done to the debris. NFPA 921, *Guide for Fire*

and Explosion Investigations, states that the responsibility of the investigator "varies according to such factors as the investigator's jurisdiction, whether he or she is a public official or private investigator, whether criminal conduct is indicated, and applicable laws and regulations."[4]

> **TIP**
>
> Public fire investigators have an implied responsibility that, for the good of the general public, they must endeavor to make an accurate determination of the cause and origin of fires occurring within their jurisdiction.

Of course, the investigator must disturb debris both to locate the area of origin and to identify the potential cause of the fire. If the investigator needs to alter the debris significantly in the search, he or she should adequately document the scene before moving the debris. Documentation enables others to see the scene before it was disturbed. Such documentation can consist of copious photos taken from every angle and direction. Videotaping can also be beneficial to show that the investigator is taking the proper actions during the investigation. In addition, it can give a better view and perspective of the evidence. Nevertheless, investigators should never rely on video alone; that method must be used in conjunction with photos.

Sometimes during the investigative phase, it is necessary to remove all the debris and room contents and then bring them back into the scene for reconstruction purposes. Some scenes may appear as just one huge pile of carbon. Dropped insulation and beams from above may prevent the true scene from being revealed in person and in photographs. To clarify the development of the fire, the item responsible for the ignition sequence may be removed and brought back into the scene. Disassembly of an item may also be necessary to identify whether it was involved in the ignition sequence. When done by a qualified public investigator in a situation where the scene needs to be processed without delay, none of these actions should be considered to be spoliation (**FIGURE 8-3**).

It cannot be stressed enough that the investigator should endeavor to photograph and document the entire scene as much as practical. In some circumstances, the investigator would be wise to photograph and document those items that were not involved in the ignition sequence, including all utility equipment and appliances capable of producing heat, to show that they were not involved in the ignition sequence.

FIGURE 8-3 The act of shoveling off the scene while searching for evidence in itself does not constitute spoliation. Courtesy of Russ Chandler.

> **TIP**
>
> The investigator should endeavor to photograph and document the entire scene.

In addition, the investigator should maintain photos and documentation in the case file that support alternative hypotheses that are discarded when the final determination of the fire area of origin and cause is made. This enables others to have the same opportunity to see as much of the scene as possible and come up with their own determination of the fire's origin and cause.

Although the first responder investigator will not often see investigators from the private sector, they should be aware that these individuals may also have an interest in the fire scene. Typically, these investigators will not arrive at the scene until hours or even days after the incident occurs. Private-sector investigations follow different procedures. For example, if private fire investigators identify potential evidence that may require destructive testing, they must notify all known interested parties, such as the full-time assigned investigator, before conducting that testing. This enables entities who have an interest in the outcome of testing to be present, if they so desire, so that they can all ascertain the evidentiary value of the item being tested.

Discovery Sanctions and Penalties

The repercussions from not following accepted procedures with evidence can be varied and far-reaching. During a trial, all parties have an opportunity to examine the evidence that will be used by both sides. This process, called discovery, is one of the foundations of a fair trial. Any evidence from the fire scene may have a dramatic impact on the outcome of the trial. Should someone, such as a first responder or an inexperienced investigator, fail to handle, secure, collect, or store evidence properly, the courts and all interested parties lose the opportunity to use that evidence in the trial.

To ensure that evidence is not mishandled, the courts may impose discovery sanctions, referring to punishment for failing to follow discovery rules. In the case of fire scenes, these sanctions can be quite harsh— taking the form of monetary fines, the exclusion of the expert witness to testify, or even dismissal of the case. Monetary fines imposed by the court are usually levied against an individual or the individual's employer. Depending on the circumstances, the employer may stand behind its staff and pay the fines. Conversely, in cases of gross misconduct, the employer may decide not to support the individual, leaving the fine to be paid by the person responsible for the spoliation. Not only can this penalty be costly financially, but it also jeopardizes the individual's credibility in future court cases as well as his or her future employment in that department or future employment in any other department.

> **TIP**
>
> To ensure that evidence is not mishandled, the courts may impose discovery sanctions, which is punishment for failing to follow discovery rules.

When the court disallows the testimony of the origin and cause expert, the perpetrator of the crime may go free, in essence allowing this person to commit other crimes. The far-reaching implications may be that others who were thinking of committing arson might now have the encouragement to carry out the act. Moreover, the dismissal of the case not only leaves the victims with a sense of injustice, but also may reduce or eliminate their possibility of any financial recovery.

Should the destruction of the evidence be intentional or negligent, tort (civil lawsuits) action can be taken against the fire fighter, investigator, or their department. In these situations, the plaintiff will most likely file the suit against the individual, the individual's immediate superior, the fire chief, the senior government official such as the county administrator or city manager, and even the political entity governing the jurisdiction. In doing so, the plaintiff is seeking the largest remedy.

The lawsuit itself can prove damaging to the credibility of the individual even if the individual is innocent. The case usually plays out in the press long before it is ever heard in court. In most cases, the locality represents the first responder or investigator in court. Nevertheless, its primary responsibility is to look out for the interest of the jurisdiction and not just the individual. Recognizing this fact, some first responders and investigators in this situation have retained their own legal counsel at their own cost. As can be imagined, there is also a psychological cost for the individual and his or her family.

Criminal charges could be levied against an individual who negligently or intentionally engages in spoliation. A charge of obstruction of justice has broad implications, and can occur during the investigation as well as during the trial. There need not be actual proof of obstruction, but any evidence that the individual attempted to obstruct the investigation or trial is sufficient for a conviction. Although penalties vary from jurisdiction to jurisdiction, obstruction of justice is always a felony, and sanctions may include both a financial penalty and prison time. It may be harder for investigators to counter such charges because they are in a position of trust and as such may face stronger scrutiny.

> **TIP**
>
> Criminal charges could be levied against an individual who negligently or intentionally engages in spoliation.

Designing Training Programs

As mentioned at the beginning of this chapter, a training program is essential to prevent spoliation incidents. Too many variables make it impossible for a single program to fit the needs of each and every department. The following subsections describe some of the considerations that must be taken into account in designing a training program.

Proper Investigative Attitude

The first step is to ensure that the investigator has a proper attitude. An investigator must believe in the

U.S. Constitution and court system, and accept that everyone is entitled to a fair and impartial trial. A fair trial constitutes one in which both sides get to see and have the benefit of all the evidence. When it comes to fire scenes, that includes *all* of the evidence from the scene—not just what the investigator thinks is important. Investigators must remember that, regardless of how certain they are about their findings, they are neither the judge nor the jury. In other words, they must remember that they are *seekers of truth and not case makers*, as stated in the International Association of Arson Investigators (IAAI) Code of Ethics.[5]

> **TIP**
>
> An investigator must believe in the U.S. Constitution and court system, and accept that everyone is entitled to a fair and impartial trial. A fair trial constitutes one in which both sides get to see and have the benefit of all the evidence.

In the past, some public- and private-sector fire investigators have had scenes bulldozed and destroyed at the completion of their scene examinations. This is spoliation at its worst, and it implies that no one would ever second-guess their findings. This leads us to the second attitude that every investigator must possess: Every investigator must be prepared to have a second expert come onto the scene and make a different determination of a fire's origin and cause. Investigators do not have to like it, but they must realize that such second opinions will happen and are not personal. Very capable and talented fire investigators with equal credibility and expertise can look at the same scene and come up with the same or different causes of the fire. With a proper attitude comes an open mind to ensure that investigators really are seekers of truth.

Training Program Design Committee

A design committee must ensure that the program includes all necessary training. At a minimum, this committee should consist of a representative from the local prosecuting attorney's office, a member of the department's training division, and a representative from the local fire and arson investigating team. The department might also consider including someone from the private sector such as a claims manager from an insurance company as well as an investigator or engineer from the private sector who specializes in fire investigations. These individuals have a different outlook on the process and can provide recommendations and insights into how to work together for the good of the public.

Creation of Agency Policies

Before a training program can be established, the fire department must review or create agency procedures and policies on actions to be taken at the scene by both suppression and investigative personnel. For example, a policy must be established on how the electrical power will be disconnected to a structure if the utility company cannot immediately provide this service.

The agency's policies should outline when the investigation will take place and what requirements are necessary to be on the scene legally. They should also address when the investigation can be conducted, in compliance with the dictates of the U.S. Supreme Court rulings on fire scene investigations, such as *Michigan v. Tyler* or *Michigan v. Clifford*. As NFPA 921, *Guide for Fire and Explosion Investigations*, indicates, these responsibilities differ for public- and private-sector investigators. It should be established that the investigation begins with the suppression crew on the scene. Based on their findings, they can conclude the investigation by entering the area of origin, material first

ignited, and heat source on the fire incident report. Just as importantly, they can document what brought the heat and ignited material together. If the first responder investigator cannot determine the fire origin and cause, then he or she should request a full-time assigned investigator.

The agency's policies should also address how evidence will be handled. Will it be maintained by the locality, and if so, which section of the law allows the locality to collect evidence of a noncriminal nature? Of course, issues of liability and spoliation should be addressed as well. Once decisions on these issues have been made, all of the relevant policies must be included in the training program.

Policy definition is not complete unless the department addresses how other entities, such as insurance companies and private-sector investigators, will address spoliation. Those entities follow a different set of rules and policies than apply to the public sector, and all fire investigators must know how to work together. For this reason, it is a good idea for the training program design committee to include representatives from the private sector. Once a policy is designed, tested, and approved, it should be included in all basic training of personnel and readdressed in ongoing in-service training.

Wrap-Up

SUMMARY

- The importance of preventing spoliation of evidence cannot be stressed strongly enough. Even first responders can be guilty of spoliation of evidence when they do not follow accepted practices and procedures. Ignorance is not a defense for the first responder or the investigator should evidence be damaged.

- The primary purpose of spoliation rules and punishments is simple: Everyone involved in criminal cases or civil litigation must ensure that there is a fair and impartial hearing of the facts. When evidence is destroyed, the ability to find all facts is diminished. Destruction of evidence can jeopardize the opportunity to seek the truth in the investigation, so every effort must be made to protect evidence. All fire fighters and investigators must resist the temptation to touch and move objects that may be potential evidence.

- To guard against spoliation, all personnel who conduct investigations must have a good ethical attitude. In addition, a good training program for all parties involved, including firefighting personnel, must be in place to prevent spoliation. Before training is delivered, the agency must create a policies and procedures manual documenting ways to avoid spoliation issues. Once a policy is designed, tested, and approved, the training program should consistently reinforce it to trainees.

- Any damage done in searching for victims, assuring extinguishment of the fire, and doing overhaul to prevent rekindling is not spoliation. Likewise, the act of shoveling out the scene and the removal and replacement of contents back into the scene is not spoliation. The disassembly of an item may be necessary to identify if it was involved in the fire; this is not spoliation.

- Negligence in protecting evidence from spoliation can lead to several types of negative consequences. In the discovery stages of a trial, sanctions and/or punishment may be levied if the fire department or the investigator is found to have committed spoliation of evidence. This punishment could take the form of fines, exclusion of evidence, or even exclusion of the investigator's testimony. This could result in an alleged arsonist going free and committing further crimes.

KEY TERMS

Discovery A pretrial device used by both (all) sides in a case to obtain all the facts about the case to prepare for the trial.

Discovery sanctions Penalties for failing to comply with discovery rules.

Overhaul The act of searching for hidden fire, ensuring that the fire is completely extinguished.

Potential evidence Something that may, or could, possibly be used to make something else evident; findings not yet used or that have yet to be used to prove a point or issue or support a hypothesis.

Spoliation The destruction of evidence; the destruction or the significant and meaningful alteration of a document or instrument.[6] It constitutes an obstruction of justice.

REVIEW QUESTIONS

1. Define *spoliation*.

2. List and describe all parties who could be affected by spoliation.

3. What protections do public investigators have during their examination of the fire scene?

4. Whom should private-sector investigators contact if they are going to conduct destructive testing on a piece of evidence?

5. Why should the private sector be involved in a training program on spoliation for the fire department?

DISCUSSION QUESTIONS

1. You are at a fire scene when you hear an investigator arrange for the debris to be bulldozed; he says that he has no intention of letting another fire investigator second-guess him. How do you feel about this investigator's actions? What should you do about his actions?

2. If you are involved in firefighting, discuss what you have seen in the past that could have been a spoliation issue.

3. A fire has claimed the life of a fire fighter. A private investigator has been hired by the defense counsel of the alleged arsonist. Why should the investigator willingly provide any and all evidence to the other investigator when ordered to do so by the courts? Should that information be provided even if it has not been ordered by the court but requested by defense counsel? Should the investigator intentionally hold anything back?

REFERENCES

1. *Webster's New World College Dictionary* (2nd ed., p. 1376) (1984). New York, NY: Simon & Schuster.
2. *Black's Law Dictionary* (5th ed., p. 1257) (1979). St. Paul, MN: West Publishing Company.
3. National Fire Protection Association. (2008). *NFPA 921: Guide for fire and explosion investigations* (pp. 921–15). Quincy, MA: NFPA.
4. National Fire Protection Association. (2008). *NFPA 921: Guide for fire and explosion investigations* (pp. 921–104). Quincy, MA: NFPA.
5. International Association of Arson Investigators. *I.A.A.I. Code of Ethics*. Retrieved from http://www.firearson.com/insideiaai/codeofethics/index, asp
6. *Black's Law Dictionary* (p. 1376). St. Paul, MN: West Publishing Company.

Evidence

LEARNING OBJECTIVES

Upon completion of this chapter, you should be able to:

- Describe the four types of evidence.
- Describe the authority of the fire department and the investigator to collect evidence.
- Describe the steps (including preliminary assessment) to take to protect the scene and preserve evidence.
- Describe the process of identifying evidence.
- Describe the proper process for collecting and preserving evidence.
- Describe how to process evidence, including tagging, analysis, reports, and documentation of the chain of evidence.
- Describe the process of protecting evidence and ways to prevent contamination.
- Describe first responders' considerations in the collection and preservation of evidence.

Case Study

It was a senseless accidental fire that destroyed 58 apartments, putting more than 100 individuals on the street. The investigation took three days, with the investigators seeking not only the cause of the fire but also the reason for the extension of the fire throughout the sprinklered building.

On scene were the local fire marshals (full-time assigned investigators) as well as investigators contracted by the insurance company. The primary investigation was done by the local authorities prior to letting the private-sector investigators come onto the scene. Once the local investigators completed their investigation, they shared the findings with the building owners, tenants, and insurance representatives.

The cause of the fire was attributed to overheating of an exterior light fixture in the common area, as the result of a tenant changing its configuration and then insulating the fixture with a shirt to block the light from entering his bedroom window. Flames extended up the exterior of the building and into the building's large attic space. This fire extension circumvented and eventually overran the sprinkler system.

This should have been a simple case, but it was not. What seemed to be a minor lapse in protocol for evidence retention ended up with an investigator, the chief fire marshal, the fire chief, and the jurisdiction all defending themselves in a civil lawsuit.

During the investigation, the investigator wanted to compare the lightbulb from the area of origin with a sample, an exemplar bulb, from the spare supplies at the complex. The fixture in question was damaged but intact and was rated for a maximum 60-W bulb. The bulb in the fixture was broken, but the stem, element, and base remained. The investigator broke the new bulb to see, in a nontechnical visual manner, if the bulb in the light fixture was truly a 60-W bulb. The outer glass of the new bulb was broken away and a comparison of the filament as well as the bulb stem proved inconclusive. The investigator was left with a new bulb with jagged glass edges on the base. For safety's sake, he set the new bulb in a small Styrofoam cup and placed the cup and bulb in a metal evidence can.

The assigned investigators completed their work and cleared the scene. Physical evidence for the accidental fire was left behind in a protected area for the insurance company's representative and the building owner; this included the bulb from the area of origin. The new bulb broken by the fire investigator was still in the small metal can; it was taken for disposal from the scene by the investigator. In error, that metal evidence can was removed from his vehicle and placed in the department's evidence locker despite not being logged in as required. All evidence being placed into or taken out of an evidence locker must be logged in and out to supplement the documentation of preserving the chain of custody of the evidence. The investigator later stated that he planned on tossing the bulb he had broken, but forgot to do so. As a result, it remained in the evidence locker.

Sometime later, a lawyer and a private investigator representing the renters from the apartment complex showed up at the investigator's office and asked to see the evidence from the fire. Various tenants did not have fire insurance and were seeking recourse to recover from their losses. They hoped to file a lawsuit against the apartment owners. Staff (not the original investigator) went to the locker and provided all of the items in the locker. They were shown the can with the sample light bulb broken by the original investigator. When the original investigator heard about this event, he realized that what the lawyer and private investigator were shown was not the actual bulb from the fire. At that point, the lawyer was contacted and told of the situation. Based on this report, the lawyer believed that the investigator was fabricating evidence.

The lawyer filed civil charges against the investigator, his supervisor, and the chief fire marshal. Immediately the press—both newspaper and TV—put the investigator and the department before the public based on the allegations of falsifying evidence. This type of situation can be a career-changing event, create major stress for families, and induce a loss of confidence. Credibility is certainly at stake.

On the day of the trial, cameras were rolling outside the court as the defendants approached. Everyone was seated, and the plaintiff, the tenants' lawyer, brought forward the complaint. All plaintiffs' evidence was presented, and then they rested their case. The defense lawyer representing the investigator, his supervisor, and the chief fire marshal asked for dismissal based on the lack of evidence. Fortunately, the judge dismissed the case, giving a summary judgment on behalf of the fire officials. Truth prevailed, as all parties were deemed innocent of any wrongdoing.

All of this drama stemmed from what might seem to be an innocent action, but was actually a breach in protocol for proper evidence handling, storage, and documentation. Together, the accused fire department members had more than 60 years of dedicated service to that community; they were highly regarded by their peers in neighboring jurisdictions and across the state. They had received numerous commendations from

their department and had risked their lives many times. Even though found innocent, their credibility suffered damage in the eyes of some people, and they faced the perception of wrongdoing in the eyes of the public.

 Access Navigate for more resources.

Introduction

Evidence is covered in several sections of this text. From the time that the call comes into dispatch, the investigator will start collecting evidence. There are four types of evidence:

- Demonstrative evidence
- Documentary evidence
- Testimonial evidence
- Real or physical evidence

Each form of evidence is addressed in this chapter. However, the primary concern here is the physical evidence that will be identified, collected, and preserved by both the first responder and the assigned investigator.

Demonstrative Evidence

Demonstrative evidence is used at trial to illustrate any particular issue to be brought up. It can include diagrams, photos, and other tangible artifacts from the scene or elsewhere that are applicable to the case. These types of evidence serve as tools to help educate the jury and judge about the facts of the case. Each item brought forward must be relevant and may be challenged by opposing counsel before being admitted as evidence.

Charts created by the attorneys or investigators are good examples of demonstrative evidence. Photographs are a staple for almost every trial, as they allow the judge and jury to see what is being described and explained by the witness. In years past, courts would require the photographer to testify about the authenticity of each photo. When digital photographs came into common use, concerns arose about the potential to alter them using computers. Today, most courts allow the witness to stipulate that the photo accurately depicts what was seen during the investigation and was not altered. Most importantly, testimony must state that the items to be used are a fair and accurate depiction of the subject at hand. The **expert witness** will be the primary user of the demonstrative evidence.

Documentary Evidence

Documentary evidence is written evidence germane to the trial. It could include bank records, business records, telephone records, insurance policies, proof of loss from the claimant, incident reports, written statements from witnesses or the accused (including transcripts from interviews), correspondence, or even the investigator's notes. The documentary evidence brought forward in a trial may need to be supported by testimony regarding its authenticity. For example, someone from the bank may need to validate that the bank statements entered into evidence are the true and accurate documents.

Steps must always be taken to assure that these documents have been legally obtained and no Fourth or Fifth Amendment violations occurred—that is, that they are a result of a legal search. Some evidence, such as written documents from a witness, must be validated as being direct statements rather than **hearsay** (facts overheard by another person).

Testimonial Evidence

Testimonial evidence is the verbal evidence given by witnesses. In a court of law, any testimonial evidence must be reliable and relevant to the case. It cannot be supposition, hearsay, speculative, unfounded, or an opinion from a non-expert witness.

The first responder investigator will not necessarily need to testify in court. However, if the first responder found a physical item that will be a key part of the trial, then he or she may receive a **subpoena** to testify as to the item's authenticity, where it was found, and so on. Similarly, if the accused made a statement to the first responder about a crime, then the first responder may be compelled to testify about that statement in court. In this situation, the first responder would be testifying as a **fact (or lay) witness** because the evidence was something that was seen, heard, or touched.

A fact witness, for the most part, is presumed to be competent. However, opposing counsel has the right to challenge a witness's competency, perhaps based on testimony contradicting previous testimony heard by the court. Opposing counsel may want to show that the witness is biased against the defendant or that the witness's perception at the time of the incident was not reliable, perhaps because of fatigue from fighting a large fire. In some situations, first responders have been challenged about what they heard because the fire scene can be so noisy from pumps, sirens, and other equipment.

Assigned investigators could testify as fact witnesses about what they heard or saw, but they will most likely be testifying an expert witnesses. An entirely different set of rules and circumstances applies to this type of witness. Any expert witness will undergo an examination to show his or her qualifications to testify as to the area of origin, the cause of the fire, and, in some situations, the identification of the responsible party. This witness will also testify as to the burn patterns and evidence collected.

The requirements for the assigned investigator and the rules of the court for this type of testimony are quite vast and complicated. For more information about expert witnesses, see Jones & Bartlett Learning's *Fire Investigator: Principles and Practices* as well as NFPA 921, *Guide for Fire and Explosion Investigations*, and NFPA 1033, *Standard for the Professional Qualifications for Fire Investigator*. Should a first responder desire to become an assigned investigator, these resources will be great assets.

Physical Evidence

The following information instructs the first responder investigator about evidence that he or she may discover and handle, but also goes beyond the actions taken by such an investigator. The first responder will find it beneficial to understand the additional steps that the assigned investigator will take to collect sufficient evidence in a more complex fire investigation. The information presented here is not exhaustive, but will help the first responder be an asset to the assigned investigator.

First and foremost, it is critical that each department or investigative division establish a **physical evidence policy** for all investigators, including first responder investigators. This policy must cover all aspects of handling and storing evidence for each incident investigated. It should also cover what type of evidence the first responder investigator should

> **TIP**
>
> First and foremost, it is critical that each department or investigative division establish a physical evidence policy.

collect, under what circumstances the first responder should collect and protect such evidence, and how to store it.

The physical evidence policy is only as good as those who have input in its creation. There are benefits to involving all fire investigative levels in the development of this policy, including experienced first responder investigators. In addition to the resources within the department or division, collect input from local law enforcement and the laboratory that will be processing the evidence. Lab personnel can specify the type of containers required, the documentation needed, and best practices in retention of the evidence to match their own policy.

Everyone within the department must be trained on the policy to the level that they will be involved in handling evidence. Everyone within an investigation unit must be thoroughly trained on all aspects of the policy. By comparison, the engine company officer may be trained on evidence recognition and ways to protect evidence from being damaged. As the first person to conduct an investigation, he or she needs to be well aware of the process of identifying evidence and under which circumstances the officer should call for an assigned investigator. Engine company officers should also know how to collect and hold evidence under extreme circumstances.

> **TIP**
>
> The engine company officer should be trained on evidence recognition and ways to protect evidence from being damaged.

Even frontline fire fighters should be trained to recognize evidence and protect it from harm. Fire fighters need to know what not to do to assure that no spoliation occurs (see Chapter 8, *Spoliation*). They must also be aware of **contamination** issues, especially given that several firefighting tools are powered by hydrocarbon fuel and should not be taken onto a scene.

> **TIP**
>
> Even the frontline fire fighter should be trained to recognize evidence and to protect it from harm.

The National Fire Academy has an older hand-off course titled "Arson Detection for the First Responder." This course highlights what to look for with an incendiary fire and which actions the fire fighter should take. The locality, or state or province training agencies, may provide a similar training course.

Photography as Part of the Collection of Evidence

Part of the physical evidence policy needs to include documenting the evidence through photography. The first set of photographs should cover the entire scene before it is further disturbed in the evidence collection process. There must be photos of the burn area as well as photos of areas not burned, which will show the extent of the damage. It is also important to photograph any appliance or device capable of being a source of ignition. Even household appliances or devices that were not burned should be photographed to show that they were not involved in the fire, thereby eliminating those items as a potential heat source.

The first responder investigator can use a camera on a personal or assigned cell phone. Today, such technology can provide high-resolution photographs that can be of great value in the investigation. Almost all of these phone devices can provide not only single photos but also panoramic photos of the area and close-up photos of tags on appliances showing model and serial numbers.

Full-time investigators may have a 35-mm digital camera with various lenses to perform close-up or distant photography. They may also have video cameras that can be set up to document the entire evidence collection process, thereby confirming use of proper technique and substantiating that evidence collection was done properly.

To prove that all investigatory work was done properly, the investigator should photograph the area where the evidence will be collected before disturbing it. It can also be beneficial to show the open empty evidence can in the area where the sample will be taken. Photograph the process of collecting the evidence, especially showing any tools used. Once the can is properly filled, leaving a head space, it should be photographed with the lid and investigator's latex gloves beside the can. The can should then be sealed, with proper labels affixed to it. This, too, should be photographed to show that the container has been properly sealed and labeled at the scene.

In years past, the ladder truck would stay on the scene to allow the investigator to not only see the scene from a higher angle but also photograph the area from above, which could be beneficial in the investigation process. Today, with new technology comes new opportunities: Many fire departments have added drones to their toolbox. These devices enable the officer in charge (OIC) to see the overall scene, safety officers to track different functions, and the fire investigator to gain a fuller picture of the fire's scope. If the fire department does not have drones yet, investigators can contact local law enforcement to see if they have this resource and if they could assist. Drones can also keep watch overhead to assist with securing the scene, letting the operator know if someone makes an unauthorized attempt to enter the area.

Authority to Collect Evidence

Who should or should not collect evidence varies. The first variable is the role of the investigator on the scene. The fire officer (first responder investigator) doing the preliminary investigation should always protect evidence he or she recognizes, but should take such evidence only as a last resort, as noted in *Michigan v. Tyler*. The full-time assigned fire investigator should collect any and all evidence that would be used as part of a hypothesis regarding an incendiary fire. If the investigator had an earlier hypothesis that was discounted, the evidence used in making that earlier hypothesis should be protected as well.

> **TIP**
>
> The fire officer (first responder investigator) doing the preliminary investigation should take evidence only as a last resort to protect and preserve that evidence.

If a fire is accidental in nature, there is no crime. If there is no crime, a question arises about the legality taking evidence of any kind for any reason. Recall that the Fourth Amendment prevents "unreasonable searches and seizures by the government." Localities should research with both local legal counsel for their jurisdiction and their local prosecuting attorney as to what they can or cannot take from a fire scene that has a clear indication of being accidental in nature. Once this determination has been established, it should be documented and placed in the department's physical evidence policy. One point remains constant across all scenes, however: Investigators can search a fire scene to make a determination of the fire cause because doing so is in the interest of the general public.

> **TIP**
>
> If there is no crime, a question arises about the legality of the investigator taking, seizing, evidence of any kind for any reason.

Arguments can be made for the fire investigator taking a faulty appliance and having it tested. However, doing so without the permission of the owner of the appliance puts the locality at risk of liability. Also, when a government employee takes any evidence for an accidental fire and wants to test it, an entirely different procedure must be followed, as recommended in the section on spoliation in NFPA 921, *Guide for Fire and Explosion Investigations*.[1] As with so many other things, there are exceptions.

In joint operations between the public and private investigators, it has sometimes been decided (usually by all parties to the case) that the public investigator can hold the evidence until testing. The locality's physical evidence policy should cover these topics as well as give examples of when it would be appropriate to take evidence from an accidental fire scene.

Technically, there is no such thing as a "suspicious fire." The investigator can be suspicious of circumstances, prompting further investigation, but fires are classified as natural, accidental, incendiary, or unknown.[2] The unknown category simply means that there is insufficient evidence to prove the actual cause. Thus, the fire is still considered accidental. Laws in the United States dictate that everyone is innocent until proven otherwise. Based on this same principle, all fires are accidental until proven otherwise.

> **TIP**
>
> Many laws, and case law in particular, dictate that fires should be classified as accidental until proven otherwise.

The actual cause of the fire may not be known until the fire scene examination has been completed. Regardless of the eventual outcome of the fire investigation, fire department personnel must make every effort to preserve the fire scene as if it were a crime scene. This is a logical process, recognized by the courts as an appropriate step to take to preserve any evidence that can identify the fire as being accidental or incendiary in nature. If any findings suggest that the fire was incendiary, the public investigator should collect any and all evidence for processing, taking all appropriate steps to protect, collect, preserve, and submit the evidence for the required testing.

Private investigators and insurance investigators have an entirely different set of rules in such situations, because they have a different focus at the fire scene. When a property owner experiences a fire and the property is insured, the insurance company may send its own investigator from the company's Special Investigations Unit or it may hire a private insurance investigator. Private-sector investigators can collect and keep evidence from a fire that is either accidental or incendiary in nature. Should the property owner refuse to allow such a private investigation, the insurance company, based on the insurance policy (a legal contract), can simply refuse to pay the claim. Given this type of incentive, property owners seldom refuse to allow the investigator to take the evidence needed. In fact, in any situation other than incendiary fires, the insurance company acts as an agent of the homeowner: Both have a common interest in the outcome of the findings. The process of collection and preservation of evidence of an incendiary nature is no different for the private investigator than it is for the public investigator. The procedures and processes are very similar, if not exactly the same.

> **TIP**
>
> Private investigators and insurance investigators operate under an entirely different set of rules in fire investigations.

> **TIP**
>
> Private-sector investigators can collect and keep evidence from a fire that is either accidental or incendiary in nature. Should the property owner refuse to allow such a private investigation, the insurance company, based on the insurance policy (a legal contract), can simply refuse to pay the claim.

Evidence of an accidental nature takes on a different light in that the insurance company may be seeking evidence to show that the fire was the result of a responsible third party. This may be in the form of a faulty appliance or recently contracted repair work. Should either of these two examples be the cause of the fire, the insurance company will pay the property owner the appropriate value of the loss up to the limits of the insurance policy. The insurance company can then recover its loss through subrogation against the responsible party. (See Chapter 3, *Legal Issues and the Right to Be There*, for more information on the interests of insurance companies in fire investigations.)

Situations in which the fire starts on one property and spreads to another can create some interesting scenarios regarding the right to be on the scene to collect evidence. For example, if the fire started in a pizza shop in a strip mall and then extended into the clothing store next door, the public investigator can look at both scenes under the auspices of determining the fire area of origin, cause, and the reason for the fire extending into the neighboring tenant space. In contrast, the private investigator from the insurance company may represent the clothing store and have no legal right to go onto the pizza shop premises without permission from the business or property owner. Even if permission is given, it can be rescinded at any time, and the private investigator will have to leave the pizza shop. If the insurance company investigator represents the strip mall owner, however, under most circumstances that investigator can look at the entire property and all tenant locations.

Protecting Evidence

First responders are the first to encounter any evidence on the scene. It is important to preserve all evidence, regardless of whether the fire is later determined to be accidental or incendiary. That evidence can often help prevent future fires through education, identifying needed recalls on appliances, or putting an arsonist behind bars. Successfully prosecuting an arsonist not only stops that one individual from setting fires, but may also act as a deterrent, causing others to think twice before committing this crime.

> **TIP**
>
> It is important to preserve all evidence, regardless of whether the fire is later determined to be accidental or incendiary.

With the proper training, suppression forces can recognize evidence and take the steps necessary to protect it. Taking extra precautions when advancing attack lines, looking at doors before forcing them to see if they already have **tool mark impressions**, and limiting salvage and overhaul are all steps that can go a long way in aiding the investigation.

Overhaul is the process of searching for hidden extension of the fire. It involves opening walls and ceilings, and even going through the debris on the floor or ground to find cinders and to ensure they are extinguished so that the fire cannot rekindle. In some instances, flaming combustion can still be present within the walls, so suppression forces must continue the overhaul process to stop the fire and to prevent rekindle.

Salvage is a process of taking items that have not been damaged or severely damaged and protecting them by either covering them with salvage covers or removing them from the structure. There is a tough

balance to strike between the overhaul and salvage process and preservation of evidence. Both suppression functions must continue, but should be limited if possible, so as to prevent loss or contamination of evidence.

The first responder who does a preliminary investigation to fulfill the requirements of the National Fire Incident Reporting System (NFIRS) may discover that the fire is not accidental but rather incendiary in nature. When this happens, two things must take place. First, the OIC must follow that jurisdiction's procedure for contacting an assigned investigator to come to the scene. Second, the fire crew must protect the scene to prevent the destruction of evidence. This means limiting access to the scene to anyone who does not need to be on the scene. At the same time, suppression forces still need to carry out overhaul and take actions to remove standing water that is adding increased weight and stress on the structure.

Securing a Fire Scene

Steps to preserve evidence include first establishing a barrier using **barricade tape** or rope to cordon off the area and prevent unwarranted entry. The barrier perimeter should encompass the entire scene. Rarely is the perimeter barricade perfectly square: Instead, fire fighters or police personnel typically tie the barricade tape or rope to existing objects, such as trees, fenceposts, or even parked vehicles. Commercial barricade tape preprinted with the words "FIRE LINE— DO NOT CROSS" or "CRIME SCENE" sends a clear message to not cross that line (**FIGURE 9-1**).

> **TIP**
>
> Steps to preserve evidence include first establishing a barrier using barricade tape or rope to cordon off the area to prevent unwarranted entry.

FIGURE 9-1 Commercial barricade tape can send a clear message to not cross the line.
Courtesy of Russ Chandler.

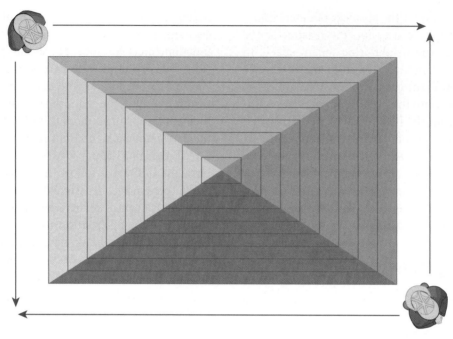

FIGURE 9-2 A simple property can be observed by as few as two or three assigned personnel. Their primary function is to make sure that no unauthorized personnel enter the scene.
Courtesy of Russ Chandler.

Because barricade tape alone will not completely deter individuals from entry, personnel must be assigned to watch the perimeter to ensure that the scene stays secure. This can be accomplished with a minimum of personnel, depending on the site. As shown in **FIGURE 9-2**, two or three people can maintain the integrity of the scene for a simple small structure. As long as all sides are under the watchful eye of an assigned person, the scene should be secure.

The department must be aware of the legal issues related to setting up barricade lines. For example, can the department prevent the owner of the property from entering his or her own property? In most jurisdictions, this can be done under the provision of protecting the person from harm during exigent circumstances such as the building still burning or being unsafe. However, in some states the department may not be able to keep out the press when they show up on the scene. Specific legal provisions might allow the press to cross fire lines provided they do not hinder the emergency operations. Researching and inserting a section in the physical evidence policy that covers securing the scene should prevent any future problems or potential conflicts.

Securing of the building is only necessary while the department is on the scene finishing extinguishment and the scene investigation. When the fire department has completed its assignment, they turn the structure over to the owner or tenant, who then has the responsibility to secure the property.

As a general rule, the private-sector investigator does not need to barricade a scene. Sometimes the barricade tape is still in place when the private-sector investigator arrives, and sometimes the structure may even be boarded up and secure. The bottom line is that there generally is no reason to establish a barricade line at that late stage, because the crowd has gone and the exigencies of the situation have long gone. However, the private-sector investigator still has a responsibility for public safety. If the scene is secure when the investigator arrives—for example, the building is boarded up and locked—the investigator must ensure that it is in that same condition when he or she leaves. This step will keep out neighborhood children and the curious and limit the investigator's liability.

TIP

If the scene is secure when the investigator arrives, the investigator must ensure that it is in that same condition when he or she leaves.

Identifying and Collecting Evidence

Before starting the process of identifying and collecting evidence, the investigator should be fully cognizant of all safety issues as outlined in Chapter 2, *Safety*. Additional safety concerns encompass actions such as cutting samples from walls, while ensuring that this step will not change the structural integrity of the building.

Evidence can take many forms—and not just the obvious, such as the gas can in the kitchen or the burned road flare on the seat of the car. Burn patterns on a wall or the structure itself may be the evidence necessary to come up with an accurate and plausible hypothesis. Once identified, all evidence must be preserved. This can often be done by picking up physical evidence, properly packaging it, logging it, and placing it in a secure area, such as the trunk of the investigator's vehicle, before it is stored in an evidence locker. But not all evidence can be preserved so easily, especially if a fire pattern pointing back to the area of origin is on an unstable wall. In such a case, the findings must be documented through photography and observations in the investigator's field notes.

Almost all patterns in a structure fire help tell the story of what happened. The interpretation of these patterns is something that is learned over time and through experience as well as from classroom learning and texts. The *burn pattern* comprises the physical effects from the fire on structural components and contents (**FIGURE 9-3**). These patterns can come from char of combustible products such as wood in the structure framing and contents. They can also be created by oxidation of metal structural components, appliances, and contents. In addition, metal may be discolored or distorted by heat. The total consumption of structural components is part of the pattern to be observed. Soot or discoloration of wall or ceiling surfaces is a type of pattern as well. These patterns are discussed in detail in Chapter 11, *Patterns: Burn and Smoke*; here, we

simply note that the observation of all patterns should be properly photographed and documented in the investigator's notes and investigative report.

Evidence can include devices and appliances, too. Smaller items could take the form of circuit breakers, switches, cords, or electrical multi-strips. Larger evidence might consist of an entire electrical panel, a clothes dryer, or even a refrigerator. In an arson case, the jurisdiction may have to lease a storage unit to hold these kinds of large items, knowing the evidence may have to be kept two years or more for the case to be tried and all appeals to be exhausted. As the investigation continues, the investigator may collect other evidence relevant to the investigation, such as papers, documents, or even signed statements.

Each type of evidence has a unique collection and preservation process. However, the typical type of evidence from the fire scene will include liquids, solids, and gases. All are essential to the final accurate determination of the fire origin and cause.

Residue Gases and Vapors

Knowing the potential gas that may have been involved in an incident dictates where to look for a sample. The vapor density of the gas determines whether the vapor will settle into low areas or rise to collect in the upper portions of a structure. Most lighter-than-air gases dissipate before a sample can be taken. This is not to say that sampling cannot be done, but the gas is less likely to be present when the investigator arrives. Gases that are heavier than air settle in low areas. In particular, they may settle in floor drains or sink drains.

When gaseous evidence might still be present at the scene, the investigator can collect it by using a vacuum container from a commercial sampling kit. These containers have had the air removed and then been sealed to create a vacuum. Using the attached tube and valve, the investigator can place the container in an atmosphere or insert the tube into an atmosphere, such as a drain pipe, and then open the valve. The pressure within the container is less than the atmospheric pressure, so the gas present at the tip of the tube will flow into the container and be trapped by the control valve. This gas can then be sent to the laboratory for examination. The obvious tests are to see whether the gas is an ignitable vapor and exactly what type of vapor is present.

FIGURE 9-3 A burn pattern can tell the story of the fire's origin and direction of travel.

Courtesy of Russ Chandler.

Another means of collecting gas and vapor samples relies on mechanical devices that pull the air sample through a charcoal trap or other absorbent material to collect the residue for testing. Other pumps pull the gas into a sample chamber, where it can later be collected and processed by a laboratory.

Liquid Samples

Liquid samples are the most common type of evidence taken at a potential incendiary fire scene. Such evidence usually takes the form of debris or flooring solids that might have absorbed a liquid **accelerant**. More often than not, the debris sent in for analysis consists of solid items cut up or scooped up from the floor and placed into an evidence container. The evidence collected is then sent to a lab to be examined for the presence of an ignitable liquid.

> **TIP**
>
> Liquid samples are the most common type of evidence taken at a potential incendiary fire scene.

When taking debris samples such as carpet, wood flooring, or concrete, the investigator is generally looking for the liquid residue trapped within the material. These types of samples are better suited to be preserved and transported in metal cans. Most laboratories prefer 1-gal cans, encouraging the investigator to collect as large a sample as possible. However, the investigator should fill the can only two-thirds to three-quarters full to allow head space at the top for sampling by the laboratory (**FIGURE 9-4**).

The investigator needs to check with the forensic lab to determine the type of container it prefers to be used for fire debris evidence. Some labs may prefer new unlined cans, as these containers assure that nothing in the can could contaminate the sample. The problem with unlined cans is that fire debris is usually moisture laden, which may eventually lead to the destruction of the can's integrity through rusting. To prevent the rusting and corrosion, some labs suggest lined cans be used for evidence collection. It is not uncommon for a criminal case to take a year, or even longer, to go to trial, so lined cans help ensure that evidence will not be compromised over this period. For the most part, lined cans have not presented a problem in the testing process. At the least, the lab may request a sample of the investigator's inventory of cans so that it can run tests to assure these containers will not present a problem in the future.

The investigator is responsible for assuring that each can or container is free of contaminants before using it. This confirmation can be as simple as just testing for the presence of hydrocarbons with a **hydrocarbon detector**. This step needs to be done each and every time for every investigation, providing consistency across the evidence collection process. It would also be a good suggestion to include this process in each fire report when this testing of the can takes place.

Each type of flooring requires specific tools to cut or break it up so that samples can fit inside an evidence can. These tools must be thoroughly cleaned and tested with a hydrocarbon detector before their use to ensure that the blades or cutting edges do not contain any oil residue that could contaminate the samples.

> **TIP**
>
> Tools must be thoroughly cleaned and tested with a hydrocarbon detector before their use to ensure that the blades or cutting edges do not contain any oil residue that could contaminate the samples.

The investigator must scrub any new saw blades clean before using them to collect evidence, because most manufacturers apply a thin coat of oil to keep the blades from rusting. This step eliminates the risk of contaminating any evidence with the oil residue. At no time should oil be used on these devices, because the oil on the tool could contaminate the sample. Although many of the saw blades may very well rust, under most circumstances they will still be usable.

The investigator can use a common utility knife (cleaned and tested) to cut up carpet. Wood flooring can be cut up and removed using hand tools or battery-operated saws. Concrete floors can be broken down with masonry chisels, a hammer, and a strong arm. Some masonry power saw blades can also make the job easier; before using them, the investigator must ensure that they will not contaminate the scene.

FIGURE 9-4 A 1-gal evidence can with debris should be filled no more than two-thirds to three-quarters full.
Courtesy of Russ Chandler.

Before using any tools, the investigator should check them again with a hydrocarbon detector to confirm that they are clean and free of any hydrocarbon contaminants. If they are used to obtain samples from more than one location, the tools must be cleaned before moving to the next sample location. Always start with clean tested tools before taking any evidence.

The petroleum sheen seen on standing water is not necessarily an indication of the presence of an ignitable liquid. Many structural contents and components are petroleum based and, when burned, leave a slight residue that may be seen as a sheen on standing water. However, if large quantities of a liquid accelerant were used, there *might* be residue left in the debris.

Any ignitable liquid containers found at the scene should be protected. If the investigation suggests that the containers might be involved in the incident, they should be marked as evidence. If the container holds any residue but cannot be sealed, every effort must be taken to recover the residue so that it can be examined by the laboratory. If the investigator plans to use an eyedropper for this purpose, it must be clean and checked before each use—although most of these types of tools should be disposable and discarded after each use. If there is not enough of the liquid to pour into an evidence container, other items, such as sterile cotton balls or gauze pads, may be used to absorb the liquid and are then placed into the evidence container.[3]

> **TIP**
>
> If the investigation suggests that containers holding ignitable liquids were involved in the incident, they should be marked as evidence.

If a liquid sample may be present in concrete, the investigator can use certain methods to extract the liquid without having to break up the concrete. For example, the investigator can use a wet broom to spread out any liquid and allow it to soak into the top layer of the concrete. Hydrocarbon fuels are lighter than water and will sit on the surface. Then, the investigator can put down a layer of an absorbent material to pick up the liquid residue. Absorbent materials such as kitty litter or even dry flour should sit on the surface for up to 30 minutes. This material can then be picked up using a clean tool and placed into an evidence container. Whichever material is used—cotton balls, kitty litter, or something else—an unexposed, clean sample of the absorbent material must be packaged in a sealed container and delivered to the laboratory as a comparison sample to show that the original sample submitted to the laboratory was not contaminated.

Solids

Solid evidence is sometimes collected for verification of the material's identity. Both accidental and incendiary fires can start when solid objects in granular form mix with other products and result in a chemical reaction. For example, oxidizers can react when mixed with an oily substance, resulting in ignition. Whether the mixing of these two components is accidental or intentional can be determined by a thorough investigation of all the facts, following the scientific methodology. (See Chapter 4, *Science, Methodology, and Fire Behavior*, for more details.)

Collecting solids requires that the investigator take a number of precautions. If more than one type of solid residue appears to be present, the investigator avoid accidentally scooping them up together, mixing them in the process. The results could be devastating should the two chemicals react and cause an exothermic reaction that results in a secondary fire. Thus, the investigator must collect both samples carefully, while ensuring that they will not come in contact with each other. Also, if the material being collected is potentially corrosive, the investigator should wear double or even triple layers of latex gloves or consider obtaining corrosion-resistant gloves to handle this type of material. For some corrosive materials, the investigator can use glass evidence containers to preserve them.

> **TIP**
>
> If more than one type of solid residue appears to be present, the investigator avoid accidentally scooping them up together, mixing them in the process.

Finding the Best Sample

Depending on the scene, the investigator may be able to identify the best sites at which to take samples. The leading edge of a floor burn or behind the baseboard that may have protected an accelerant product are both good candidate locations. But is the accelerant actually present in those locations? Many investigators resort to the use of a hydrocarbon detector to select samples. The many models on the market carry quite different price tags. Some are single-sensor units, whereas others are dual-sensor units. Multiuse units can find the trace evidence as well as measure the lower explosive range. Each investigator must decide which product best fits his or her needs and budget.

One of the best "tools" to use to look for evidence is a properly trained **K-9 accelerant dog** (**FIGURE 9-5**). The Bureau of Alcohol, Tobacco, Firearms, and Explosives

FIGURE 9-5 K-9 accelerant dogs are a great help in finding the best sample to submit for laboratory analysis. (Photo courtesy of Karl Mercer.)
Courtesy of Russ Chandler.

FIGURE 9-6 One-gallon metal paint cans are the evidence containers of choice for most investigators and laboratories.
Courtesy of Russ Chandler.

has taken a lead role in the training and use of accelerant detection dogs. The insurance industry—in particular, State Farm Insurance—has provided many of these animals to the investigations community across the United States. The trained dog's ability to detect the location of an accelerant can enable the fire investigator to find the best sample to submit to the laboratory for analysis.

> **TIP**
>
> The Bureau of Alcohol, Tobacco, Firearms, and Explosives has taken a lead role in the training and use of accelerant detection dogs.

Of course, using a K-9 at the incident scene does not guarantee that investigators will obtain a positive sample. Although a dog's ability to detect accelerants is far more sensitive than any technology presently available, investigators should not rely totally on the K-9's ability, but should use all of their knowledge, skills, and resources to do a complete and thorough investigation.

Evidence Containers

The appropriate type of evidence container to use depends on the situation, the instructions from the laboratory, the investigator's preferences, and fiscal resources. Metal and glass evidence containers have both pros and cons. Regardless of the type of container, all evidence containers must be new and unused.

Glass does not corrode or rust, can last indefinitely, and allows everyone to see the contents. Glass can be sealed adequately if done properly and with care. Do not depend on the rubber seal that comes with glass jars—these seals can react with vapors, contaminating the sample. They can also fail, resulting in the loss of

vapors from the sample. A laboratory can recommend the proper sealing method. As a general rule, placing a layer of aluminum foil between the glass edge and the rubber seal can create a proper seal while preventing vapors from reacting with the rubber seal. Drawbacks to glass containers are their fragility and the need for precautions to make sure containers are not damaged.

Alternative evidence containers on the market include nonpermeable heat-sealing bags made of nylon. These bags will keep the evidence indefinitely, and one can see what is inside the container. However, if the lab is not set up to process these bags, it makes no sense for the investigator to stock such items.

Metal cans are easy to handle, seal properly, and are durable (**FIGURE 9-6**). However, all of these containers—even lined cans—may eventually rust. Plus, one cannot see the contents. Labs tend to ask for lined cans so as to prevent the degradation of the can from oxidation with the liquid it contains. Even when a can is new, investigators should always check each and every can with the hydrocarbon detector just before use on the fire scene to prevent any contamination of the evidence placed in the container.

> **TIP**
>
> Metal cans are easy to handle, seal properly, and are durable.

Traditional Crime Scene Forensics

Many full-time fire investigators receive training in collection of all criminal evidence. This training is necessary because arson has been used many times to destroy evidence of other crimes such as burglary or homicide. If investigators have not been fully trained in collecting this other evidence, they will have to call local police forensic technicians for assistance.

Along with looking for evidence of the fire's origin and cause, sometimes investigators need to collect evidence such as blood or other body fluids. Weapons such as knives, guns, or blunt objects such as a baseball bat or heavy stick may also need to be recovered and protected. Fingerprints are a concern in case the incident turns out to be a crime scene. New research in fingerprinting can enable the investigator to recover fingerprints, even though the area is heavily sooted.

TIP

Along with looking for evidence of the fire's origin and cause, sometimes investigators need to collect more traditional evidence.

Tool mark impressions need to be documented, and then carefully collected. In some cases, fire investigators have been known to take an entire door for analysis and eventual presentation in court. Along with the impressions, the tools that may have been used to make these impressions must be collected and preserved, if they are found at the scene. The laboratory can do a comparison to see if a specific tool was used; metal flakes or other characteristics could match the tool with the damage.

If the scene is criminal in nature, and the investigator finds pertinent documents at the scene, these documents may be collected under certain circumstances. One reason to collect them is to protect them from being damaged by the elements. Should these paper artifacts already be wet, it may be necessary to collect them to be dried, so as to prevent mold and mildew that could damage them. Documents could show motive, and if they are left behind by the investigator, the perpetrator of the crime might find and destroy them.

However, if a fire started in the kitchen of a house and looks incendiary in nature, the investigator cannot go immediately to an undamaged area of the structure and search for documents. If there is probable cause to believe that evidence could be in files in the undamaged home office, for example, a guard should be left on the premises and the investigator should go to the proper legal authority, such as a magistrate, to obtain a search warrant. This ensures the rights of all concerned are protected and protects the developing case so that any evidence found can be legally admitted into court.

Tool Mark Evidence

Investigators must keep an open mind and look at all evidence. For example, a restaurant was having financial difficulties, so the owner decided to burn it down and collect the insurance money. In the process of setting up the fire scene to look like an intruder broke in and lit the fire, the owner made tool mark impressions on the exterior kitchen door with a crowbar. The door was open when the fire department arrived, a sharp-eyed first responder realized that the tool marks clearly showed that the perpetrator was forcing the door into the frame rather than prying the door open (**FIGURE 9-7**). This finding, along with other evidence, prompted the insurance company to deny the claim, and the individual eventually was arrested for the arson.

FIGURE 9-7 Tool mark impressions on doors can indicate a forced entry or an attempt by someone to make it look as if there was forced entry.
Courtesy of Russ Chandler.

Not-So-Traditional Evidence

Whenever a first responder encounters a victim within a burning structure, every effort must be made to save the individual by getting the person out of the fire scene atmosphere. Once the victim is removed and immediate life-saving actions have been taken, the first responder should start thinking about evidence issues. Any clothing removed during the victim's treatment should be kept separate from any other evidence to prevent contamination. In some circumstances, the patterns on clothing can provide valuable evidence. Trace evidence may also be present, such as an accelerant. If the fire was set to cover up other crimes, the clothing may contain evidence of the other crimes, such as assault.

Burn injuries on victims also need to be investigated and documented. However, this should be done only after the victim has received proper medical care. Documentation should include photographing the burns; these patterns are evidence.

If the first responder encounters a victim who is clearly dead, the body should be left in place, undisturbed. Burn patterns on the body and the associated clothing can also tell a story if properly documented and preserved. The location of the body and its relationship to the surrounding area are important pieces of evidence, which is why it is vital not to move a body if possible. Of course, the first and most important task is always to save victims in fire situations; this is a far greater priority than anything else, including collecting evidence, at the fire scene.

Accidental fires have been associated with appliances since their inception (**FIGURE 9-8**). Causes have varied from bad design and poor manufacturing to improper use. Sometimes misuse can be obvious, such as storing open containers of paint thinner adjacent to a gas furnace. Various appliances have been subjected to many recalls over the years when they were found to malfunction and create a heat source sufficient to ignite parts of the appliance or nearby combustibles.

FIGURE 9-8 Appliances can be involved in the ignition sequence and, as such, may need to be collected as evidence. If the fire is determined to be accidental, the appliance can then be turned over to a responsible party. If the appliance was used as part of an incendiary device, the appliance should be kept and processed as criminal evidence.
Courtesy of Russ Chandler.

> **TIP**
>
> Accidental fires have been associated with appliances since their inception.

Some people have taken advantage of these recalls and used the implicated appliances to set a fire intentionally. Properly operating appliances can be used to set a fire as well. The placement of cans or jugs of gasoline around a furnace could indicate a totally ignorant act or an intentional design to start a fire and make it look accidental. In some fires, an appliance was intentionally altered by the intended arsonist so that it malfunctioned and caused the fire.

> **TIP**
>
> The placement of cans or jugs of gasoline around a furnace could indicate a totally ignorant act or an intentional design to start a fire and make it look accidental.

When large appliances are involved in the ignition sequence, they pose a problem of keeping them as evidence. Most evidence lockers are not large enough to store a large appliance. Wherever the evidence is placed, the space must be secure, so there is no chance that the appliance will be tampered with or altered in any way. This control of the appliance is important for both accidental cases when the private sector will be seeking subrogation and when the public investigator has determined that the appliance was used as an incendiary device.

Sometimes the entire appliance need not be collected, but the switch, control, or circuitry must be taken for examination. In such a case, the investigator should put the item in an evidence envelope and pack it so that it is protected from movement and kept safe until an examination can be conducted. Smaller appliances such as microwave ovens, coffee makers, and toasters might also be part of the ignition sequence. These smaller items are easy to collect and protect. Placing these items in the evidence locker will protect them from damage.

Comparison Samples

When there are indications of the presence of an accelerant, the assigned investigator will take samples to the laboratory for analysis. These samples can consist of materials from the structure that have absorbed the liquid residue of an accelerant. Such materials could include flooring such as carpet and padding, wood planking, or concrete.

If possible, it is also a good practice to find another portion of the unburned flooring that does not contain the accelerant. Samples taken from unburned, uncontaminated areas are called **comparison samples**. Unburned areas may have been protected because they were under furniture or on the opposite side of the room. When collecting comparison samples, investigators should follow the same process used for other evidence, ensuring that no contamination occurs and properly documenting the sample. This enables the laboratory to compare the two samples and provides for a more reliable analysis.

Comparison samples for other types of evidence, such as electrical or mechanical devices, are commonly referred to as **exemplars**. Electrical switches, electrical circuit breakers, and even electrical panels can all benefit from the collection of comparison samples for analysis. The investigator should be able to see how the item appeared before the fire and all components that make it work properly.

Improper Installation

The rear of the house was fully involved in the fire when the fire department arrived. The enclosed porch was gone, the rear area of the basement was destroyed, and the rooms above were completely consumed. The fire was knocked down quickly, but there was not much overhaul to be conducted because most everything was gone except metal and concrete.

The investigator looked at the back of the structure. What stood out was a tee connector for a stovepipe, still hanging on the outside wall where the pipe came through the basement wall. On the inside, a wood stove still contained hot coals. The chimney had been enclosed by a wooden chimney chase; the remains of a 2 × 4 stud were found on the wall adjacent to the tee connector. On the ground all around the tee connector were sections of double-walled pipe. The tee connector (**FIGURE 9-9**) was designed to have the smoke enter and flow upward. The bottom of this connector was designed to be a cleanout; it was supposed to have a cover secured with three screws. However, the investigator found the cover on the ground, under the tee connector (**FIGURE 9-10**). The screws were in place on the cap, but had pushed

FIGURE 9-9 Tee connector as seen at the fire scene. The 2 × 4 stud on wall adjacent to the connector is the only remaining part of the chimney chase. Further evidence at the scene indicated that the homeowner used the chimney chase for storage. Poles for a badminton set are sitting on the tee support posts.
Courtesy of Russ Chandler.

FIGURE 9-10 The bottom of the tee connector does not have a cover plate to keep smoke, fire, and soot from escaping into the chimney chase. This cover was found directly under the debris, indicating that it fell off before the fire. Notice the indentures on the connector where screws did not penetrate the steel but bent it in. The base plate is held in position by tension of the points of the three screws.
Courtesy of Russ Chandler.

FIGURE 9-11 Base plate and the tee connector.
Courtesy of Russ Chandler.

FIGURE 9-12 A new tee connector showing the bottom plate.
Courtesy of Russ Chandler.

in at the base of the tee instead of penetrating and securing the cover (**FIGURE 9-11**). The only thing holding the cover in place was the pressure of the three screw points against the tee connector.

Because the investigator was not sure whether this was a design flaw or an installation error, he ordered and tested a new tee connector (**FIGURE 9-12** and **FIGURE 9-13**). The screws that came with the new plate were self-tapping screws. In contrast, the screws on the failed plate were ordinary sheet metal screws. During an interview with the installer, the installer admitted that he lost the original screws and just used three old screws he had in his toolbox. The insurance company paid the homeowner the limits of the policy and sought subrogation from the installer's insurance company.

FIGURE 9-13 Base plate (shiny surface reflecting wood decking) in position.
Courtesy of Russ Chandler.

Contamination

There is a very thin line between contamination and spoliation. Spoliation is a legal concern; it is defined as the act of destroying evidence, whether intentional or unintentional. For example, in the examination of a burned electrical outlet, suppose the investigator adds pressure sufficient to crumble it to dust. This is not contamination, but spoliation. Keep in mind that in some instances, both contamination and spoliation can occur from the same action. This distinction between the two will depend on how the action is described, such as with the use of an oiled saw blade when cutting up flooring: The oil from the blade may contaminate the floor samples, causing spoliation.

Contamination is a concern at almost every fire scene. Some contamination occurs during the suppression activities. Fire streams tend to move objects, materials, and residue from one point to another—and this outcome cannot always be avoided. Some sources also claim that fire fighters walking around the fire scene,

tracking in contaminants, can create positive samples for accelerants. However, a 2004 report published by ASTM International showed that footwear does not transport (track) identifiable ignitable liquid residue throughout a fire scene during an investigation.[4]

Avoid Contamination

The fire service has made great strides in sharing the message about the necessity of cleaning all parts of the fire fighters' personal protective clothing. The overall objective is to prevent contamination from carcinogenic particles. However, the same thorough cleaning of firefighting gear, including the boots, assures that no contamination from carcinogens will also remove any potential hydrocarbon residue. This avoids any possibility of contamination—or perception of contamination—of future fire scenes from residues from earlier scenes.

Fire crews should avoid using any suppression appliance such as a fan or saw in and around the area of origin, especially equipment powered by a gasoline engine. Any presence of such an appliance, even if it did not contaminate the scene, may cast doubts on the validity of any hydrocarbon evidence recovered.

TIP

Fire crews should avoid using any suppression appliance such as a fan or saw in and around the area of origin, especially equipment powered by a gasoline engine.

Under most circumstances, first responders need to concentrate on protecting what they find. An unknown footprint in the mud can be covered with a bucket, road cones can surround a tire impression in the ground, or a road cone can cover a bottle, can, or cigarette butt found in the backyard. The most important step that suppression forces can take to

protect evidence is to limit overhaul until the investigation is completed either by the officer in charge or an assigned investigator.

Water can be a contaminant. Fire fighters should limit their water flow to avoid destruction of evidence, such as knocking down drywall that may have had a pattern of interest to the investigator. Excessive flow of water can also dilute a liquid sample that may go to the lab.

Contamination can also come from the weather. Tire marks or shoe prints could be destroyed by rain, so covering that evidence could make a huge difference in its preservation. The same problem can arise in colder climates: Tire tracks in the snow could be just as important, so they also must be protected to maintain their evidentiary value.

Use of Gloves and Safety Gear When Collecting Evidence

It is essential that all safety equipment be worn all of the time. If a fire fighter's helmet falls off or becomes repositioned when the fire fighter bends over, the helmet

Snow Imprints

The family homestead was now unoccupied, but the family had made arrangements for a contractor to maintain and plow the driveway after each snowfall. On this particular night, the snow was pushed into a turnaround area, creating a snowbank a little more than 3 ft tall. When plowing stone driveways, blades are raised a bit to avoid damaging the driveway. In doing this, the plow driver left approximately 1 in. of snow left behind on the driveway.

Later that night, a perpetrator went to the house and set a small vandalism-type fire in the structure. In the snow on the driveway were the tracks from the plow and a second set of tracks from the perpetrator's vehicle. The perpetrator's tracks were visible, but not sufficient to identify any specific vehicle.

The fire was seen by a neighbor, who thought the glow was an exterior fire. In response, dispatch sent only a single engine. The fire was small enough for the engine crew to handle. After the engineer had set up, he walked around the engine to view the rest of the property. The crew had already suspected that the fire was intentionally set. The engineer saw the three sets of tire tracks in the snow: one from the engine, one set obviously from the truck that most likely had a plow, and the third from a much smaller car (narrow tire tread).

Close to where the engine was parked was the driveway turnaround space. Looking at this area, the

engineer noticed that the smaller tire tracks appeared to have backed into the turnaround—all the way into the snowbank. There in the fresh plowed snow bank was the imprint of the bumper and a perfect reverse print of the vehicle's license plate. The engineer notified the lieutenant, who called for the full-time investigator.

When the full-time fire investigator arrived, he photographed both the tire tracks and the license plate number after examining the scene. He called the dispatcher and had her run the tag; dispatch came back with the vehicle owner and address. When the investigator arrived at that address, a middle-aged man answered the door and confirmed the car with the tag in question was his, but stated he had been home all evening. He then said that his 18-year-old son had used the vehicle that evening to go out with friends.

The investigator interviewed the son with the father present. After being confronted with the evidence, the son confessed to setting the fire. Eight months later, the judge sentenced the young man to one year in jail. Because the fire caused only about $1000 in damage, the judge suspended the one year sentence and gave him three years of probation, ordered the young man to make restitution, and gave him 200 hours of community service.

should be readjusted to make sure it is secure on the head. Gloves protect the hands and help avoid contaminating any evidence handled. Wearing heavy gloves is critical to the safety of the investigator, as many sharp objects in fire debris can easily penetrate lightweight gloves. Because leather gloves may contain contaminants from other areas or other scenes, they should not be worn when handling evidence that will be subjected to laboratory analysis. Instead, the investigator should wear latex gloves when handling the evidence and then discard the gloves after collecting the sample. New gloves should be used for the next sample to ensure that no cross-contamination occurs. Obviously, investigators must have an adequate supply of latex gloves before starting the process of collecting evidence.[5]

The investigator should never place gloves inside an evidence container. If the lab needs the gloves for some reason, tape them to the outside of the container or place them in a plastic bag and attach that bag to the outside of the container.

The fire investigator must ensure all safety gear and tools are clean before entering the scene. All boots and gear must be thoroughly cleaned after each incident to prevent cross-contamination. After cleaning, the investigator should run a calibrated hydrocarbon detector over the gear to confirm the absence of contamination. If disposable coveralls are used—the operative word is *disposable*—the investigator should use a new set of disposable coveralls at the next fire scene.

TIP

All boots and gear must be thoroughly cleaned after each incident to prevent cross-contamination. After cleaning, the investigator should run a calibrated hydrocarbon detector over the gear to confirm the absence of contamination.

Tools require special care. Thoroughly scrub shovels, trowels, and other hand tools with soap and water after each incident. When they are pulled out at the next event, the investigator should test them with a hydrocarbon detector before using them. The investigator must document this test in the field notes to confirm that this task was accomplished before the investigation began.

First Responder Considerations

In areas where there is no readily available assigned investigator, a first responder can be trained in evidence collection in conjunction with the training received in fire origin and cause. A small evidence kit can be established to collect debris samples or containers, enabling the first responder to tag and document the evidence properly. Of course, there must be a procedure for the first responder to lock up the evidence, ensuring that no one else has access and that the evidence remains in the possession of the first responder the entire time before it is turned over to an investigator for processing (**FIGURE 9-14**). This maintains the **chain of custody**. If an arrest is eventually made in the case, the first responder can testify in court about the evidence collection and the chain of custody. In the case of an accidental fire, the first responder might testify in civil court should a civil suit result in a product failure or if the insurance company chooses to subrogate.

Of course, other measures can be taken to provide for evidence collection at a fire scene. Local law enforcement officers, even if they are not fire investigators, can assist in evidence collection, documentation, and storage. The key consideration is to have a plan in place as part of the department's policy, well in advance of the incident.

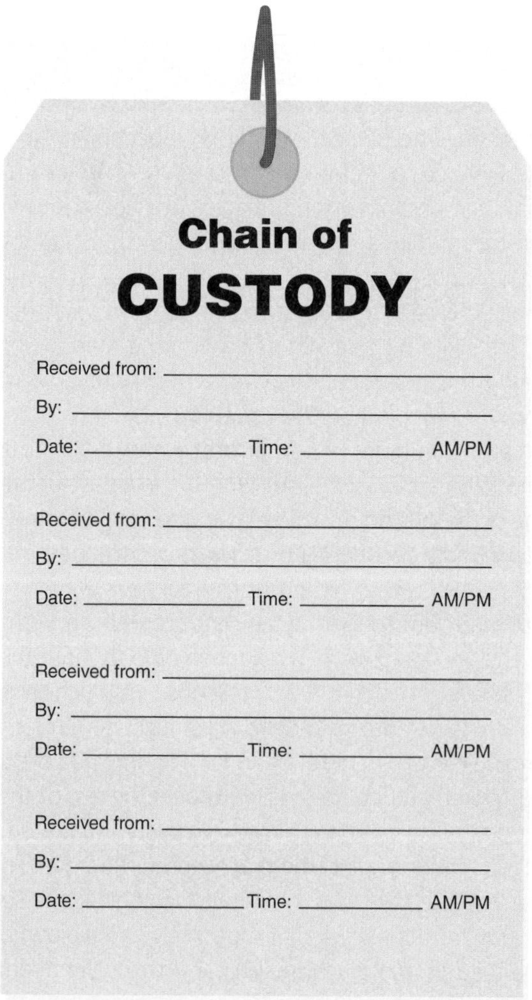
Chain of
CUSTODY

Received from: ____
By: ____
Date: ____ Time: ____ AM/PM

Received from: ____
By: ____
Date: ____ Time: ____ AM/PM

Received from: ____
By: ____
Date: ____ Time: ____ AM/PM

Received from: ____
By: ____
Date: ____ Time: ____ AM/PM

FIGURE 9-14 Typical Chain of Custody tag allowing each handler to log the transfer of possession of the evidence.

Documenting, Transporting, and Storing Evidence

Any piece of evidence is only as good as the way it is handled. The evidence must be properly and fully documented so that it will maintain its value for a future trial. Each piece of evidence must be labeled or tagged. Then, the evidence must be properly stored from the time it is collected until the investigator receives a court order to destroy the evidence.

> **TIP**
>
> Any piece of evidence is only as good as the way it is handled.

Documenting Evidence

Every piece of evidence collected by the investigator must be tagged or labeled. The information on the evidence label should include the date and time the sample was taken, the location from which it was taken, and the person who took the evidence. In addition, the label must have the case number of the fire location, along with a description of the evidence itself.

> **TIP**
>
> Every piece of evidence collected by the investigator must be tagged or labeled.

Another item to be documented with every piece of evidence is the chain of custody. When the evidence is transferred from one person to another, the name of the person receiving the evidence must be placed on the chain of custody label, including the date and time that the transfer occurred.

All of the information about the evidence collection and the chain of custody must also be included in the investigator's file. This document should be used to track each and every piece of evidence for that case.

Generally, public investigators use **commercial evidence tape** to seal evidence containers, and then place the investigator's initials on the tape. In the private sector, evidence tape is rarely used; instead, the evidence cans are sealed and then boxed for shipment via a common carrier. The sealing of the box is proof enough that the evidence was not tampered while en route to the lab.

Shipment of Evidence

If the investigator is fortunate enough to work close to the laboratory, he or she can personally deliver the evidence. Upon arrival at the laboratory, the investigator hands over the evidence and the **evidence transmittal** as part of the receiving process. The **evidence transmittal form**, which is usually provided by each state or local forensic laboratory, describes the evidence, the testing desired, and all the contact information of the investigator. It also includes a section on chain of custody. Once the evidence is turned over, the investigator receives a copy of the transmittal for the investigation file.

Because not everyone works near a forensic laboratory, some states set up collection points using their staff to transport the evidence to their lab. Some remote localities and most private-sector investigators have to rely on a common carrier to transport evidence. Evidence needs to be packed so that it will not be damaged in shipment; even metal cans can be damaged and lose their integrity if not packaged properly. Once the evidence is packaged, the investigator must remember to insert the evidence transmittal form. With the package properly sealed, it is sent by registered mail, requiring a receiving signature, via a common carrier such as the U.S. Postal Service, FedEx, or United Parcel Service (UPS).

> **TIP**
>
> Evidence needs to be packed so to that it will not be damaged in shipment; even metal cans can be damaged and lose their integrity if not packaged properly.

Once the laboratory receives the evidence, the carrier sends the signed receipt to the investigator. Staff at the laboratories will also sign the evidence transmittal, which continues the chain of custody. A copy is then sent to the investigator.

Evidence Storage

Evidence should be stored in a secure location at the state or local lab, where the investigator or lab personnel are the only persons with access. Evidence rooms must be climate controlled to keep the environment dry and cool. They should contain ample space for racks and shelves to hold the evidence. Some evidence lockers contain freezers to hold evidence scheduled for the detection of accelerants. However, care must be

> **TIP**
>
> Evidence rooms must be climate controlled to keep the environment dry and cool.

taken to ensure that the moisture within a container is not of such a volume that it could freeze, expand, and rupture the container.

If the evidence is not volatile and does not need laboratory testing of accelerants, other care issues should be observed. Items that are damp or moist may eventually mildew and mold if not properly dried out. This evidence must not be placed in sealed containers until it is completely dry.

If the evidence might potentially break down, oxidize, or corrode, special precautions should be taken. Place the evidence in plastic boxes or other containers so it does not contaminate other evidence stored nearby.

Laboratory Testing

The most frequent examination requested by fire investigators is testing of debris for the presence of an accelerant such as hydrocarbons. The evidence transmittal form usually asks the lab to test for the presence of ignitable liquid residues. The laboratory then decides which process to run on the evidence provided to garner the best results. Generally, the laboratory runs two tests—gas chromatography and mass spectrometry. Although this has been the standard for many years, some research institutions are searching for better ways to improve the science, even in the field of fire investigations.

In fire investigations, it is common to run the two tests together in a method called gas chromatography–mass spectrometry (GC/MS). The **gas chromatography (GC)** process separates the sample into individual components. The GC instrument provides a graphic representation of each component, quantifying the amount within the sample. If a large enough sample is available, this information, along with the further analysis provided by **mass spectrometry (MS)**, can enable the lab personnel to identify any ignitable liquids present in the contents of the evidence.

Of course, a forensic laboratory can provide many other services. In addition to those commonly expected, such as recovering fingerprints or doing DNA testing, a lab can perform more obscure jobs such as putting glass fragments back together from bottles or glass panes. Lab personnel can identify tire marks or shoe impressions, and even the source of small paint chips. Thanks to recent movies and television shows on crime, the public is more aware of the potential capabilities of forensic labs. Early in their careers, investigators should visit the local forensic lab that will process their evidence. It is always good to meet the members of the forensic team and to discover the true

potential of each lab along with its recommendations on evidence collection and preservation.[6]

The best tool for success with testing at the laboratory is an open line of communication between lab personnel and fire investigators. Forensic scientists can identify the correct amount of debris to collect for obtaining the best test results. They can also make recommendations on most aspects of evidence collection.

> **TIP**
>
> The best tool for success with testing at the laboratory is an open line of communication between lab personnel and fire investigators.

The laboratory personnel conducting the tests are responsible for writing the report and for testifying in court about the testing process and the results of their testing. The investigator may have to testify about how the evidence was discovered, where it was discovered, and which method was used to collect the evidence. Opposing counsel will be meticulous in examining this testimony to ensure that the investigator followed proper procedures.

Release of Evidence

Evidence for either a criminal case or a civil suit may have to be kept for years, awaiting trial. Then, it may need to be stored even longer, to a date beyond any potential for an appeal. In criminal cases, much of the evidence submitted in the trial is kept by the courts. However, lab samples, debris, and volatile liquids are usually returned to the investigator. When the evidence is no longer needed, the first step should be to contact the owner of the evidence to see if that person may need it for any reason. Even a can of debris may be of value. Following the criminal trial, the property owner may want to file civil suit, where such evidence may be submitted as part of the litigation. If the owner wants the evidence, it should not be returned until the investigator petitions the courts to release the evidence to the owner.

If the property owner has no need for the evidence, it is advisable for the investigator to get a signed release stating that fact. The next step in most jurisdictions is to petition the courts to have the evidence destroyed. The manner of destruction will be a local decision. After the evidence is destroyed, policy may require the investigator to return the destruction order to the courts, stating the date of the destruction. Common sense dictates that copies of all documents should be kept in the investigation file.

Not all evidence collected is used at trial. Usually, the prosecuting attorney decides what will be admitted in court. The investigator must maintain any remaining evidence not used in the trial just as if it had been used in court. This is necessary in case a retrial occurs or the evidence is needed for an appeal. When petitioning the court to release or destroy the evidence, all of the evidence should be listed in the request and on the judge's order.

A similar process applies in civil cases. If the investigator for some reason is the holder of the evidence and has been subpoenaed for the civil trial, all evidence must be kept and maintained as if it were a criminal case. However, some civil trials involve multiple litigants. In such a case, the evidence must be made available to all parties. In all such situations, advice from the jurisdiction's attorney or the local prosecutor should provide direction for how the evidence is to be viewed and who can examine it. A key consideration is that no destructive testing of the evidence should be conducted without all known parties having a chance to be involved or at least to observe. This is clearly a situation in which spoliation of the evidence is of prime concern to all parties. See Chapter 8, *Spoliation*, for further information on spoliation.

Wrap-Up

SUMMARY

- There are four types of evidence: demonstrative, documentary, testimonial, and physical evidence. Even though first responders may be only fact witnesses, they should be aware of each type to understand all aspects of a full investigation.
- Evidence consists of a multitude of items, ranging from objects to patterns. Investigators must have the training and expertise necessary to identify potential evidence and its value so that they will know what should be collected. First responder investigators should have knowledge of how the full-time investigator will handle evidence so they may be of assistance in this process.
- All investigators must know the legal parameters regarding taking and securing evidence. If findings suggest that the fire was incendiary in nature, the investigator can—and more importantly, should—collect any and all evidence as to the fire's area of origin and cause. Any evidence that may indicate who might have caused the fire should be collected and preserved as well.
- When a fire is declared accidental at the time of the fire scene examination, the assigned investigator does not need to collect evidence to prove that hypothesis. Under most circumstances, the investigator has no legal authority to collect and secure evidence from such a scene. However, if evidence may be of value to an absent property owner, and leaving it behind may lead to its damage or disappearance, the investigator should consider taking and securing the evidence. Evidence can then be provided to the property owner once contacted. The physical evidence policy should address this action, and it should be cleared by the local prosecuting attorney or jurisdiction before taking on such a liability.
- Instead of securing and keeping the evidence, it may be appropriate to find a relative or neighbor and turn the entire structure over to them, including any evidence, until the property owner arrives. If no one is available to take on this responsibility, it may be necessary to post security on the structure until a responsible party can be located.
- Once evidence has been collected by the fire officer or assigned fire investigator, it should be locked up and secured. These actions must be documented, and a list of who the evidence is turned over to and when should be retained to maintain the chain of custody.
- Once collected, the evidence should be processed for testing by a forensic laboratory as soon as practical. This improves the chances of getting the best results from the test process. Testing can range from identifying whether a liquid is ignitable to determining the presence of an ignitable liquid in a debris sample. Forensic testing of fingerprints, tool mark impressions, and so forth may also occur.

■ After the evidence has been used in court and all appeals have been settled, investigators must arrange to dispose of the evidence. This can be done by returning it to its owner or by petitioning the courts for permission to destroy the evidence. The evidence may still have the potential to be used in civil cases, so this possibility must be explored before the evidence is released or destroyed.

KEY TERMS

Accelerant An item or substance used to ignite, spread, or increase the rate of fire growth.

Barricade tape Brightly colored tape used to warn people not to enter an area. Lettering may include "fire line," "keep out," "police line," "caution," and so forth. Tape comes in various colors and in large rolls of 500 ft, 1000 ft, or more.

Chain of custody A means of documenting who had control of the evidence from the time of collection through the trial and up to its release or destruction.

Commercial evidence tape Tape with a sticky side used to seal evidence containers. Tape contains wording such as "evidence" or "do not tamper." When tape is removed, it leaves a visible indication that clearly shows the package or container has been opened.

Comparison sample An uncontaminated sample of what is being tested that helps examiners identify any contaminants.

Contamination Anything introduced into the fire scene or into the evidence that makes test results unreliable or raises doubts about the results of the laboratory testing.

Demonstrative evidence Evidence in the form of the representation of an object, in contrast to physical (real) evidence, documentary evidence, or testimony evidence.

Documentary evidence Evidence in the form of media such as paper records, photographs, videos, and voice tape recordings.

Evidence transmittal form A form designed to identify the evidence to be submitted to the laboratory and the tests to be conducted on that evidence; it can also serve as the chain of custody of the evidence.

Exemplar A comparison sample that enables the investigator to compare the damaged item to an undamaged version of the product.

Expert witness A person who is permitted to testify in court due to his or her special knowledge and expertise in a particular area relevant to the case.

Fact (or lay) witness A person with knowledge about what happened, what he or she saw, or what he or she heard that is firsthand and relevant to the case.

Gas chromatography (GC) A laboratory test method that separates the recovered sample into its individual components; it provides a graphic representation of each component along with the amounts of each component present.

Hearsay Information that is heard or received from other people than the originating party. This information cannot be substantiated and most times is considered rumor.

Hydrocarbon detector An electronic device intended to be used in the field to identify the location of ignitable liquids; it is used to improve the probability of obtaining the best sample to submit to the laboratory for testing.

K-9 accelerant dog A dog specifically trained by a reputable organization to identify the presence of an accelerant and give the appropriate signal of its location.

Litigant Someone who is involved in a lawsuit.

Mass spectrometry (MS) A test used to identify the presence of an ignitable liquid in a submitted sample; it indicates the composition of a physical sample by generating a spectrum representing the masses of sample components.

Overhaul Process carried out by fire suppression personnel to examine and uncover any remaining heat source that could rekindle the fire. This often includes the term *salvage,* where the fire fighters take action to protect contents of the structure from further damage by removing or covering with tarps referred to as salvage covers.

Physical evidence policy A locality's policy that guides and directs the proper collection, storage, handling, use, and disposal of all evidence.

Salvage A suppression activity used to protect the contents of a property from smoke or water damage by removing objects from the structure, covering materials within the structure with tarps, and/or removing water with squeegees, pumps, water vacuums, and so forth.

Subpoena The command of a court to have a person appear at a certain place at a specific time to give testimony.

Testimonial evidence Testimony offered to prove truth of the issue at hand, usually from a witness.

Tool mark impressions Potential identifiable marks from tools used by the perpetrator of the crime.

REVIEW QUESTIONS

1. Why should the fire investigator not seize evidence from any accidental fire?

2. Under what circumstances might the fire investigator want to collect evidence at an accidental fire? How can this be done properly and legally?

3. What incentive does the insured have to allow the insurance company investigator to conduct the investigation and take any evidence?

4. Why should the investigator test tools with a hydrocarbon detector upon arrival at the scene of an incendiary fire?

5. What is a fact witness, and what can he or she testify to in a trial?

6. What is a comparison sample, and why is it so important?

7. Can an investigator continue using the same tools, such as shovels, at different locations at the fire scene? If so, what needs to be done (if anything) to the tool to prevent contamination?

8. Define the four types of evidence.

9. What conditions must be met to destroy evidence?

10. What is an evidence transmittal form?

DISCUSSION QUESTIONS

1. What various items might be found around a house that could be used to mark or preserve a shoe-print impression, a 2-ft-long tire impression, or a 1-gal gasoline can found in the side yard? For those already in the fire service, what might be found on the apparatus on the scene that could be used to perform the same tasks?

2. Under what circumstances might a volunteer fire officer decide to take evidence found on an incendiary fire scene, especially if the only full-time assigned investigator is days away from visiting the scene?

3. The fire investigator identified a portable heater with a manufacturer defect as the cause of the fire. Does the investigator have the right to take that heater to use in fire prevention efforts? If in doubt, under what circumstances might this be acceptable?

REFERENCES

1. National Fire Protection Association. (2008). *NFPA 921: Guide for fire and explosion investigations* (pp. 921–103). Quincy, MA: National Fire Protection Association.

2. National Fire Protection Association. (2008). *NFPA 921: Guide for fire and explosion investigations* (pp. 921–158). Quincy, MA: National Fire Protection Association.

3. National Fire Protection Association. (2008). *NFPA 921: Guide for fire and explosion investigations* (pp. 921–137). Quincy, MA: National Fire Protection Association.

4. Armstrong, A., Babrauskas, V., Holmes, D. L., Martin, C., Powell, R., Riggs, S., Young, L. D. (2004). The evaluation of the extent of transporting or 'tracking' an identifiable ignitable liquid (gasoline) throughout fire scenes during the investigative process. *Journal of Forensic Science, 49*(4). Paper ID: JFS2003155.

5. National Fire Protection Association. (2008). *NFPA 921: Guide for fire and explosion investigations* (pp. 921–136). Quincy, MA: National Fire Protection Association.

6. Daeid, N. N. (2004). *Fire investigation* (pp. 160–181). Boca Raton, FL: CRC Press.

CHAPTER 10

Sources of Ignition

LEARNING OBJECTIVES

Upon completion of this chapter, you should be able to:

- Describe various heat sources capable of being a source of ignition.
- Describe how the forces of nature in the form of lightning, escaping gases, and radiant heat from the sun can provide sufficient energy to be potential heat sources.
- Describe how smoking materials can be a potential heat source.
- Describe friction heat as a potential heat source for a fire.
- Describe how chimneys and associated appliances can contribute to a fire as sources of ignition.
- Describe chemical reactions and their involvement as a potential heat source.
- Describe and understand the concept of spontaneous ignition and where it can and cannot occur.
- Describe the role of appliances, large and small, as sources of ignition.
- Describe how animals and insects can become a potential heat source.
- Describe how lightbulbs can be a potential heat source.

Case Study

A homeowner in Virginia was faced with another year of high fuel oil costs. He had a floor furnace; the chimney for the furnace was cement block, attached to the outside wall of the structure. To save money, the homeowner decided to purchase a woodstove and stove pipe to connect the wood stove to the existing chimney with a chimney thimble. A thimble is a terra-cotta ring that accepts the pipe from the woodstove, directing vapors, smoke, and gases up into the chimney.

The plan was simple: Cut a hole in the wall adjacent to the chimney, chisel a hole into the cement block chimney, attach the thimble, put the woodstove into position, and connect the pipe between the thimble and the woodstove.

The homeowner chose a position on the wall for the thimble. He cut a hole in the wood paneling, through the sheet rock, and then cut away the exterior sheathing, exposing the cement blocks of the chimney. Next, the homeowner took a chisel and carved his way through to the interior of the chimney. The opening he created was a little less than 6 in. in diameter. The interior of the chimney was only 9 in.^{2} (3 in. across), which was not to code. in consequence, the diameter opening from the woodstove was 8 in., the opening into the chimney was less than 6 in., and the interior of the chimney was 3 in. (**FIGURE 10-1**).

The homeowner put mortar on the face of the cinder block (**FIGURE 10-2**), pushing the thimble up against the block. The mortar could not have held the thimble in place, so the wall sheathing, sheetrock, and wood paneling—two of which were quite combustible—served to hold the thimble in position (**FIGURE 10-3**).

With the woodstove in place and the mortar just barely dry, the homeowner put kindling and wood into the stove, then lit the wadded paper to

FIGURE 10-2 The thimble was 8 in. across, with an interior that was 6 in. in diameter.
Courtesy of Russ Chandler.

FIGURE 10-3 The thimble was intended to be held in place with the mortar, shown where it is being positioned. The hole cut in the paneling, sheetrock, sheathing, and siding held the thimble in place until they ignited.
Courtesy of Russ Chandler.

start the fire. Warmth started spreading through the room. The burning kindling had a good draft, pulling the smoke from the woodstove burn chamber into the pipe, through the thimble and into the narrow chimney flue. But as the larger pieces of wood caught fire and started giving off smoke, it proved too much for the chimney—and the smoke, gases, and heat started coming out the front of the woodstove. Closing the doors did not help because the smoke started coming out the damper vents, from around the pipe as it came into the stove, and from the pipe as it entered the thimble. The family rushed to open windows and doors. Soon they noticed more smoke from the area of the thimble, and then the wall paneling surrounding the thimble burst into flames. Finally, they called 911, while

FIGURE 10-1 The interior of the flue is shown; when measured, it was only 9 in.^{2}: 3 in. by 3 in.
Courtesy of Russ Chandler.

they filled pitchers, pots, and other vessels to throw water onto the wall.

Unbelievably, the family stayed in the house until the fire department arrived. Without a doubt, they kept the fire from spreading on the paneling, but they put themselves at an extreme risk of losing their lives. The home's occupants were treated for minor smoke inhalation at the scene, but refused to go to the hospital. Fire damage to the home was limited to around the thimble. However, soot was deposited throughout the house.

Starting the fire in the woodstove that night most likely saved their lives. Had they not used the woodstove, the floor furnace would have been used that night. There is no doubt that carbon monoxide from the floor furnace would have seeped from the chimney into the thimble, piping, and woodstove, escaping into the home. The family members very well could have died in their sleep.

A question arose when filling out this fire report: Was the first material ignited the paper used to start the woodstove or the paneling holding up the thimble? Both answers could potentially be correct. When someone flicks a thumb across the wheel of a lighter, the first material ignited could be considered the fuel from the lighter itself, but the fire report will most likely cite the material that the lighter ignited. With that point in mind for the woodstove case, many would report that the first material ignited was the paneling. The cause of the fire was an improperly installed woodstove, but some might protest that the actual cause was a total lack of knowledge by the installer.

 *Access Navigate for more resources.*

Introduction

This chapter, in focusing on sources of ignition, devotes little attention to chemicals and incendiary devices. Discussions of these items should be limited to organized training events, where instructors pay careful attention to who attends and how the information is disseminated. That said, the ready availability of the Internet means that many people, including teens, have an abundance of information on how to make chemical incendiary devices. Sometimes even Hollywood films share this potentially dangerous information, prompting impressionable individuals to experiment with chemicals and improvised devices.

Sources of ignition are all around us. We frequently use them in everyday life—at home, in the office, in the factory, and even at play. Throughout history, the primary igniters used by people have changed. For example, when the Europeans started to inhabit the New World in the 1600s, the open fire used for cooking and heating kept them alive, yet, when not controlled, took their lives as well.

Consider the case of Jamestown, one of the first colonial settlements. More than 400 years ago, English settlers came to this site in Virginia, where they built stockades and homes. They built their structures with thatch roofs, with chimneys made of sticks coated with mud. Almost the entire settlement burned not once, but twice. The source of ignition for these fires was most likely the fireplaces: The mud coverings might have failed or embers escaped from chimneys and landed on the thatch.

Since the match was invented, it has been a primary igniter of hostile fires, both accidental and deliberate. In 1827, John Walker invented a match that could be ignited by striking it on a rough surface. The danger associated with the **strike-anywhere match** eventually prompted the invention of the **safety match** by Johan Edvard Lundstrom, in 1855 in Sweden. This match could ignite only if struck on a specific surface, rendering the match safe in most circumstances.

Even today, matches and lighters often play the role of primary igniter, along with associated smoking materials. However, many small appliances, chemicals, and other heat-producing items are also sources of heat for uncontrolled fires. Heat causes fire—though the devices that create the heat have changed over time.

> **TIP**
>
> Heat causes fire—though the devices that create the heat have changed over time.

In every fire investigation, the area of origin is the first point determined. In the area of origin, the investigator should find the first item ignited. That ignition was caused by a heat source, which may also be found in the area of origin or in close proximity. Sometimes the heat source is missing from the scene because it was destroyed as a result of the fire or was moved, possibly by suppression activities. Armed with a clearly defined area of origin and other supporting evidence, the investigator can infer the culpable heat source, even if it cannot be found. Nevertheless, the heat source must be established to make a final determination as to the fire's origin and cause.

Competent Ignition Sources

A heat source must be able to deliver enough heat to cause ignition. For this to happen, sufficient heat must be present, at or in excess of the ignition temperature of the first material ignited. The heat must also be in contact with the first material ignited for a sufficient duration to heat the fuel to its ignition temperature.

The investigator will need to be able to establish the temperature capability of heat sources as well as the ignition temperature of most fuels; he or she will need access to reliable resources to make these determinations. The investigator can then determine if that heat source could have ignited that specific material. It will be beneficial to have such competent sources documented to be used in the report just in case it may be needed in legal hearings in either criminal or civil court.

Lightning

It seems fitting to first address the one heat source that we can do little to control. Lightning is a static discharge between two differing potentials, negative and positive. It can discharge from one cloud to another cloud, or from ground to a cloud, or from cloud to ground.[1] The discharge can even occur within the same cloud when the lower surface of the cloud has a potential different from the upper surface of the cloud.

Over the period 2007–2011, an estimated 22,600 fires were started by lightning each year in the United States, which resulted in an average of 9 civilian deaths, 53 civilian injuries, and $451 million in direct property damage annually.[2] Worldwide, more than 8 million cloud-to-ground discharges of lightning occur every day.[3]

TIP

Over the period 2007–2011, an estimated 22,600 fires were started by lightning each year in the United States, which resulted in an average of 9 civilian deaths, 53 civilian injuries, and $451 million in direct property damage annually.

In a study done in the northern Rocky Mountains, researchers discovered that 1 in 25 lightning strikes are capable of starting a fire.[4] The 18th edition of the National Fire Protection Association (NFPA) *Fire Protection*

Handbook states that the average strike is 24,000 amperes, but strikes have been measured as high as 270,000 amperes.[5] According to the National Weather Service, however, lightning strikes can peak at an average of 30,000 amperes and can have as much energy as 300,000 amperes.[6] Clearly, lightning contains a lot of energy—more than enough to start a fire if conditions are right.

TIP

According to the National Weather Service ,lightning strikes peak at an average of 30,000 amperes and can have as much energy as 300,000 amperes.

Almost everyone has seen the result of a lightning strike on a tree. The lightning energy is so massive that the moisture in the tree is vaporized violently, literally exploding and ripping open the trunk of the tree. Even with this much energy and heat, however, a fire may not necessarily occur. As discussed in Chapter 4, *Science, Methodology, and Fire Behavior*, all conditions have to be just right for a fire to start. Lightning bolts contain unimaginable amounts of energy, but they last for only a very brief amount of time. This is the key question: Is the target mass large enough to absorb the heat? The answer in most circumstances is yes. In fact, if a tree (dead or alive) contains an appreciable amount of moisture, a lightning strike will not cause a fire.

When lightning strikes structures, it can cause other damage as well as fires. According to the NFPA, 5300 structure fires from lightning were reported in the United States in 2014. Of those fires, 3900 affected homes, resulting in $408 million in damage along with 35 civilian injuries. Nonhome structure fires in 214 resulted in $65 million in damage, with 3 civilian injuries.[7]

Depending on the amount of damage experienced, there may or may not be obvious indications of a lightning strike on the structure. Clues to lightning's role may include items that would usually not melt in a fire, such as copper, but were vaporized as the result of lightning energy. The clues at the physical scene along with witness accounts may provide enough evidence to prove that lightning was the heat source for the fire.

Seismic Events and Escaping Gases

A seismic event is not a source of ignition. However, the event itself changes the dynamics of the environment, so what typically would not have been a source of ignition can become responsible for starting many fires. Earthquakes are perfect examples.

The most famous fire associated with a seismic event was the Great San Francisco Fire of 1906. The fires associated with the initial earthquake were the result of many events. Primarily, the earthquake brought together open flames and destroyed buildings, providing lots of fuel. Fires are seen in many earthquakes when available heat sources are brought in contact with fuels as the result of the movement and destruction of buildings.

Smaller seismic events can create fractures in rock that enable the release of **methane gas**, which typically would not have made it to the surface. Although not a source of ignition, this gas is fuel that can be easily be ignited from the energy released from someone flipping a simple light switch.

Deep in the ground, in various areas, are pockets of natural gas. This gas is colorless and odorless—as well as deadly and ignitable. Natural gas may escape to the surface from time to time, but seismic activity can also move the layers of rock, sometimes creating a path for the gas's egress. Natural gas is lighter than air and will migrate upward. As this gas moves toward the surface, it may also find a human-made path, allowing a more rapid escape to the atmosphere. Trenches around large pipes that are backfilled with stone to allow water drainage and provide stability, for example, may create a path for the gas's migration. Wells, both deep and shallow, can create a conduit for escaping gases. Because most wells are capped, the gases may escape along the water pipe trench leading to the structure and terminating in a basement or crawlspace. Once inside a structure, because the gas is lighter than air, it collects in higher locations. If a source of ignition is present, the gas may detonate, creating an explosion and potentially a subsequent fire. Such events can come as a surprise.

Methane Gas Surprise

The insurance investigator received a strange assignment to go to the western, mountainous part of the state to do a preliminary investigation. Getting to the site required a 6-hour trip, one-way. The strange part was that the investigator was asked to deliver a check made out to the insured. Usually, the claims manager handed out the checks, not the investigator. The investigator's instructions were to look at the structure, listen to the insured, and if the facts matched the scene, hand over the insurance check; if not, he was instructed to return the check to the main office.

At the insured's home, the investigator found debris all around where the home used to sit. It had been a small square home, and each of the four exterior walls were just laid out as if they had been pushed over. The roof was almost intact, but it lay behind the house. There were signs of small fires here and there where items such as curtains had burned and been consumed, but the fire did not appear to have spread. The only other structure on the small property was a small well house that was built as a replica of the original home with white siding and a peaked roof.

The investigator made a cursory investigation by walking around the structure and noticed the presence of a full basement. The homeowner arrived, and he and the investigator talked in the side yard about 20 ft from the well house. The homeowner explained that he got up early in the morning and went to the basement to get some tools he planned to use on a project. The only light in the basement was a ceiling-mounted ceramic base light fixture with a single bulb and a pull string to turn it on and off.

The homeowner reached for the string—and that was the last thing he remembered. When he returned to consciousness, he looked around and found that the house was pretty much gone. The string from the light was still in his hand, and he had burns on his head and face. Neighbors had called emergency services, and the homeowner was transported to the hospital, where he was treated and released. He experienced first-degree burns on his face, and his hair, including his eyebrows, were gone.

The homeowner had lived in this community his entire life, and he worked in the nearby coal mine. He said he knew exactly what had happened: Methane gas from an abandoned mine had seeped up through his well and then into his house. The investigator, not being from this part of the country, knew little about mines or methane problems. The investigator felt some skepticism as he listened to the story.

Suddenly, there was a loud explosion. The investigator saw the roof of the pump house hit the ground about 4 ft away. All four walls of the well house were laid out, exposing the above-ground pump and pressure tank. There was a slight whiff of smoke but no fire.

The investigator gave the check to the homeowner, had him sign a receipt, shook his hand, and headed back to the main office, better educated, adding more to the sum of his experiences.

Radiant Heat

Sunlight is the ultimate radiant heat, but by itself is not intense enough to ignite common combustibles. However, when the rays of the sun are concentrated, the focal point can produce enough energy to become a competent ignition source. Bouncing the sun's rays off a shiny concave surface, such as the bottom of a metal can, can focus a beam of light sufficient to create heat to ignite a fuel source.

> **TIP**
>
> When the rays of the sun are concentrated, the focal point can produce enough energy to become a competent ignition source.

Focusing the sun's rays can be done with glass such as a magnifying glass, a glass sphere, or pieces of broken glass. Cut glass, broken glass, bottles, or decorative glass pieces can refract light, creating a focused beam that can then be applied to an ignitable fuel. The focal length of a focused beam of light tends to be short, so the first material ignited must be in close proximity. This mechanism should not be confused with the phenomenon seen with prisms. A prism will scatter light, projecting a rainbow effect rather than concentrating a beam.

> **TIP**
>
> The focal length of a focused beam of light tends to be short, so the first material ignited must be in close proximity.

The focused rays of the sun are capable of starting a fire, though such events are rare. In years past, this type of heat source was usually associated with wildland fires. Today, more products sold as decorative items for homes are capable of this phenomenon, and fires associated with concentration of radiant heat by these products are occurring more frequently in residential structures. As with all fires, if glass is suspected of being a potential heat source, a systematic and thorough investigation should be able to prove a reliable hypothesis.

Smoking Materials

Cigarettes, cigars, and pipes are commonly used smoking materials. Although fewer people smoke today than did 10 years ago, a sufficient amount of tobacco is still used to consider smoking materials a frequent source of heat for an uncontrolled fire.

Cigars generally go out if left unattended, so they present fewer fire hazards. To keep a cigar burning, the smoker needs to draw on the cigar, pulling air across the coals at the end of the cigar to keep it burning. When placed in an ashtray, cigars eventually self-extinguish.

> **TIP**
>
> Cigars generally go out if left unattended.

Pipes work in the same way: The user pulls air across the burning leaf in the bowl, supplying more oxygen, causing the tobacco to glow more intensely and burn. Left unattended, pipes also tend to go out. Consequently, smokers have to relight both cigars and pipes from time to time. However, pipes can be involved in the ignition sequence if a bowl is emptied into or onto potential fuel. When fanned by a breeze, the coals may heat up the same way they do when someone draws on the pipe. The air passing the coals can increase the burning rate, release more heat, and potentially ignite the combustible fuel. Again, this is not a common occurrence.

> **TIP**
>
> Pipes can be involved in the ignition sequence when the bowl of the pipe is emptied into or onto potential fuel.

By comparison, cigarettes are the main ignition culprit as far as smoking materials are concerned, and have been blamed for many fires across the United States. If no other heat source can be found, it can be easy to blame smoking materials. Statistics compiled between 2012 and 2016 estimated that, on average, 5 percent of all structure fires were started from smoking materials (1 in 20 homes). These fires resulted in 590 fatalities and 1130 injuries annually. Thus, 1 in 4 home fire deaths over this period was related to smoking material fires.[8]

> **TIP**
>
> The common cigarette has been blamed for many fires across the United States.

A cigarette will continue to burn when left unattended or when carelessly discarded. A cigarette by itself may not always have sufficient heat to cause ignition. Studies have shown that a cigarette placed on top of a sheet will char the sheet but will not ignite it. However, when a cigarette is placed between the folds of a sheet or between the cushions of a couch, it is insulated. The heat can then build up because it is not dissipating. This may lead to enough heat to ignite combustibles.

As the number of smokers in the United States declines, the number of fires associated with cigarettes is decreasing as well. In the future, this downward trend and the possibility of new legislation requiring self-extinguishing cigarettes may decrease the number of smoking-related fires and associated deaths and injuries even further.

In a fire scene investigation, cigarette butts or cigar stubs may be found on the scene. If the investigator suspects that they may have come from the perpetrator of a crime, they can be collected and submitted to the laboratory for forensic testing. Even if the sample is insufficient to show DNA, the evidence may reveal the user's blood type. Science is improving every day. For example, new research indicates that a cigarette butt tossed into a hydrocarbon fuel for ignition is capable of providing DNA data that could lead to the identification of the smoker.[8]

Matches

Two basic types of commercial matches are used today. The safety match is the most prevalent and, as its name implies, is the safest type to store, carry, and use because it relies on the match striking the chemical surface on the matchbook or box to ignite. The strike-anywhere match, which may be found in camping or survival stores, will ignite by striking on almost any surface. Both matches share a similar construction: a head containing a chemical to initiate the fire; tinder to keep the fire burning, which is the wood or cardboard that the chemical is attached to at one end; and the handle, which in this case is the same as the tinder.

The safety match became a necessity as more and more fires were started accidentally by the strike-anywhere match and its predecessors. Earlier matches contained chemicals such as white phosphorus and yellow phosphorus. One type of match could ignite when the user bit into a small glass ball of sulfuric acid attached to the head of the match, so that the acid mixed with chemicals coated on the head of the match; the ensuing chemical reaction resulted in ignition. Perhaps not surprisingly, these devices were just as hazardous to the user as they were for starting unintentional fires. A third type of match on the market today is the wind and waterproof match. These matches have a higher chemical content and are treated to resist moisture (**FIGURE 10-4**).

Matches as Evidence

Because of its size, the residue of a match is not always easy to locate at a fire scene. The investigator should certainly look for it, however, and must always consider a match as well as any other heat source once the fire's area of origin has been located. When residue of a match is found on the scene, the investigator must consider whether suppression activities placed it there from water flow, or whether the match was there before the fire actually started.

FIGURE 10-4 The wind and waterproof match has a higher chemical content, as can be seen in comparison with the safety match in the bottom of the photo. The strike-anywhere match may have two colors, indicating the two chemical components necessary for the reaction.
Courtesy of Russ Chandler.

The design of the match allows it to continue to burn for an extended time. Some cardboard matches are impregnated with paraffin to allow burning and to provide moisture proofing. However, as a safety feature, cardboard and wood matches are also chemically treated to retard the afterglow when the match is fanned or blown out. When the match is blown out and dropped, it should not have a glowing ember that can reignite and ignite other combustibles.

> **TIP**
>
> Cardboard and wood matches are chemically treated to retard the afterglow when the match is fanned or blown out.

A cardboard match is torn from the base of the matchbook. This action might sometimes create a characteristic pattern, allowing the laboratory to prove that a specific match was torn from a specific matchbook. Today, most matchbooks have manufactured perforations—small cuts at the base of the match where it attaches to the matchbook. This limits the amount of area that must be torn so that the match can be more easily removed from the matchbook. It also reduces the surface area that actually tears, in turn limiting the potential for—but not preventing—a comparison between match and matchbook.

The laboratory can also compare the cardboard match against the matchbook, looking at width, color, thickness, and composition of the cardboard. Such a limited comparison by itself may not sound like much, but as with any case, it is the accumulation of all the facts that leads to knowing the truth. It is just as important to eliminate suspects and innocent parties as it is to find the culprit.

Lighters

Two types of liquid fuel lighters are on the market today. The older style is usually metal and has a fuel cavity that can continually be refilled with liquid lighter fluid. Lighter fluid is usually naphtha and sometimes benzene. The fuel cavity is filled with a fiber material that absorbs the liquid, which limits leakage. The lighter uses a wicking process to bring the liquid to the top of the lighter, where it can vaporize near the ignition device. The cap of the lighter keeps the vapors from totally escaping, allowing enough vapor to be present when the lighter is opened. Should the lighter be unused for any appreciable time, the lighter fluid may vaporize and escape, requiring it to be refilled on

a regular basis. The leakage dissipates in small quantities over time, so that it rarely constitutes a hazard.

> **TIP**
>
> Lighter fluid is usually naphtha and sometimes benzene.

The igniter is a sparking device that consists of a piece of flint and a rotating steel (a wheel with a rough surface). The flint is held against the wheel with a spring. The friction of the steel across the flint creates a series of sparks that fly into the area where the vapors are present and ignition occurs. To extinguish the fire, the user simply closes the lid, smothering the fire.

> **TIP**
>
> The igniter is a sparking device that consists of a piece of flint and a rotating steel (a wheel with a rough surface).

The predominant lighter in use today is the **disposable butane lighter**. Its plastic body serves as a pressure vessel for the fuel (butane). Under pressure, butane becomes a liquid; when a valve is opened on the top of the lighter, the vapors escape and become the fuel for the lighter's operation. Because the gas continues to escape when the lighter is activated, it is essential that immediate ignition take place; otherwise, the collection of the vapors would grow to an unsafe volume. Although some devices on the market still use the flint and steel igniters, science has provided a better and more reliable igniter in the form of **piezoelectric ignition**.

The pressure on the activator that opens the fuel valve also strikes a quartz crystal. This crystal releases an electrical discharge when it receives an impact or pressure. The electrical discharge takes the form of an arc, which provides sufficient energy to ignite the escaping gas. Many of these lighters also enable the operator to adjust the amount of fuel escaping, thereby increasing or decreasing the flame height.

Although most of today's butane lighters are disposable, some refillable models exist. These units can be refilled from a larger cylinder of butane with a simple pressure-filling device. Because these units are not disposable, they tend to be sturdier and are usually constructed of metal instead of plastic.

Because most, but not all, butane lighters are disposable, they may be left at the scene of a fire for a

Matching the Match

The investigator was at the scene of a vehicle fire. The late-model car had been parked beside the owner's house. According to the owner's statement to the fire officer in charge, the owner had gone out first thing in the morning and tried to start the vehicle. He stated that he could smell gasoline, so he assumed that he had flooded the engine. Knowing that a flooded engine would not start, he went back inside to get a cup of coffee, thinking he would try to start it later. He looked outside about 10 minutes later and found the vehicle burning. Fortunately, the fire department's engine company was nearby and was able to extinguish the fire in short order.

Working from the least damaged area to the most damaged area, the fire investigator looked into the trunk and found no problems. There was little fire damage and no indication that the fire started in the trunk area. Next, he examined the engine compartment. The exterior of both the trunk and the engine compartment showed little damage; the same was true for the engine. The only real damage was to the insulation under the hood, which was damaged from the outside where the flames had escaped from the windshield, reflecting heat downward on the top surface of the hood.

In the engine compartment, the investigator noted an obvious odor of gasoline. This was an older vehicle with a V-8 engine and a carburetor. This configuration created a small-depressed area under the carburetor that was capable of holding a small quantity of liquid. Upon close examination, the investigator found approximately ¼ in. of a liquid, later identified as gasoline. Within this liquid were a dozen cardboard matches, with the heads of the matches showing signs of carbon as if they had already been burned. The liquid and the matches were recovered, placed in appropriate containers, and labeled for identification and tracking of the chain of custody.

Looking in the passenger compartment, which was severely damaged, the investigator found all of the front seat's combustible materials were consumed, leaving only the metal frame and wire springs. In the center of the seat on the driver's side was a white residue. It was about ½ in. in diameter and about 1 ft long, somewhat intact. This was collected and properly sealed and tagged.

The evidence clearly suggested that the fire started in the passenger compartment, most likely by a road flare that was dropped onto the seat, as indicated by the white residue found on top of the seat springs. The owner, when questioned, advised that he was not smoking in or around the vehicle, that the vehicle was locked when he went out to start it first thing that morning, and that he had no road flares in the passenger compartment.

The last question the investigator asked was whether the owner had any matches in his possession. The owner reached into his pocket and came out with a matchbook. Thinking that the investigator wanted to light a cigarette, he apologized that there were no matches left. The investigator then asked if he could have the matchbook; producing a small evidence bag, he asked the homeowner to drop the matchbook into the bag, which he did. The investigator then asked the homeowner to sign a release for the matchbook, which he did.

The laboratory was able to prove that the matches in the engine compartment came from the matchbook surrendered by the vehicle's owner. The white material in the passenger compartment was consistent with the residue from a road flare. Faced with these facts in an interview, the homeowner confessed to setting the fire. He further stated that the vehicle had just came back from the shop with an estimate that it would take more than $1000 to repair. The insurance company would not pay for the repairs but would pay if the vehicle was burned. The owner stated that he poured gasoline onto the engine, stepped back, lit each match from his matchbook, and tossed the matches toward the engine compartment. Each match went out and he ran out of matches. He then confessed that he went to the trunk, took out a road flare, ignited it, and tossed it onto the driver's seat, and then closed the door.

In addition to demonstrating a good scene examination, proper collection of evidence, and good interview skills, this fire investigation resulted in an arrest and successful conviction. A key lesson from this case is that any question that goes unasked is a missed opportunity. When the investigator asked for the matchbook, it was given freely. It is surprising how often a perpetrator will give up evidence willingly when asked so as not to appear guilty.

variety of reasons. If not consumed by the fire, the surface of the lighter may contain latent fingerprints. If the fire consumes the lighter, the plastic pieces will be destroyed, but the metal parts may survive. Searching for these small items may require a more detailed search. If found, they may identify a particular brand that can be traced to locations where they are sold in the area.

Hobby Micro Torch

A device similar in operation to a lighter is a hobby micro torch (**FIGURE 10-5**). This device is usually sold to hobbyists to braze and solder small particles in hobby-type projects. These torches are capable of reaching fairly high temperatures, in the range of 2400°F, which is much higher than the average lighter. Both lighters and micro torches use the same fuel and sometimes have the same type of igniter. The difference that enables the higher temperature in the micro torch is the head assembly, which allows the introduction of air into the burn chamber to achieve a more efficient flame. Larger models may be as much as 6 in. long, whereas the smaller units may be only 3 in. tall. A device of this size can easily be concealed.

TIP

Micro torches are capable of reaching temperatures as high as 2400°F.

Micro torches are not very expensive, but it is unlikely they will be left at the scene. Of interest to the investigator would be finding a suspect at the scene and finding such a unit on that person or in his or her

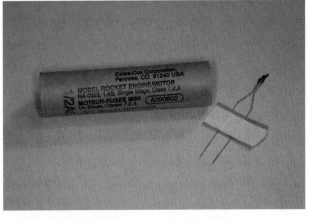

FIGURE 10-6 Model rocket motors and their igniters can be found in hobby stores.
Courtesy of Russ Chandler.

vehicle. By itself a micro torch may mean little, but it can be just one more piece of the puzzle that requires further investigation.

Other Hobby Ignition Sources

Every investigator should take a trip to the local hobby store just to see the type of materials available to the curious juvenile, careless hobbyist, or clever arsonist. In addition to micro torches, rocket motors (**FIGURE 10-6**) are available for small model rockets. They come in a variety of styles for different models, with burning durations depending on the rocket's load or anticipated flying height.

Hobby shops may even carry electric matches. These remote igniters use low voltage, but their energy is sufficient to ignite common materials. Some even look like a matchbook with wires attached. Other electric matches are used for fireworks displays and explosives; they all work using the same principle of delivering an electric charge with sufficient heat to ignite an explosive charge.

Friction Heating

Friction heat is created when two items rub together to produce heat. Not all friction results in fire or in measurable heat. For example, the slow rubbing of tree branches may eventually rub off bark, but will not lead to combustion. The movement (kinetic energy) of an object with sufficient friction (resistance from that movement) can translate kinetic energy into thermal energy; the key variable is the roughness of the surface. Items forced together with pressure create more friction and, in turn, more heat. The opposite is true when liquid is involved: The friction of objects wet from water or oil is decreased and less heat is produced.

FIGURE 10-5 This micro torch, which uses butane, is only 4 in. long and can easily be carried in a pocket.
Courtesy of Russ Chandler.

TIP

Friction heat is created when two items rub together sufficiently to produce heat.

Rotors

A **rotor** can take many forms—from the mathematical rotor, a convex figure that rotated inside a polygon and that touches every side or inside face of the polygon, to the blades of a helicopter. In the fire investigation context, a common heat source is the rotor of an electric motor or alternator. The rotor is the rotating portion of the electric motor; the stationary portion is called the **stator**.

By their very nature, these electric motors and alternators get warm when working properly. However, when the bearings that hold the rotor in place fail, a massive amount of friction occurs that can heat up to the point of becoming a competent heat source. The availability of a nearby fuel source dictates whether a fire will occur. With the advent of plastic parts in and around the motor, such as fan blades, the fuel could be part of the assembly. A motor may also produce sufficient heat to become a heat source because of oxidation as the result of moisture. Rust on the surface can create more friction, which in turn creates more heat.

TIP

When the bearings that hold the rotor in place fail, a massive amount of friction occurs that can heat up to the point of becoming a competent heat source.

When a motor is found in or near the area of origin and the investigator suspects that it may have been involved in the ignition sequence, every effort must be made to ensure that there is no spoliation of the evidence. Take several photographs to document the condition of the motor, and then collect it as evidence. An engineering laboratory may be able to examine the unit to tell whether the heat was internal or external, adding to the data being collected to make a final hypothesis.

Brakes

Motor vehicles have rotors associated with the braking system. However, the source of heat in most braking system failures usually involves the brakes themselves, rather than the rotor. When brakes fail, the pads sometimes lock in place. As the vehicle continues to roll, it creates a massive amount of friction, which in turn creates massive amounts of heat. This energy can then ignite grease, plastic, rubber, and other materials. The brakes themselves can come apart, sending hot fragments flying and igniting nearby combustibles such as grass and brush. This can go on for miles before the driver realizes what has happened. In some incidents, the brakes on tractor trailers have overheated so much that the heat was transferred upward into the trailer, igniting the contents of the trailer.

TIP

The brakes themselves can come apart, sending hot fragments flying and igniting nearby combustibles such as grass and brush.

Muffler Systems

Mufflers of motor vehicles or other metal parts hanging down sufficiently low to touch the ground can become a source of sparks. This can include hanging metal bumpers or safety chains used when pulling trailers. As the vehicle increases in speed, the volume of sparks increases as well. Each of these sparks is a glowing metal ember that is fully capable of igniting small pieces of fuel found along the road, such as grass, leaves, or paper and plastic litter. The road surface need not be asphalt for friction to create sparks. Indeed, rocks on dirt roads can create sparks—a source that has been the cause of many wildland fires.

Muffler problems are not limited to passenger vehicles, recreational vehicles, or freight trucks. Multitudes of vehicles that are used for off-road purposes, including bulldozers, logging vehicles, and farm machinery, have their own characteristics that could provide a heat source.

Logging vehicles and other vehicles designed to be used in wildland terrain have screens around the engine compartment to keep out leaves, pine needles, and other combustibles. All too often these screens get clogged up; then, instead of crews cleaning them on a regular basis, they are removed. When this happens, leaves and other items can collect in the engine compartment, resting up against the exhaust manifold. The heat from the engine is sufficient to eventually ignite these materials. The probability of a fire is increased by the air flow created by the vehicle's fan, which helps any smoldering items break into open flame.

Often, when dealing with a fire that may have been caused by sparks from a motor vehicle, the investigator will have to depend on eyewitness accounts. If sparks were seen coming from a vehicle but the witness did not see the fire actually start, it may be hard to propose this as a hypothesis; it will be conjecture, not fact.

Sparks from Catalytic Converters

A catalytic converter is a device designed to reduce the toxicity of emissions from the exhaust of an internal combustion engine. Catalytic converters can be found on cars, buses, and trucks. Some construction equipment and forklifts also use catalytic converters. The basic principle is to allow exhaust gases to flow through a ceramic honeycomb (some models use ceramic beads) that is coated with metals that first remove the nitrogen oxides and then separate out the oxygen molecules, thereby creating oxygen and nitrogen. The second stage then burns off any residue hydrocarbons or carbon monoxide, combining these molecules with the oxygen from the first stage. With an internal operating temperature of 1300°F, catalytic convertors provide for perpetual ignition.

Most manufacturers provide adequate shielding around the catalytic converter to dissipate as much heat as possible. The average exterior temperature of 600°F is not a problem under most circumstances.

The average road provides sufficient clearance of combustibles. However, parking a vehicle on the side of the road, where vegetation, including fallen leaves, can come in close proximity or contact with the catalytic converter, creates a fire hazard.

> **TIP**
>
> The average exterior temperature of 600°F for catalytic convertors is not a problem under most circumstances.

Certain conditions can cause a catalytic converter to be hazardous. For example, impact damage to the converter could break up the ceramic interior. As the converter then heats up and exhaust passes through, small pieces of the ceramic, heated to incandescence, could be blown through the exhaust system. These ceramic pieces, when coming out of the exhaust pipe, would be hot enough to serve as a heat source for leaves, grass, and other vegetation.

An improperly running engine can sometimes create problems with the operation of a catalytic converter. Anytime that the vaporized fuel injected into the cylinders does not ignite, it is forced into the

New Insurance Policy

The local farmer had asked his insurance agent to reevaluate his farm property to make sure he had enough insurance coverage should something go wrong. The agent checked all the structures with the farmer. They needed to check one more remote barn on the property to get the square footage of that building. The farmer rode out to the building with the insurance agent because the agent had air conditioning in his vehicle.

The barn was rarely used because of its condition and remote location. As such, the farmer had plowed all the land around the barn, planting hay for his livestock. He left no path to the barn. It had been exceptionally dry, and the weather was hot. Rather than walk the distance, because the hay was only about 18 in. tall, the agent and the farmer drove the car to the barn. Parking just outside the door to the barn, the insurance agent left the vehicle running to keep the car cool for their return trip to the farmer's house.

By the time the rural fire department arrived, the vehicle was fully involved. About half of the barn was

engulfed in flames and at least four acres of young hay was burning. The fire officer (first investigator) interviewed both the insurance agent and the farmer separately. Both explained that while taking measurements inside the barn, they had smelled smoke. Outside smoke was coming from under the vehicle, and through the smoke they could see the glow of flames. The glow was exactly in the area of the catalytic converter.

With no cell towers in the area, the pair started to run for help. Because of the long distance back to the farmhouse and the hot weather, that trek turned into a brisk walk. Before they could get to a phone to call for help, a neighbor saw smoke in the distance and called the fire department.

Months later, the new barn was almost complete. The fire officer had seen the farmer at lunch in town and asked about his insurance claim, curious as to the final adjustment on the property. The farmer just smiled and said all was taken care of by the insurance company.

exhaust system. When this occurs, that fuel enters the catalytic converter where, at 1300°F, it is ignited. Evidence of this action is a rough-running engine and a glowing red (orange) catalytic converter. Eventually, the operator may notice a burning smell within the interior of the vehicle, caused by the transfer of the heat from the converter to the carpet padding just above the converter. This can eventually lead to the ignition of the carpet material, along with any combustibles sitting on the carpet.

Signs of this problem are usually first reported by witnesses such as the vehicle's driver and occupants, who can verify after the fact that the motor was running rough and that they smelled smoke. The driver may even indicate that the floor area was warm or hot. A witness exterior to the vehicle may have noticed the glow from under the vehicle in the area of the catalytic converter. Burn patterns will show low burning above and adjacent to the converter. The converter itself may show degradation and damage from the presence of the vehicle's fuel.

Chimneys and Fireplaces

The chimney has been a staple of the American home since the time of the early settlers. The common chimney today is constructed of brick with a liner or steel, and ideally it is at least double-walled steel. A steel chimney may be placed within a chimney chase that is usually wood framed, with siding that matches the structure.

Properly used and maintained, today's chimneys provide few problems and can be an asset to any home or business. It is only when construction is improper or chimneys are not used properly that they can be involved in an unwanted fire.

Metal Chimneys for Fireplaces

Metal chimneys are designed for specific applications, ranging from hot water heaters to crematories. Historically, those dedicated to appliances have provided few problems. Metal chimneys for fireplaces and wood-stoves may be involved in residential fires, but usually because the user failed to follow proper precautions and burned the wrong fuel or installed the metal pipes improperly.

When the examination of a fire scene points back to the metal chimney as the area of origin, it is beneficial to obtain the instructions for assembly from the manufacturer. Under most circumstances, the appliance will come with the proper chimney. The instructions will indicate the amount of heat that the device and the chimney can handle. Many of the current metal

fireboxes (fireplaces) with metal chimneys can easily handle a wood fire from logs and kindling. In contrast, they may not be able to handle the heat from a pressed log consisting of sawdust and paraffin.

> **TIP**
>
> When the examination of the fire scene points back to the metal chimney in the area of origin, it is beneficial to obtain the instructions for assembly from the manufacturer.

Some firebox assemblies are advertised and installed as zero-clearance units. This descriptor implies that wood assemblies can be installed in close proximity to or against the exterior of identified metal portions of the firebox. Under most circumstances, this design feature holds true as long as the unit is used according to the instructions provided.

On rare occasions, investigators might pull onto the scene and be in awe of the pattern greeting them as they step from their vehicle. The burn pattern shown in **FIGURE 10-7** seems too obvious to be real—but it was real. Chapter 11, *Patterns: Burn and Smoke*, discusses this particular pattern. Following the sides of the V downward, the investigator will find that it points to the area of origin in many circumstances. In this case, the area of origin was the firebox of the metal fireplace (**FIGURE 10-8**).

In this fire, the metal fireplace, as seen from the interior (**FIGURE 10-9**), was heavily damaged as the result of the first fire in the unit. What is difficult to see is that the metal was heavily warped and pitted—not what would be expected from usual use of the unit.

FIGURE 10-7 Burn pattern of a fire that started at the firebox of a zero-clearance stove with a metal chimney. Notice that the wooden chimney chase, with the exception of the base, has been consumed.

Courtesy of Russ Chandler.

FIGURE 10-8 Close-up of the rear outside surface of the firebox. This is the area of origin.
Courtesy of Russ Chandler.

FIGURE 10-9 Interior of a new firebox after being used just once to burn paraffin logs, contrary to the recommendations of the manufacturer. Instructions against using pressed logs were found on the coffee table in the same room.
Courtesy of Russ Chandler.

FIGURE 10-10 Corner of the fireplace insert showing proper zero-clearance installation. Fire extended beyond the fireplace because the owner used paraffin logs against recommendations of the manufacturer and installer.
Courtesy of Russ Chandler.

FIGURE 10-10 shows the base of the fireplace unit. The clearances and installation all met manufacturer requirements, but the owner used a paraffin log, which was not recommended.

Brick Chimneys for Fireplaces or Woodstoves

A brick chimney appears to be mortar and brick from the outside. When looking at the top of the chimney, under most circumstances, one will see a portion of the chimney liner protrude. Codes require this liner to be installed through the entire length of the chimney, from the firebox to the top of the chimney. The liner provides a conduit for the hot fire gases and smoke to continue their upward rise to the top of the chimney to be released into the atmosphere. The liner is essential to the proper operation of a brick chimney.

> **TIP**
>
> The liner provides a conduit for the hot fire gases and smoke to continue their upward rise to the top of the chimney to be released into the atmosphere.

A brick chimney without a liner is a hazard. The mortar that holds the bricks in place is not designed for or capable of containing the hot gases or smoke from the fire. In time, gases and smoke can permeate the mortar, leading to a potentially uncontrolled fire outside of the chimney.

> **TIP**
>
> A brick chimney without a liner is a hazard.

A common residential brick fireplace chimney liner is clay or terra-cotta. However, many other materials are used, such as steel, ceramic, concrete, and other combinations of specialized materials. All of these liners, when properly installed, do what they are designed to do: withstand the heat from the firebox and safely remove hot gases and smoke.

When the examination of the fire scene indicates that the fire started in the close vicinity of the chimney, examine the chimney closely. The first question to ask is whether there was a fire in the fireplace or woodstove. If so, the occupants may be able to indicate the maintenance history of the fireplace, such as when the chimney was last cleaned.

Conduct an exterior detailed examination to look for cracks and missing mortar. When doing so, take care not to damage the chimney in this examination—do not pry or prod. **FIGURE 10-11** shows a chimney in relation to the adjacent burn patterns. The presence of a flue liner coming out of the top of the chimney usually indicates that a liner was in place, implying the chimney would be safer than one without a liner. The use of a high-resolution digital camera can aid in the investigation. From the fireplace damper or through a thimble, insert the camera and use auto focus to take a series of photos, changing the angle slightly for each photo. **FIGURE 10-12**

FIGURE 10-11 The liner extending above the brick usually indicates the chimney is lined.

Courtesy of Russ Chandler.

FIGURE 10-12 Photo of the interior showing that the liner was just one piece of tile held in place with metal hangers. No other liner pieces were installed in the chimney. The metal strap was clearly installed by the brick mason at the time of construction. This cost-saving measure resulted in an extension of the fire into the home.

Courtesy of Russ Chandler.

shows the findings from such an examination: No flue liner was installed below the top portion of the chimney. In this case, the metal strap was clearly installed by the brick mason at the time of construction—a cost-saving measure that resulted in an extension of the fire into the home.

If it is safe, also take photographs from the roof, looking down through the chimney. See whether a truck company can help you take these photos safely from the platform of an aerial apparatus.

It is critical to use extreme caution when approaching any chimney. If the structure surrounding the chimney has sustained any damage, assume that the chimney is damaged as well. Chimneys that experience degradation from a hostile fire can lose their structural integrity. Even just leaning a ladder against the chimney may cause it to collapse, risking the investigator's life and the lives of others. If there is any question, the use of an aerial ladder may be the only platform safe enough from which to continue the investigation.

The full-time investigator will know some local resources that may be available to assist in the examination of the chimney. Many chimney sweeps have camera systems that they use to inspect chimney conditions. Some of these systems even have recorders for future reference, which could prove to be a beneficial tool in the examination.

The first thing to look for is proper construction and the presence of a flue liner. Check the condition of the liner; see whether it is broken up or pieces are missing. Is there creosote in the chimney? How much creosote has adhered to the chimney? How much has it restricted the size of the chimney opening?

When conducting interviews, find out whether anyone saw fire coming from the chimney top. This may indicate a chimney fire, which could have been the cause of the fire in the structure. Sometimes

FIGURE 10-13 A fire starting in the chimney chase in a subdivision consumed all material in that area. The examination of a neighboring house built by the same contractor and crews revealed that they failed to remove the boards for a temporary casting to create the concrete top.
Courtesy of Russ Chandler.

occupants might have heard a roaring sound, which could indicate a chimney fire. Such events may be caused by the ignition of a buildup of creosote in the chimney but can also result from other items in the chimney burning, such as birds' nests or large quantities of cobwebs, sticks, and leaves.

Also, check the construction of the chimney in both the living spaces and concealed spaces such as attics. On rare occasions, the brick mason may have chosen to brick around wood beams, leaving them inside the brick assembly, exposing them to heat, which breaks down the wood and eventually leads to ignition of the beam. A proper installation has the beams headed off with a wood faceplate; essentially, a box is made around the opening for the fireplace to go through the roof. To state the obvious, if the area has been consumed and the chimney is still intact, a telltale opening in the chimney where it was built around the wood beam will exist.

Another type of masonry chimney is made from concrete blocks that are then boxed in and covered with a wood-frame chimney chase. The chase is covered with the same siding material as is used for the rest of the house. The top cap usually consists of metal. Sometimes the top surface of the chimney surrounding the flu, the crown, is concrete; if so, it requires a wood temporary framing. Fires have resulted from the construction crew not going back in and removing this framing (**FIGURE 10-13**), which rests against the flue liner. As much as a year or two later, depending on the frequency of use, the wood will eventually break down and possibly ignite.

Wind-Blown Sparks

Wind can cause a burning object to become more incandescent, raising its temperature. In addition, if the

burning particles are small enough, the wind can pick them up and deposit them elsewhere. This movement may place a burning ember (spark) in contact with combustible material. Under the right conditions, the mass of the fuel and its ignition temperature might start an unwanted fire.

Chimneys and Wind

The wind can have an amazing effect on both chimneys and sparks. Occasionally, downward-blowing winds can create a downdraft in a chimney, possibly stirring the sparks in the fireplace and even blowing them into the room. Sparks landing on a hardwood floor most likely will not have the energy to ignite the wood into flaming combustion, but can char it. Most modern carpets char as well, but with enough energy they may eventually ignite. Combustible items such as newspapers, dried or fake flower arrangements, and kindling can be ignited under the right circumstances by a spark from a fireplace.

Another phenomenon is the **Bernoulli effect**, in which wind blowing across the top of the opening of the chimney creates a strong updraft. The negative pressure created by this effect pulls the air out of the chimney faster than the usual upward draft. The void of the air leaving the chimney is replaced by air from the room. As this air passes over the burning fire and ashes at a faster-than-normal rate, it fans the fire, increasing the burning rate and possibly picking up sparks and embers that usually would not rise up the chimney. Because the sparks are larger, they have more mass and burn longer. In fact, sometimes they may burn long enough to travel up through the chimney and out into the atmosphere. Sparks, if still incandescent, can land on the roof and could have sufficient energy to ignite leaves or vegetation debris on the roof. If the roof is constructed of combustible materials, such as cedar shingles, sparks could even ignite them over time. The fact that the fireplace was in use, the location of the area of origin, and documented high winds may support this theory of the heat source.

> **TIP**
>
> In the Bernoulli effect, wind blowing across the top of the opening of the chimney creates a strong updraft.

Placing Christmas or birthday wrapping paper in the fireplace may seem like a good way to get additional heat and dispose of the paper at the same time. The balled-up paper is easily ignited. It already has little weight and as it is consumed, becomes even lighter. However, the draft created by the fire, along with the additional draft created when the paper ignites and rapidly burns, may potentially be sufficient to pull some of the charred paper into the chimney with the draft. It then might be capable of igniting the creosote on the chimney walls; it might also leave the chimney and land on the roof or in the yard, where it can ignite dead vegetation, creating an unwanted and hostile fire. Evidence for this type of ignition includes the residue or wrapping paper at or near the area of origin. The presence of other pieces of carbonized wrapping paper at the scene may support the theory. As with all situations, a thorough interview and witness statements may help make the determination as well.

Burning Leaves

When a leaf-burning event gets out of control, human error is usually involved. Mistakes such as starting a fire with gasoline can result in a fire instantly being too large to handle. Pouring the gasoline and delaying the ignition can result in the vapors migrating away from the leaf pile, possibly igniting other combustibles as well as injuring the person igniting the vapors.

Not staying with a pile of burning leaves can allow the fire to get out of control, endangering structures and wildlands. It takes only a small breeze to stir embers, pick up burning particles, and transport them to other vegetation. The burning leaves themselves can blow away in a heavier wind. When the burning pile has been reduced to ash, it is still burning inside of the pile. What may look like an extinguished fire can flare up with a fresh breeze, spreading embers into unburned areas.

> **TIP**
>
> What may look like an extinguished fire can flare up with a fresh breeze, spreading embers into unburned areas.

When conducting an investigation, any pile of debris and ash may serve as an indication of a potential heat source. The wind at the time of the fire may also indicate a direction. However, the wind direction at the time of the investigation may be different from at the time of the fire. An eyewitness account of the smoke travel direction may be a better indicator.

Trash Burning

The burning of household, office, and some commercial trash is hazardous because of the release of toxic gases from burning plastics and other chemically impregnated items. As such, trash burning is usually outlawed, but it still takes place in some rural communities. The different materials burned can lead to the fire flaring up from time to time as well as the fire releasing sparks that can fall in unburned vegetation or other debris, resulting in an extension of the fire.

When trash is placed in an open burn barrel, the sparks can rise in the hot air currents. They travel upward until the hot air currents stop providing enough updraft; then, the spark falls back down. Should sparks have sufficient energy to remain burning when they fall, they could ignite nearby vegetation, spreading the fire away from the barrel. The placement of a small-weave screen over the barrel usually prevents the sparks from leaving. Regretfully, not everyone takes such a precaution.

In fire investigations, both burn piles and burn barrels, as well as leaf piles, may show a path of char leading right back to the source. Again, it is good to document whether there was a fire in the barrel in the first place.

> **TIP**
>
> Burn piles and burn barrels, as well as leaf piles, may show a path of char leading right back to the source.

Credible Proof of Wind Direction

Two fire investigators and an engineer, all hired by different insurance companies, were debating the burn patterns of an accidental fire. The conflicts all came down to the wind direction at the time of the event. One of the investigators had interviewed the night guard at a neighboring car dealership, and the watchman said the wind was to the northwest. Yet the engineer had called the local television station whose weather department said the wind was consistently to the southeast. The debate would have continued except that the investigator revealed that the guard was an off-duty fire investigator who discovered the fire because the smoke filled the car lot, which was situated just northwest of the fire location. Based on a reliable first-hand account, it was agreed that the wind was from the southeast blowing to the northwest.

Chemical Reactions

Some chemicals react when combined with others, releasing heat at sufficient levels to ignite nearby combustibles. Chemicals such as organic peroxides and metals may ignite when in contact with moisture or air—some with explosive force. Although the average investigator may not run into too many igniting metals, he or she may encounter other chemical processes that produce heat. Something as simple as fiberglass repair operations require a catalyst to be mixed with the resins. This chemical reaction, under the right conditions, can give rise to surprising temperatures.

Essentially, many chemicals can react and produce heat that can ignite nearby combustibles. These chemicals are in use in almost every community. It is up to the investigator to be diligent and resourceful in identifying these hazards as heat sources that could have started a fire.

Spontaneous Combustion

Some materials are susceptible to spontaneous heating, which in turn may lead to spontaneous combustion. These materials produce heat under various circumstances. If the heat is not allowed to dissipate, it can continue to increase in temperature until the heat is sufficient to spontaneously ignite or ignite nearby combustibles. The NFPA's *Fire Protection Handbook* provides examples of different materials and their capability to spontaneously combust.

For a spontaneous reaction to occur, there must be an exothermic reaction to take place at a normal temperature and in the presence of oxygen. This reaction must accelerate rapidly as the temperature increases. The arrangement of the material must be such that it does not allow, or drastically limits, the dissipation of the building heat. Lastly, the material undergoing this process must be capable of smoldering.

> **TIP**
>
> For spontaneous reaction to occur, an exothermic reaction must take place at normal temperature and in the presence of oxygen.

Drying Oils

Drying oils such as linseed oil, tung nut, or fish oil are all susceptible to spontaneous heating. The reaction depends on the percentage of the oil in contact with oxygen, along with its ability to dissipate heat. A crumpled rag soaked with linseed oil left on a work bench will build up heat within the rag because the folds do not allow the heat to dissipate, possibly heating to ignition. That same rag hung on a clothesline can dissipate the heat as the reaction occurs, not allowing heat to build up, until it has safely dried.

Upon locating and examining a fire's area of origin, the fire investigator may find traces of such a rag possibly contaminated with a drying oil. Even after being fully involved in fire, the rag residue may still hold its shape enough to be recognized. The lack of any other potential heat source may also be an indicator—but not proof—that the rag was involved in the fire ignition. Interviews can be conducted to identify whether and when the occupants were using the rag and what substance was on the rag. The laboratory analysis may be able to confirm the presence of an oil. Finding the container and the project that the oil was being used for can help to support or refute a hypothesis.

Hay and Straw

As discussed in Chapter 4, *Science, Methodology, and Fire Behavior*, the self-heating of hay or straw is one of the more well-known spontaneous ignition events. Although this type of event does not happen frequently, it is important for the investigator to understand how it occurs.

The conditions must be just right for self-heating to begin. First, moisture must be present. Different studies have found varying moisture percentages in different types of hay and straw. The process of condensation creates heat, but this by itself usually is not enough to cause ignition. However, the presence of microorganisms in the moist bales can start the fermentation process, which creates even more heat release. If that heat has no means of dissipating, it continues to build and eventually reaches the ignition temperature of the baled material. If the reaction within the mass slows down, the farmer may find charred hay in the pile. If the heating process reaches its peak and extends to the outer layer of the hay mass, enough oxygen is present to enable the process to proceed to flame production.

> **TIP**
>
> The presence of microorganisms in moist bales can start the fermentation process, which creates even more heat release.

Because of the nature of this event and the ready supply of fuel, all too often the fire results in major destruction, limiting the opportunity for the investigator to find any evidence. However, in a structure that was secure and in the absence of other heat sources, the investigation may turn up evidence to prove what happened. Should there be sufficient remaining debris, it may be obvious whether the fire started inside or outside the stack. Spontaneous ignition starts on the inside, moving upward. In contrast, fire that starts on the outside is the result of other heat sources.

One piece of evidence may help the investigator. Sometimes in the process of spontaneous combustion in hay, a **clinker** is created. This glassy, irregularly formed mass might be found in the center of the area of origin. It can be gray to green in color and range in size from small particles to chunks that are 2 and 3 in. long. The size is not an indicator of anything other than the basic components in the hay at the time of the fire, but the presence of a clinker does suggest that the first material ignited was hay and the heat source was spontaneous combustion.

> **TIP**
>
> Sometimes in the process of spontaneous combustion in hay, a clinker is created.

Plywood and Pressboard

Plywood and particleboard are made by mixing wood particles with a resin compound, placing the mixture in a form, and pressing it into shape using high heat and pressure. These products, if allowed sufficient ventilation and placed in smaller stacks with the ability to cross-ventilate, usually do not pose any problem. However, if plywood or particleboard sheets are stacked in such a way as to prevent the dissipation of heat, self-heating will take place that could result in spontaneous ignition. Like hay and straw, the location of the area of origin at the center of the plywood stack is the first indicator that the fire event involved spontaneous ignition.

> **TIP**
>
> Like hay and straw, the location of the area of origin at the center of the plywood stack is the first indicator that the fire event involved spontaneous ignition.

Appliances

Appliances come in various sizes, small and large. Larger appliances can be categorized as electric or gas, based on their power source. In most homes, small appliances are powered by electricity; some require simple motors or servos for movement. The fire service frequently encounters small appliances that use heat for its intended purpose, including toasters, toaster ovens, electric skillets, and coffee makers, to name a few.

When appliances are suspected of playing a role in a fire event, investigators must first determine whether the appliance was involved in the ignition sequence. If so, what was the reason? Did the unit itself malfunction, was there misuse or abuse by the user, was the appliance improperly installed, or was there a design or manufacturing defect? Next, investigators must determine whether the fire was set by an intentional act and intended to look accidental.

Small Appliances

Misuse and abuse of small appliances often cause failures that result in fire. Any bagel too thick for the toaster may jam the appliance in the cooking position, continuing to cook the bread and eventually igniting the bagel. If the toaster is under the cabinets, flames could impinge on the underside of the overhead wood, spreading the fire.

> **TIP**
>
> Misuse and abuse of small appliances often cause failures that result in fire.

Toaster ovens have similar problems when food is left in them too long or when paper and plastic materials are left on top of the units, which creates a potential fire hazard. Cooking grease-laden items also poses problems. Foods such as bacon that is cooked in a small, flat pan too close to the burner is a potential source of fire. As the grease heats up and spatters, it comes in contact with the heating elements, creating the potential for ignition.

Coffee makers had a rough start when first introduced, being plagued by failure after failure in both design and manufacturing. The thermostats and thermal fuses in these new devices often malfunctioned, which caused overheating and subsequent ignition of the plastic housing and nearby combustibles. Today, recalls might occur occasionally, but the safety problems with coffee makers are much less significant than those noted in the past. Many units have automatic shutoffs that limit the potential fire hazard.

Portable heaters are a serious problem, especially in the workplace. Trying to keep warm, workers often place the heater under the desk, next to a plastic trashcan or a cardboard box of papers. In a fire investigation, the burn pattern may help narrow the ignition source candidates down to portable heaters as a potential fire cause.

> **TIP**
>
> Portable heaters are a serious problem, especially in the workplace.

When investigators find a small appliance at the area of origin, it is imperative that they do not alter or experiment with the device at the scene. The movement of a knob could constitute spoliation, making it impossible to determine whether the device was on or off at the time of the fire. If any indicators suggest that the device was involved in an incendiary act, the device must be protected and saved for a full-time investigator to collect and send it to a laboratory for testing. If there are no indications of any incendiary nature, because the appliance actually still belongs to the homeowner, it should not be seized. The fire investigator might obtain the owner's permission to take the appliance. However, the insurance company may have a vested interest in examining the appliance, which outweighs the homeowner's interest (see Chapter 8, *Spoliation*). Most insurance companies will share their findings with the local investigator upon request, so this may be the more logical path to follow to examine the appliance.

> **TIP**
>
> If there are no indications of any incendiary nature, because the appliance actually still belongs to the homeowner, it should not be seized.

Small appliances can also be used to set fires intentionally. Timers can turn on appliances when the perpetrator is far away from the property and has a solid alibi. Heaters and combustibles can be left in close proximity. Was it a foolish action or an intentional act that created this situation? The answer can be uncovered only with a complete analytical investigation.

Larger Appliances

As a general rule, a properly operating refrigerator or washing machine is infrequently involved in the ignition sequence of a fire. Some gas-operated refrigerators are available, but there have been few problems with them. Appliances involved with heating do have a history of fire-related incidents. Like smaller appliances, the majority of fire-related problems with large appliances involve misuse or abuse.

The primary energy source coming into the appliance is the investigator's first consideration. Gas lines on stoves usually include a loop or enough line to allow the occasional movement of the unit, such as for pulling out the stove for cleaning. However, constant movement of the appliance could create a leak resulting from loosened fittings or eventually lead to a small stress hole in the pipe. Electric appliances are not immune to damage from this type of movement. Most larger electrical appliances have an electrical pig tail, which is a 2- to 4-ft electrical connection made with flexible (stranded) wire that can withstand movement. Constant movement can create problems, but there have also been instances in which the appliance has abraded or nicked the insulation on the wire, creating a hazard. In other cases, the appliance has ended up sitting on the cord, although this is usually obvious because the appliance no longer sits level.

When a larger appliance is present in the area of origin, all aspects of the appliance's use must be explored. A careful removal of debris may indicate what was happening at the time of the fire? Is there cloth debris from dish cloths or towels near or on the burners of the stove? Were there clothes in the dryer? Document in writing and in photographs the positions of all knobs on the appliances, along with proof that the appliance was or was not hooked up to power or to a fuel line.

> **TIP**
>
> When a larger appliance is present in the area of origin, all aspects of the appliance use must be explored.

A fire resulting from food left on the stove or in the oven too long is something almost every fire fighter has encountered. The electric burner element can leave an impression on a steel pan if the pan is left on the burner too long. Although investigators might tend to think that is the source of the fire (and it might be), they should also consider the possibility that the pan was damaged in a prior incident and had nothing to do with this fire.

Both the fittings for the gas appliance and the electrical cord insulation may be consumed in the fire. This may limit the investigation, but there are other things to examine. Signs of electrical shorting on the wire may be evident. However, investigators need to exercise caution in assuming which occurred first—that is, whether the short caused the fire or the fire caused the electrical short. A flattened-out area of the electrical feed or a kink or hole in the gas line may provide valuable information. During the interview phase, ask questions about appliance movement and whether any problems with the appliance were experienced prior to the fire.

Gas problems may exist in other areas in addition to fittings and piping. Gas coming into an appliance must be regulated down to a usable pressure. The gas control houses the regulator and sends the fuel when needed to heat the water or dry the clothes. Should the regulator fail, it could allow too much gas to flow, creating a larger-than-wanted flame and leading to a fire. However, most appliances have thermal-limiting switches that detect any high-heat situation and cut off the regulator, shutting off the gas; in electrical appliances, the limiting switch shuts off the electricity.

FIGURE 10-14 A small heater in the closet area became a hazard when the occupant stuffed dirty clothes in and around the water heater, forcing combustible materials up against the unit. The area of origin was the access area for the heating element where insulation had been removed; this allowed heat to be transferred to the clothing.
Courtesy of Russ Chandler.

> **TIP**
>
> Most appliances have thermal-limiting switches that detect any high-heat situation and cut off the regulator, shutting off the gas; in electrical appliances, the limiting switch shuts off the electricity.

Larger appliances are insulated to prevent the unwanted escape of heat, as both an energy-saving measure and a safety feature. In older appliances, the insulation may settle, exposing upper areas and allowing them to radiate more heat than desired. Sometimes the insulation is taken out, such as in the access areas for hot water heaters. The insulation behind the panel allowing access to the heating elements is intended to be removed when replacing the element, and the insulation should be replaced when the job is done. Not putting this insulation back in place generally will not cause a fire problem except if the water heater is improperly used and combustible materials are crowded against uninsulated areas to the point that the residual heat cannot dissipate. **FIGURE 10-14** shows just such a fire where the tenant used the space around the hot water heater for storing dirty clothing until laundry day.

Gas appliances also rely on proper ventilation, with air entering the unit to allow for proper combustion and for the removal of the gas residue following combustion. Blocking the air intake of a gas furnace prevents proper combustion and may cause flame to roll out or allow radiant heat to heat up nearby combustibles. Materials such as sticks and leaves that fall into a gas furnace or hot water heater flue can block the escape of gases, endangering the occupants and also forcing heat back into the appliance area, which creates a potential problem.

Lightbulbs

Typically, a proper-wattage incandescent lightbulb in a light fixture will not cause a problem. When bulbs of higher wattage than recommended are used, especially in enclosed fixtures, heat can build up to the point of potentially igniting nearby combustibles.

A 60-W lightbulb can produce temperatures as high as 255°F, and a 100-W bulb can produce temperatures of 300°F. If these bulbs are insulated and ventilation is restricted, they can reach higher temperatures.

> **TIP**
>
> A 60-W lightbulb can produce temperatures as high as 255°F, and a 100-W bulb can produce temperatures of 300°F.

Over the years, more types of bulbs have come on the market. Fluorescent bulbs were popular in the early 1920s; they had low heat output and were economical, but the electrical ballast needed to operate the light often overheated and started fires. Halogen and mercury vapor lights produced tremendous lumens but also produced sufficient heat to ignite combustibles coming in contact with the surface of the bulb.

Over the years, fluorescent lamps have been improved through the introduction of new technology. The new compact fluorescent lamps (CFL) are both safer and more economical. Also, standard fluorescent tube lamps no longer have a large bulky ballast, making them less likely to become a source of ignition. More recent technology has given us light-emitting diode (LED) bulbs, which have a low heat output and are not a common source of heat for a hostile fire.

Also available are high-intensity bulbs that are capable of reaching even higher temperatures. Some of the smaller units can easily reach temperatures of 500°F. Quartz bulbs can reach temperatures in excess of 1000°F. In case of fires caused by this source, close examination of the area of origin will reveal the residue from the bulb as well as the fixture, which will aid in the investigation.

Animals and Insects

Animals and insects are not frequently involved in an ignition sequence, but it can happen. For example, a light fixture that is not used for some time and then turned on can sometimes be involved in a fire situation, especially if it turns out to be the base of a hornet nest. Hornet nests are the consistency of paper, and the insulated bulb, in time, could ignite the material. Other reports cite slugs climbing into ground-level outside timing devices and shorting out the circuitry, resulting in a fire.

Larger animals can also be involved in an ignition sequence, usually as the result of eating through insulation on electrical wires. Some in the scientific field have explained that such events should not happen because the wire insulation would not taste good to the animal. Nonetheless, when investigators find a rodent at the exact area of origin, its body charred and stretched out, with its teeth still in contact with the wires, they might never know its motive, but one thing will be clear: The animal did chew through the wires.

The call came in as "smoke in a residence." The engine company arrived at the home and found just a haze of smoke throughout the second floor of the house. It appeared to be coming from the 5-year-old boy's bedroom. The boy wanted to explain what he was doing before he smelled the smoke. He said that he threw his small plush elephant, Dumbo, into the air; it disappeared and then there was smoke. The family, it seemed, had watched the Disney movie *Dumbo* that evening.

The boy insisted that his elephant had disappeared. There was no sign of it on the floor, under the bed, or anywhere else in the room. The lieutenant saw a free-standing halogen pole lamp in the corner. The bulbs for these units were halogen (R7S base) and capable of reaching temperatures as high as 500°F.

These lamp stands are approximately 5 ft tall and direct the light upward, allowing it to be reflected from the ceiling. Sure enough, there was Dumbo, the small stuffed toy; it had fallen into the top of the lamp, coming in contact with the halogen bulb. The cloth of the stuffed toy had charred; some of the small Styrofoam beads had melted. Fortunately, the bulb had failed, preventing the toy from going to flaming combustion.

The following year, it was noticed that all the halogen pole lamps for sale now had a wire screen over the opening of the top to keep items from coming in contact with the bulb. It would appear that data had convinced the lighting industry that a new safety measure had to be taken.

Wrap-Up

SUMMARY

- We are surrounded by heat sources every day, both natural and human-made. Even on the warmest day of the year, people may use electric blow dryers or want a morning cup of hot coffee or tea. Smoking materials still abound and add to the number of accidental fires each year. A crystal shade pull may concentrate the sun's rays, heating the curtain to the ignition point. This chapter covered only a few of the available heat sources that first responder and full-time fire investigators may encounter; it would be impossible to identify every heat source in all possible situations.

- As products change, so do the dynamics of heat sources involved in unintentional fires. The strike-anywhere match remains a leading cause of fire, but new science and technology created the safety match, which has most likely prevented a countless number of fires. As lightbulbs increased in lumens, their heat output also increased, causing these devices to become potential heat sources. New technology now has maintained those lumens with less heat by-product. First responder and full-time investigators must stay current on such technology changes that affect fire risks.

- Common everyday items can be a potential heat source. For example, various types of matches, lights, micro torches, and model motors can serve as heat sources. Building components such as woodstoves and chimneys can also be competent heat sources.

- The innocent burning of trash can create a situation in which the controlled fire gets out of hand due to escaping embers. Most state statues require an attendant be present at any open fire to prevent the controlled burning of trash, leaves, or brush from becoming an uncontrolled fire.

- Appliances for the most part work perfectly well. Sometimes, however, an appliance—perhaps through faulty design, manufacturing failure, or inappropriate use—can be the source of heat for a fire in the home or workplace.

- Identifying a heat source must be done scientifically and accurately. This is accomplished through a systematic and thorough fire scene investigation. In addition to keeping an open mind, it is essential that the investigator is able to research and uncover information about all aspects of the potential heat source. In this manner, future fires may be prevented by better understanding and controlling potential sources of heat.

KEY TERMS

Bernoulli effect As it applies to chimneys, a phenomenon in which wind blowing across the opening of a chimney decreases the pressure in the flue, increasing the updraft.

Catalytic converter A device installed as part of the exhaust system of an automobile; designed to reduce emissions by burning off pollutants.

Chimney chase A decorative hollow covering around a cement block or metal chimney, designed for aesthetic purposes; it is usually framed in wood and covered with the same material as the siding of the structure.

Chimney liner The covering inside the chimney that enables the flow of hot gases and smoke from the chimney and keeps them from seeping through the mortar and into the structure.

Clinker A solid glass-like object found as residue in a large hay fire.

Disposable butane lighter A device filled with a pressurized liquid ignitable gas; friction or an electrical arc is used to ignite the escaping gas as designed. Such a lighter is also capable of putting out the flame, cutting off the fuel and making the device safe enough to be carried by the user.

Drying oils Organic oils used in paints and varnishes that, when dried, leave a hard finish.

Electric matches Electrical devices that can act as an igniter from a remote distance.

Kinetic energy Energy as the result of a body in motion.

Methane gas Colorless, odorless, flammable gas created naturally by decomposing vegetation; it can also be created artificially.

Micro torch A miniature device that uses butane as a fuel to create flame; typically used by hobbyists. The mechanism can inject air into the chamber, allowing for a high heat output.

Piezoelectric ignition Certain crystals that generate voltage when subjected to pressure (or impact).

Rocket motors Hobby propulsion kits that provide thrust through the burning of an ignitable mixture.

Rotor The part that rotates in an electric motor.

Safety match A match designed to ignite only when scraped across a chemically impregnated strip.

Spontaneous combustion A process in which a chemical reaction takes place internally, creating an exothermic reaction that builds until it reaches the ignition temperature of the material involved.

Stator The part of an electric motor that stays stationary, housing the rotor.

Strike-anywhere match A match where the head is chemically impregnated to ignite when scraped across any rough surface.

REVIEW QUESTIONS

1. Define *competent ignition source*.

2. How much energy can a lightning bolt produce?

3. How could methane gas enter a house? Would it be detectable by the occupants?

4. Why are cigars less likely to be an ignition source than cigarettes?

5. What type of evidence can a paper match and matchbook provide?

6. Describe the different types of matches.

7. Name three types of ignition devices that can be found at a local hobby shop.

8. What would cause sparks to come out of the exhaust from a catalytic converter?

9. What keeps hot gases, smoke, and sparks from coming out through the mortar of a brick chimney?

10. Define *spontaneous combustion*.

11. Give three examples of spontaneous heating, and explain how each occurs.

12. How can small appliances be involved in fires as potential heat sources?

DISCUSSION QUESTIONS

1. When you look around your classroom (or home), what potential sources of heat do you see?

2. Some states have passed laws allowing only fire-safe cigarettes to be sold in that state; other states are considering similar legislation. Why will the tobacco industry not switch to manufacturing fire-safe cigarettes on its own instead of waiting for government mandates?

REFERENCES

1. National Fire Protection Association. (1997). *Fire protection handbook* (18th ed., pp. 1–66). Quincy, MA: National Fire Protection Association.
2. Ahrens, M. (2013, June). *Lightning fires and lightning strikes.* Quincy, MA: National Fire Protection Association. Retrieved from https://www.nfpa.org/News-and-Research/Data-research-and-tools/US-Fire-Problem/Lightning-Fires-and-Lightning-Strikes
3. National Fire Protection Association. (1997). *Fire protection handbook* (18th ed., pp. 10–78). Quincy, MA: National Fire Protection Association.
4. National Fire Protection Association. (1997). *Fire protection handbook* (18th ed., pp. 1–69). Quincy, MA: National Fire Protection Association.

5. National Fire Protection Association. (1997). *Fire protection handbook* (18th ed., pp. 3–53). Quincy, MA: National Fire Protection Association.

6. National Weather Service. *What is lightning?* Melbourne, Florida: Forecast Office, Lightning Center. Retrieved from https://www.weather.gov/mlb/what_is_lightning

7. Ahrens, M. (2017, April). *Structure fires started by lightning.* National Fire Protection Association. Retrieved from https://www.nfpa.org/-/media/Files/News-and-Research/Fire-statistics-and-reports/US-Fire-Problem/Fire-causes/oslightningstructurefires.pdf; (2019, January). https://www.nfpa.org/News-and-Research/Data-research-and-tools/US-Fire-Problem/Smoking-Materials.

8. National Fire Protection Association. (1998). *Principles of fire protection chemistry and physics* (pp. 243–247). Quincy, MA: National Fire Protection Association.

CHAPTER 11

Patterns: Burn and Smoke

LEARNING OBJECTIVES

Upon completion of this chapter, you should be able to:

- Describe and understand the concept of normal fire growth, direction, and the patterns that fires create on building components.
- Describe the examination of burn patterns on ceilings and wall surfaces.
- Describe the examination of burn patterns on wood framing and doors.
- Describe the examination of protected areas in and around other burn patterns.
- Describe the impact of suppression activities on fire growth and direction, and explain how suppression can affect the resulting burn patterns on the structure and contents.
- Describe unique patterns associated with concrete, glass, and steel springs.
- Describe various burn pattern indicators that point to where the fire originated.

Case Study

The restaurant was located in a typical strip mall, where the storefronts were mostly glass with glass doors. The floor plan showed the seating area in the left half of the occupancy going from front to back. The right half of the structure included the restrooms at the front and the kitchen at the back, with these areas being separated by an office.

The business owner had recently experienced employee problems and had changed the locks on the day of the fire. He was the only person with keys, all of which were in his possession at the time of the fire. The structure had been checked by the owner as he left the building, and he was adamant that there were no persons left in the building when he locked up. He also affirmed that he had personally locked both the front and back doors before leaving at nearly midnight.

The owner had parked his car near the road, quite some distance from the restaurant. As he was getting in his car, he heard glass breaking. He looked up and saw three young men in front of his restaurant, where the front glass had been broken out. He then saw one of them throw something else in the restaurant, and he immediately saw flames. He yelled and the three young men took off. Instead of giving chase, the owner stopped and called the fire department and the police.

The police conducted a search for the three males. Several young people were pulled over and questioned that night based on a weak description from the restaurant owner.

An engine crew arrived and extinguished the fire, which was predominantly burning the carpet from the front window across the dining room floor and into the kitchen. Based on burn patterns, the engine officer (first responder investigator) suspected that the fire involved a liquid accelerant and immediately called the full-time fire investigator on the radio. Upon arrival, the fire investigator did a walkthrough of the scene with the engine officer. Next, he conducted a scene examination, and interviewed the owner. The investigator then called law enforcement and halted the search for the three young men who had allegedly attacked the business.

The fire consisted of the burning of a fuel, later identified as gasoline, along the dining room carpet and kitchen tile floor. The tables, chairs, and other contents had not yet reached their ignition temperature, thanks to a quick extinguishment of the fire. A brick was found just inside at the large broken picture window at the front of the restaurant. However, the only fuel container found on the property was a 5-gallon red plastic container located on the floor just outside the restroom door, near the front door, which was some distance from the window. Both front and back doors were still locked; there was no other access to the property except through the broken front window.

The ignitable liquid pour pattern on the floor extended from the window, across the patron seating area to the restroom, and then made a left turn and went toward the back of the building. The pour pattern then took a right-hand turn, through the kitchen door, around the food preparation table in the middle of the kitchen, and back out the second kitchen opening, looping back to the path leading from the restroom.

The owner, by his own admission, had exclusive opportunity to gain entry into the restaurant and pour the liquid. He had the keys and was in the vicinity when the fire started. Faced with the facts, the owner confessed. A number of scenarios could have played out in this particular case had the owner been more resourceful in his planning and had the response from the initial engine taken longer. Ultimately, the burn patterns clearly refuted the owner's statement about the brick and fuel being thrown through the window.

Further investigation showed that the restaurant's profits were down; the business had employee problems, which might have been the reason for changing the locks; and the owner was in other financial difficulties. But the property was insured, and the owner also had business interruption insurance. Based on the fire investigator's report and its own investigation, the insurance company denied the owner's insurance claim. After a brief trial, the owner was sentenced to 10 years in jail.

 *Access Navigate for more resources.*

Introduction

No single text or training class can fully educate someone on the interpretation of burn patterns and normal fire progression. The sum of their experiences will help investigators become proficient in this field, including studying fire behavior from watching fires burn and observing how fires react under various conditions. The experience of seeing how ventilation of a burning structure changes the direction of fire travel can be invaluable to the investigator. Ventilation can be natural, as when high heat causes a window to break, or it can be caused by a fire fighter cutting a hole in the roof. Watching these types of events and then seeing the smoke and fire patterns created on structural surfaces and contents are an education all in themselves.

> **TIP**
>
> The experience of seeing how ventilation of a burning structure changes the direction of fire travel can be invaluable to the investigator.

> **TIP**
>
> Seeing the smoke and fire patterns created on structural surfaces and contents can educate the fire investigator about how these patterns form.

Through these types of experiences, investigators can learn to recognize the patterns left behind by fire and smoke on structural surfaces and contents so that they can subsequently interpret the patterns and make a presumption about the fire direction that will lead back to the fire area of origin. This statement is not meant to imply that only an experienced fire fighter can be a proficient fire investigator—but it certainly makes sense that fire fighters' experiences in the field may help them transition from fighting the fire to exploring the cause of the fire. Other members of the police community and engineers who investigate fires also have their own experiences and specialized knowledge that can help in many other aspects of the investigation. To learn to interpret burn patterns, they will need to undergo additional specialized training and have opportunities to see fires burn and to watch burn patterns develop. Likewise, an individual with just a fire background would be well advised to obtain training on conducting interviews and observing human behavior.

> **TIP**
>
> Fire fighters' experiences in the field may help them transition from fighting the fire to exploring the cause of the fire.

Burn and smoke patterns comprise the indicators left behind after the fire has been extinguished. By examining these patterns, the investigator can interpret the direction of fire travel, follow the growth of the fire, and eventually determine the area where the fire originated.

The burn patterns identified and used in the analysis of the fire scene must also pass a test before being considered reliable evidence. Is the scenario proposed by the investigator based on sound scientific principles? If this is not the case, then the analysis may be flawed. Science tells us that fires move upward and outward. It also tells us that with other influences, fires can increase dramatically in volume, can change direction, and can move downward as well. All of these factors must be taken into account, and further examination may be required if conflicting patterns are encountered—that is, patterns that are out of place from what one would expect in a normal fire growth and spread. The final conclusions of the interpretation of the burn patterns should be supported by science.

> **TIP**
>
> Fires move upward and outward.

This chapter focuses on burn patterns, discussing basic concepts about plumes, fire travel, and the impact on fire patterns of fire suppression actions. It addresses both the simple patterns and the impact of flashover on existing fire burn patterns. You will learn about the fuel load impact on patterns and the patterns on different types of materials.

Patterns

Fire patterns are the markings on a structure and its contents left by fire and smoke. This damage can be inflicted by flames, hot gases, or radiant heat. Such damage can be seen on combustible and even on noncombustible surfaces in the form of heat marks or soot. The investigator examines these patterns with the intention of determining the direction of fire travel and eventually identifying where the fire started.

> **TIP**
>
> The investigator examines fire patterns with the intention of determining the direction of fire travel and eventually identifying where the fire started.

Fire patterns can involve the charring of combustible materials such as wood or plastics. In some cases, the material may have been consumed, and its absence forms part of the burn pattern as well. Metal surfaces may show charring of the paint. Once the paint has burned away, or if the item had no paint, the metal

FIGURE 11-1 Patterns on furnishings may include sooting, heat changes, and oxidation.
Courtesy of Russ Chandler.

surface may then oxidize. This oxidation pattern can also help the investigator. Some metals, such as pewter, lead, and pot metals, have a low melting temperature and may melt if the items are in the path of the fire. This, too, can show the direction of fire travel or point the investigator toward the area of origin.

In addition to the partial burning, oxidizing, or total consumption of materials, other indicators can help the fire investigator. In the less-damaged areas, materials may simply become discolored from heat, and some items may be warped or distorted as well. Of course, the smoke and sooting can also aid in the investigation (**FIGURE 11-1**). All of these indicators must be taken into account when creating the hypothesis of the origin and cause of the fire.

However, these patterns may not always be straightforward indicators of the fire's origin and progression. Perhaps the fire started in an area with limited combustible materials. As the fire grew and moved from one room to the next, it might have encountered materials capable of great heat release. Such materials might also produce high volumes of smoke and soot, creating a more devastating appearance than that found in the room of origin. Moreover, if the fire breached a window, air currents might have entered the structure and changed the direction of fire travel, creating deeper char as the fresh air fed the fire.

> **TIP**
>
> Fire patterns may not always be straightforward indicators of the fire's origin and progression.

The ultimate goal when examining burn patterns is to find the least damaged area. The investigator then works toward the most damaged area, which may point toward the area of origin. During this process, the investigator will compare like surfaces, so as to determine where the char or smoke damage is the heaviest. Then the investigator must take into account, and eliminate, any patterns created by ventilation of a larger fuel load in the path of the fire.

Basic Fire Patterns

Under most circumstances, normal fire growth is upward and outward. In this process, the initial patterns on the wall will have the appearance of a V if the fire occurs near a vertical surface. The V is created by the flames and heat of the fire as it impacts the wall. The width of the V depends on the size and width of the fire that is burning: As the fire increases in size, so will the pattern created by that fire. The longer the fire burns in any area, the more damage that will occur in the area. Not all fire damage is directed upward and outward, however. Because of direct fire impingement, radiant heat, and other factors, a fire will also burn downward, albeit at a much slower rate. This brings us to one of the tools every investigator needs: common sense.

> **TIP**
>
> Under most circumstances, normal fire growth is upward and outward.

> **TIP**
>
> Because of direct fire impingement, radiant heat, and other factors, a fire will also burn downward, albeit at a much slower rate.

> **TIP**
>
> Every investigator needs the same tool: common sense.

If the fuel has a low heat-release rate, which creates a fire with shorter flames, an inverted V pattern may be seen on the wall. As the fire grows in volume and intensity, this pattern may end up being masked by new patterns, which may potentially destroy any evidence of the initial inverted V.

Investigators must ask themselves whether the patterns left behind makes sense. Is this what one would expect to see under the current known circumstances?

If something does not seem right, it is the investigator's duty to search until all questions can be answered. This is all part of the analytical process.

After a fire begins, its natural progression may create an hourglass shape (**FIGURE 11-2**). This occurs when the bottom of the pattern is formed by the natural shape of the flames created by air entrainment, as the flames lean inward. This air is not always equal from all sides, so the fire may appear skewed to one side or the other depending on the air currents. As the fire intensifies, it releases more heat and larger flames

FIGURE 11-2 The inverted V can indicate a small fire that burned itself out or one that continued to burn, possibly forming an hourglass shape, which was partially created as the result of air entrainment. The third type of pattern, the V pattern, points back down to the area of origin.

that produce the upward and outward flow. This in turn creates an hourglass shape—a natural pattern shape that may be readily seen by the investigator.

This same type of pattern can sometimes be seen in areas of the structure other than the area of origin. These patterns could be caused when burning material falls on other combustibles, possibly at the floor level. An example would be curtain material that is ignited at ceiling level as a result of the buildup of the heat layer in that area. As the curtains start to burn, they then fall off the curtain rod, dropping to the floor. This creates a second floor-level pattern on the wall. The location and the fact that the burn pattern is less intense may indicate the presence of a drop-down fire, rather than a second, deliberately set fire.

Some patterns suggest that multiple fires were set in the structure. The investigator must be cautious and not jump to the conclusion that the fire was incendiary in such cases. Instead, following the analytical approach, the investigator should search for a reason for patterns that imply multiple areas of origin. As an example, consider an industrial occupancy in which a large electrical wire shorts out and the overcurrent protection does not activate. The wire will heat evenly across its entire length, igniting similar materials in proximity to the wiring.

The actual patterns to be interpreted are usually two-dimensional, but the investigator should try to visualize the fire in a three-dimensional form. The initial fire, with no outside influences, will resemble a cone—that is, narrow at the bottom with a wide section at the top. Depending on the distance from objects, this cone may eventually intersect and come in contact with walls or contents. When the cone comes in contact with the wall, as shown in **FIGURE 11-3**, it creates a two-dimensional pattern on the wall. This pattern may resemble a U shape. Based on the height of the pattern on the walls and marks on the contents, the investigator can envision the entire cone pattern. This, in turn, may help the investigator narrow the search for the area of origin.

FIGURE 11-3 A cone pattern from the fire intersects and leaves marks on the wall and contents.

TIP

The actual patterns to be interpreted are usually two-dimensional, but the investigator should try to visualize the fire in a three-dimensional form.

The ceiling area directly over the cone, if still intact, will show its own pattern (**FIGURE 11-4**). It may be circular in nature or slightly skewed if any winds or air movement affected the area. The amount of fuel can also influence a pattern. The fire could start in a limited-fuel area, but then extend to an area with an abundance of fuel; there, it can create a large pattern, even sometimes masking the original pattern.

When fuels have a high heat-release rate, they burn with more vigor, giving off gases and flames that create more of a vertical pattern, straight up or out, thereby producing the V pattern. This type of fuel creates more heat at a faster rate. The heat builds up in confined spaces in the upper areas near ceilings, and banks down and leaves distinct marks on walls and furnishings. This upper heated area also affects other items in the room by radiating the heat downward, creating patterns on the top surface of contents.

Generally, the banking down of the heat layer creates a solid line, a clear delineation along the walls and furnishings. However, the heat layer can be somewhat fluid, creating occasional wavy lines. This fluidity is evident when the thermal layer of hot gases, flowing across the ceiling, comes in contact with an obstacle such as a wall. At this point, the hot air and gases hit the wall and roll down and then back up. The result is a unique pattern with a slightly lower burn than the level of the rest of the thermal layer. In the example shown in **FIGURE 11-5**, the fire started in the room to the right; the upward movement of the heat can be seen on the trim work of the opening to the room. On

FIGURE 11-5 These patterns show the lower burn in the room to the right. The impact of the flowing gases from right to left created a rollover pattern on the wall to the left as the gases and heat hit the wall, and then rolled down and back up, causing a slight dip in the pattern.
Courtesy of Russ Chandler.

the left, the result of the rolling effect on the wall can be seen. This slightly lower mark is where the thermal layer hit the wall and rolled down and then back up, staying buoyant.

TIP

Generally, the banking down of the heat layer creates a solid line, a clear delineation along the walls and furnishings.

As the fire spreads, it will go through openings. One way to analyze the flow of this layer of hot gases is to think of it as a fluid. However, unlike liquids that collect at the lowest possible point, the buoyant gases accumulate and flow upward. Consider the fluidity of the gases as they collect at the ceiling and bank down. Once they reach the door opening to the next room, they will roll over the upper frame of the door and spill into the ceiling area of the next room. If the door is closed, the hot gases and smoke leak through the cracks and small openings, much as a liquid would do at floor level. As these hot gases move out of the room of origin, they will continue to damage the ceiling area and upper portions of the room, creating patterns on the wall and on the doorframe.

Such patterns can lead the investigator back to the area of origin—whether in the basement, a living area, or the attic. That origin can be at floor level, on the wall (e.g., at an electrical outlet), about 3 ft off the floor (e.g., on a range or countertop), or at the ceiling (e.g., at a failed light fixture). An open mind and an analytical methodology will lead the investigator to a sound conclusion.

The fuel load in a given area may show the heaviest char. The investigator needs to examine this area to

FIGURE 11-4 A circular pattern directly over the area of origin. What was initially a sooted area was then burned clean as the fire continued to burn and grow.
Courtesy of Russ Chandler.

determine whether this area was subject to a longer burn time or increased char due to increased fuel. Ventilation plays a role in the char patterns as well. Where ventilation occurs, an abundance of air will enhance the combustion process, creating more heat and, in turn, more char. This area may therefore need to be eliminated as the area of origin. The culmination of all the facts leads to a determination of an area of origin and an eventual cause of the fire. The following sections discuss some areas where patterns can be found and examined.

Ceilings

Patterns on the ceiling are a result of heat rising, which is stopped only by an obstacle, which is the ceiling. The ceiling area may or may not be intact. Because the ceiling may be the first area to be affected by the buildup of heat, it also may be the first area to fail. Most ceilings are made of **drywall**. As the gypsum in the drywall breaks down, it weakens and may fall on its own accord, or, with the addition of suppression water, it may get heavy and drop to the floor, taking most of its pattern with it. However, the debris on the floor from the ceiling may hold clues, as will the structural elements that held the ceiling drywall. First and foremost, the heaviest area of damage may be in the area of origin.

It is rare, but sometimes the patterns are limited to an oval pattern on the ceiling when the area of origin is directly below that pattern. Most scenes will have patterns across the entire floor of origin. The original direction of fire travel may have been down a south-facing hall, only to have the structure ventilate itself and strong winds push the pattern back on itself, so that it takes another direction.

> **TIP**
>
> Most scenes will have patterns across the entire floor of origin.

> **TIP**
>
> The investigator needs to observe the patterns on the ceiling and compare them to the patterns on the walls.

As an example, an investigator might find the following fire pattern evidence at a fire scene:

- The investigator needs to observe the patterns on the ceiling and compare them to the patterns on the walls. Together they may show heavier damage nearer the kitchen.

- Working toward the kitchen, the patterns on the wall may be lower to the floor.
- Just outside of the kitchen, the investigator finds a breakfast nook with massive damage. The large amount of debris in this area shows that the bench seats were padded and there were lots of newspapers on the table. Evidence indicates that the window failed early, allowing the fire to burn more intensely.

Realizing that this may be a fuel load and ventilation effect, the investigator must observe the patterns in the rest of the room, which show heavy damage in the kitchen cabinets. A V pattern goes up the cabinets, the ceiling directly overhead is missing drywall, and the ceiling rafters have fire damage.

By itself, this evidence does not allow the investigator to make a final determination. Other issues must be addressed, such as potential source of heat and first material ignited. Interviews and a thorough investigation will solidify the final determination.

Wall Surfaces

Many types of wall surfaces exist, both commercially installed and self-designed. For commercial establishments, fire resistance requirements may differ from those for residential structures. However, occupants of both commercial and residential structures can create some unusual designs that may change the fuel load dynamics. This could range from a concrete wall with no covering to an office in an airplane hangar where old silk parachutes are draped as wall coverings. Investigators should not rely solely on what the code mandates for wall or ceiling furnishings, but instead should go into every structure with an open mind—they may find some very interesting wall surfaces. The next few sections address specific issues that may be encountered with various construction features and finishes on walls.

> **TIP**
>
> Investigators should not rely solely on what the code mandates, but instead should go into every structure with an open mind.

Drywall

The most commonly encountered wall material is drywall, also referred to as *sheetrock* or *gypsum board*. Drywall consists of gypsum plaster in a core wrapped with strong paper. The drywall surface can be painted, covered in stucco or wallpaper (some wallpapers have a plastic laminated surface), and even covered with veneered wood paneling or any other finish. The addition of these coverings over the required sheetrock changes the way the wall will react in a fire situation. Sometimes wall coverings retard the fire progress, but quite often they increase the fuel load.

In general, drywall is fire retardant. It will resist burning but will eventually fail as the fire progresses. The paper surface will be the first to burn away, and then the gypsum will start to break down as the moisture is driven off in a process called calcination. As this takes place, the gypsum will allow the heat to transfer through the drywall and burn off the paper on the other side. When the moisture is driven off, the gypsum takes on a whitish appearance. In this state, it may start to lose its cohesiveness and will weaken. All it might need to fail completely is the application of a water stream from suppression forces. The drywall in the ceiling may fall from its own weight or from the added weight of moisture from the fire streams.

> **TIP**
>
> Gypsum starts to break down as the moisture is driven off in a process called calcination.

The degree of damage to the drywall paper or the gypsum around the room may vary, with more damage occurring in one area than in another. The difference may reflect either a localized high fuel load or a longer burn time, thereby indicating the direction of fire travel (**FIGURE 11-6**). Commercially available gauges allow the investigator to measure the depth of the calcination to determine the amount of damage at any given point in the room. When conducting the examination of the calcination of the drywall, it is critical to ensure that like surfaces are being compared.

> **TIP**
>
> When conducting the examination of the calcination of the drywall, it is critical to ensure that like surfaces are being compared.

Suppression water applied to areas of drywall calcination may cause the drywall to fail and fall to the floor. These fire streams may have been necessary to stop the fire. Although fire fighters are trained not to flow water unless necessary, a less experienced fire

FIGURE 11-6 Sheetrock with the paper burned away, leaving the gypsum exposed to the fire and fire gases. Comparing these walls to adjacent walls may help the investigator determine the direction of fire travel. In this case, the fire came from the room to the left and flowed into the hall on the right.
Courtesy of Russ Chandler.

fighter may not give this a thought. Such fire fighters may need to be educated by the fire officer or fire investigator about how unnecessary water flow can destroy evidence.

Manufacturers produce some drywall products designed to meet certain fire-resistive ratings. To allow for longer fire protection, additional bonding fibers may be added to the gypsum during the manufacturing process. Drywall also comes in different thicknesses and lengths. The length is intended to make the hanging process easier, creating fewer joints to tape and seal. The thickness of the drywall varies according to the need for fire resistance and soundproofing. In some industrial occupancies, more than one layer of drywall may be installed to give the wall a better fire protection rating.

Wood Walls

Paneling is wood-based but may have a plastic surface. The plastic and the wood of similar pieces of paneling can be compared to determine the direction of fire travel or to show where secondary burning of a large fuel load in the path of the fire took place. Be sure to compare like surfaces in the comparison, just as for drywall.

Paneling has a tendency to curl as it burns, and because of its combustible nature, it may disappear altogether. When the paneling burns away, it exposes the material underneath. If this is a concrete wall, such as in a basement or in commercial structures, the installer may have used furring strips attached to the concrete to provide spacing and a surface to attach the paneling. Unlike in a wood stud wall, where the 2 × 4's go from floor to ceiling, the furring strips may be spaced at the top half; matching that vertical line, a short strip may be placed on the bottom part of the wall. This allows scrap wood or less expensive lumber to be used as individual strips. The investigator, in turn, should be cautious about interpreting the gap in the strips as being burned away; in fact, they may have always been two separate pieces of wood (**FIGURE 11-7**).

Wood boards may be placed either horizontally or vertically on the walls to give a rustic appearance. Different types of wood may also be present. Oak will not char and blister in the same way that pine or ash will. The different types of char blisters are of no significance to the investigator. Instead, the important point is whether there is a line of **demarcation** between an area of char and an area of no char, or in some circumstances whether there is deep char with much lighter char. This demarcation can show the fire or heat path.

> **TIP**
>
> Oak will not char and blister in the same way that pine or ash will. The different types of char blisters are of no significance to the investigator.

FIGURE 11-7 Missing sections of furring strips on concrete walls require close examination. They may not have been burned away; instead, shorter pieces may have been used in construction.
Courtesy of Russ Chandler.

Wallpaper

Wallpaper can be made of actual paper, cloth, or plastic, or a combination of any of those materials. The charring of the paper needs to be compared with the same type of charring found elsewhere in that room or area. Some paper with plastic coating will bubble or blister, leaving an unusual pattern. Some cloth coverings may have raised areas in the design that will be more prone to ignition. When interpreting the patterns on a wall to find the fuel load or the direction of fire travel, the investigator must be sure to compare like surfaces for a more accurate interpretation.

> **TIP**
>
> Wallpaper can be made of actual paper, cloth, or plastic, or a combination of any of those materials.

A kitchen is one area where differing burn patterns are often observed on walls. In one area, the wall surface may be covered with wallpaper that has a plastic film—a composition intended to keep grease and spattered oils from staining the walls. The adjacent wall in the dining room may be just drywall with a coating of paint. The plastic in the kitchen's wall covering serves as an additional fuel, which may create a visibly different char than any burning covering on the drywall in the dining room, even though the dining room may have been exposed to more heat or for a longer period of time. **FIGURE 11-8** illustrates this char effect from the plastic film on wallpaper.

FIGURE 11-8 Wallpaper with a plastic film on the surface will burn differently from other types of paper surfaces. Here, the baseboard has been removed, revealing a pour pattern where an ignitable liquid ran down the wall and collected behind the baseboard. A sample was collected and came back from the laboratory as positive for gasoline.
Courtesy of Russ Chandler.

Concrete Walls

When examining burn patterns on concrete walls, the investigator must first ensure that he or she is dealing with true concrete. New design techniques enable builders to create architectural designs with Styrofoam, which is then covered with a cloth material and sprayed with a water-resistant material or concrete. The resulting finish looks like stucco. This construction is used predominantly on exteriors but has sometimes been used in interiors as well.

Concrete and stucco on walls will pop off when exposed to heat. This effect, called spalling, results from moisture within the concrete being heated to its boiling point, 212°F, at which point it turns into steam. In the transformation from a liquid to a gas (steam), water expands to 1700 times its volume. That means 1 in.^{3} of water becomes 1700 in.^{3} of steam. This action can be forceful enough to pop off the surface of the concrete. The ability of the moisture in concrete to reach the boiling point depends on the amount of captured heat; thus, reflected heat can cause spalling under stairwells or in small areas where heat builds up. In consequence, spalling may not necessarily indicate the presence of additional fuel or proximity to the area of origin. Depending on the structure, the concrete may not contain any appreciable moisture, so spalling may not always occur.

Some concrete walls are sealed with a finish to limit moisture absorption or for decoration. The sealer in some cases can become a fuel as well. Investigators are advised to find out as much about the construction and post-construction changes to the structure as possible if they observe an unusual burn pattern on a concrete wall.

As with any other wall surface, investigators should examine the concrete wall for heat and soot patterns. Any finish applied to the concrete walls may have burned off in some areas and not others. All of these factors provide clues and indicators, allowing investigators to work toward the area of origin.

Doors

As the fire grows and the thermal gas layer expands, moving outward from the area of origin, it will eventually impact at least one door. Doors can exhibit the same burn patterns as walls; in some cases, there may be more damage on doors because the walls are made of fire-resistive sheetrock and the doors are made of combustible wood. The fire patterns may also indicate whether a door was open or closed at the time of the fire.

When a door is closed, the gap between the door and the doorframe may be tight enough to limit escaping gases and heat. Frequently, this is truer on the door edge with the hinges. As such, the surface of the door may show more damage than the edge if the door was closed at the time of the fire. By examining the patterns on the door, the investigator should be able to tell whether the door was open or closed at the time of the fire. The most telling clue will be a lack of char on the inside edge of the door at the hinges, which indicates that the door was closed at the time of the fire.

If the door is closed at the time of the fire, the top of the door will most likely experience heat. As the fire burns, air is pulled in through any space at the bottom of the door. Thus, the top will be burned and the bottom may show a lesser degree of fire damage. However, as with most issues in fire investigations, there is no one answer for everything. Should drop-down fire occur near or at the base of the door, the investigator may find fire damage along the bottom of the door. Likewise, if the fire was incendiary in nature and the arsonist poured an ignitable liquid through the doorway, then the bottom of the door will be burned as well. By looking at the scene indicators, the investigator should be able to determine which scenario may have caused the charring under the bottom edge of the door.

If the door was open at the time of the fire, it should have the same level of damage on the face of the door as on the wall. Depending on the distance of the door from the area of origin, the door may even be part of the V pattern showing the upward and outward flow of the fire (**FIGURE 11-9**).

FIGURE 11-9 This door was open at the time of the fire; the V pattern is evident on the surface of the door. The edge of the door and doorjamb where the hinges are located are charred, another indicator that the door was open at the time of the fire.

Courtesy of Russ Chandler.

Flashover within a room where the doors were open or closed may create so much charring as to obscure and eliminate all previous patterns. This does not necessitate that the investigator should discard any patterns found or stop looking for patterns in a room that experienced flashover.

> **TIP**
>
> Flashover within a room where the doors were open or closed may create so much charring as to obscure and eliminate all previous patterns.

Exterior doors should be checked to determine whether they were open or closed at the time of the fire. A close examination should be conducted to look for any signs of forced entry. During the follow-up meeting with fire suppression personnel, the investigator should verify the condition of the doors when the fire fighters arrived on the scene. Did the suppression forces have to force any doors or windows when carrying out their duties? Did they try the door to see if it was unlocked prior to forcing it? This may seem like a foolish question, but in the heat of the moment when given an order to force the door, a fire fighter may do just that—force it before trying the doorknob.

> **TIP**
>
> Check whether the suppression forces had to force any doors or windows when carrying out their duties.

The investigator also needs to check with other emergency services personnel about their actions. A police officer or rescue squad personnel may have attempted a rescue if they arrived before fire suppression personnel, forcing the door. Or they may have found the door open or unlocked—both situations are of interest. Neighbors and even strangers have been known to attempt a rescue when they see flames. Check with witnesses to see if this was the case.

Any marks on doors that might indicate forced entry should be examined closely, documented, and photographed. If the investigator is not experienced in this area, he or she should check with local law enforcement for assistance. Most police agencies have experts in this area and should be able to provide someone who has expertise in taking tool mark impressions for later use in the investigation. During this process, additional evidence may be discovered in the form of paint chips or metal fragments that can help identify the tool used to force the entry.

Windows and Glass

Windows have a lot to tell the fire investigator. Patterns on windows close to the area of origin all too often are heavily damaged by fire because of the availability of air and the fire burning through the window. Glass will fail relatively quickly in a fire.

When exposed to pressure from an explosion of any type, the glass may break, sending shards into the yard. When working from the least damaged area to the most damaged area, at some point in the process the investigator should look at the yard and surrounding area of the structure. Glass shards found in this area need to be examined closely. If they are sooted, it may indicate that the fire was burning, soot collected on the glass, and then an explosion occurred, sending the broken, sooted shards of glass to the outside of the structure.

If the glass found in the yard came from a window of the structure and is not sooted, that finding may indicate that the explosion came first, which in turn may have started the fire. The investigator should wait before rushing to make such an assumption because more investigation will need to be done. Could the fire have been burning in another area of the structure in a room being protected by a closed door? In such a case, the fire could have eventually reached a fuel that caused an explosion, causing the door to that room to fail and in turn impacting the unsooted glass and sending it into the yard. Granted, that scenario may seem far-fetched, but the investigator must address these kinds of possibilities to formulate a proper and accurate hypothesis of the fire area of origin and cause.

> **TIP**
>
> If the glass found in the yard came from a window of the structure and is not sooted, that finding may indicate that the explosion came first, which in turn may have started the fire.

The glass may not have always traveled far out in the yard. In the process of a detailed investigation, the investigator will want to look directly below a window that has failed as the result of the fire. By looking on the ground under the window on the outside, and on the floor under the window on the inside, the investigator can remove the fire debris in layers until the floor or the grass/ground is reached. If the glass under the debris in both circumstances is sooted, then the window most likely failed as the result of the fire. If the investigator finds clean glass at the bottom, the window may have been broken prior to the fire—a finding that suggests a whole new set of questions about what or who may have done this and why. Most important, it is critical to keep an open mind and let the evidence solve the puzzle, not speculation.

> **TIP**
>
> If the investigator finds clean glass at the bottom, the window may have been broken prior to the fire—a finding that suggests a whole new set of questions about what or who may have done this and why.

In any given structure fire, varying types of damage to glass can occur. Sometimes glass will develop half-moon-type cracks from heat. When more heat is applied, the glass might melt, falling down on itself in multiple folds. It can also **craze** (crazing)—that is, multiple small cracks may form that do not quite go all the way through the glass. This same effect occurs in pottery when the pots are pulled from the hot kiln and sprayed with a mist of water. This crackling or crazing of the glaze on pottery has been used for centuries. Thus, crazing of glass in a structural fire is most likely the result of rapid cooling, possibly from the application of suppression water spray, and nothing more.

Reconstruction of broken glass bottles may provide valuable evidence in a fire investigation. Such bottles can be cracked, broken, or even crazed. If enough of the glass can be collected and if there is good evidentiary reason for the reconstruction, then a laboratory may have the expertise and ability to undertake the task. However, tempered glass, which is found on glass doors, car windows, and television screens, cannot be reconstructed. When it breaks, it shatters into many small fragments rather than sharp shards, making it safer than regular glass (**FIGURE 11-10**).

Burn patterns on windows also give some clues as to the fire's development. If a window has been mostly consumed, its remains have only limited value. If it was a wood sash window, the burn pattern may indicate whether the window was open or closed as the flames impinged on the window. An open **single-hung window**, in which only the bottom sash moves up and down, will have a protected area where the sash was sitting at the time of the fire. If there

FIGURE 11-10 Tempered glass broken but still remaining in the door frame. This is not crazing, but rather the result of its design, which is intended to make the glass safer should it break.
Courtesy of Russ Chandler.

was no damage from the flames, soot patterns may give the same information. **Double-hung windows** allow both the top and bottom sashes to operate independently, moving up and down. Both the upper and lower sashes should be checked for a protected area, which indicates the position of the window at the time of the fire.

> **TIP**
>
> Burn patterns on windows give some clues as to the fire's development.

Other types of windows may offer clues as well. Several models require a crank handle to open the window. On **jalousie windows**, slats of horizontal glass all tilt open as the crank handle is turned. On **awning windows** and **casement windows**, one, two, or more sections of glass open independently when the handle is turned.

Jalousie and awning windows can have either wood or aluminum framing. Aluminum frames melt easily and can leave a pattern on the outside showing that the window was pivoted out at the time of the fire. The handles themselves are made of a metal, such as aluminum or pot metal, and may melt early in the fire. Sometimes the residue from the handle may be present, but in mass destruction of that area it may be impossible to tell whether the handles were off or on at the time of the fire.

The presence or absence of the window handles may yield valuable clues. Some families have taken the handles off to prevent young children from opening windows, especially on upper floors. A more interesting finding would be open windows with the handles missing—that could be the action of the arsonist who wants to assure plenty of oxygen is available to feed the fire.

Wood Framing

Once the wall coverings have burned away, they expose the structure's wood framing to the fire. The wood framing may provide a pattern that can help the investigator determine the direction of fire travel or the fire origin. NFPA 921, Guide for Fire and Explosion Investigations, uses the term arrow patterns to describe the patterns left on wood structural elements. When examined, arrow patterns point toward the area of origin. Wall studs farther from the area of origin have damage at a higher level than do those studs that are closer to the area of origin. As the patterns being observed get lower and lower, their arrangement

resembles an arrow pointing back toward the area of origin (**FIGURE 11-11**). The studs themselves may show more rounded edges on the side facing the area of origin.

TIP

As the patterns being observed get lower and lower, their arrangement resembles an arrow pointing back toward the area of origin.

When subjected to flames and heat, the wood studs start to char. There is no definitive way to determine that a fire burned for a certain length of time with a certain amount of char. Nevertheless, a comparison of the depth of the char on wood framing can help the investigator determine a direction of fire travel. The

An Engine Operator Observation

The house was large, even for the neighborhood. The two-story, L-shaped structure had a two-car garage with rooms over the garage on the shorter wing. The front of the house was classic in design, with evenly spaced casement windows on both the first and second floors. In contrast, the back of the house consisted of mostly sliding-glass doors looking out over a short backyard and into a scenic bay, with houses lined up on the other side of the bay.

The first 911 call came in for smoke in the neighborhood, and a single engine was dispatched. Before the engine arrived, 911 calls started coming in for a structure on fire with fire coming from the roof. As the dispatcher reported this information, the first engine arrived and reported fire through the roof. A supply line was laid and an attack line was stretched to the front door. Because the door was locked, forced entry was required, which took time for the three-man crew on the engine to accomplish.

The operator of the first engine had everything lined up and took a moment to look around. It was then that he noticed that fire was coming from the room on each end of the L-shaped structure, but no fire was evident in the middle of the structure. The flames that appeared to be coming through the roof were really flames coming from the large windows of two rooms on the second floor level, which were on opposite ends of the structure. The roof was still intact. This structure was more than 6000 ft², so a considerable distance separated the two ends of the building.

Something just did not look right. Then, it hit the engine operator: He was looking through the windows of the house at the lights from the homes on the other side of the small bay. He took further stock of the situation and saw that the casement windows were all open, on both the first and second floors. This positioning was unusual because it was December and the temperature was just above freezing. Even more important, the engine operator could see into the windows in the middle of the second floor, in the center of the house, and there was no smoke. He pointed this observation out to another fire fighter who arrived with the second alarm. After getting attack lines in place, the fire fighter made a quick check and discovered that all the casement windows on the first floor had no handles and could not be closed.

Before additional attack lines could be laid, the entire second floor became involved in fire and the officer in charge decided to do an exterior attack. No one was able to confirm whether handles were in place on the second-floor windows, and there was too much fire damage to verify their presence one way or another.

Needless to say, the observations of the engineer on the first engine proved invaluable to the assigned investigator. All evidence pointed toward the homeowners being involved in setting the fire. Although an arson charge could not be established, the insurance company denied the insurance claim based on evidence of insurance fraud.

FIGURE 11-11 Arrow patterns pointing back toward the area of origin in the corner. As shown here, more than one arrow pattern may be found in any given area.
Courtesy of Russ Chandler.

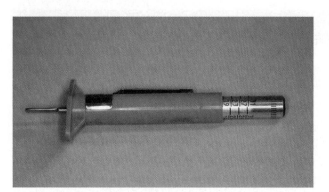

FIGURE 11-12 A depth gauge showing the nomenclature. This device is more commonly used in the automotive tire industry to measure tread depth.
Courtesy of Russ Chandler.

depth of the char can be compared from stud to stud and from wall to wall, based on the premise that the deepest char will be found at or near the area of origin. The investigator should be aware, however, that the fuel load may cause heavy charring in an area that is not the area of origin. Also, ventilation can create a situation where more char occurs because the fire burned more intensely near an opening such as a failed window or door.

The depth of the char can be measured with a simple, inexpensive device that can be found at most automotive hardware stores. A tire-depth gauge in increments of 64ths or 32nds of an inch is an ideal tool, given its measured increments and blunt blade on the end (**FIGURE 11-12**). One person should take all measurements because the blade must be pushed into the char with the same pressure each time a measurement is taken. If the studs still maintain their original

shape and size, then the tool can be placed with the face against the char on the stud; the measurement can then be read on the upper side of the device.

> **TIP**
>
> The depth of the char can be measured with a simple, inexpensive device that can be found at most automotive hardware stores.

Some studs may have experienced some partial consumption as a result of the fire, so the investigator must compensate for that missing material. The investigator should cut an unburned section of a stud, and then place it flush (even) with and parallel to the stud being measured. This will then show where there is missing char. Take the measurement from the edge of the char to the edge of the sample stud; this measurement is the missing char. Then, push the depth gauge into the char as before, and take the reading. Adding the two measurements together will give the total char. Using this same method, the investigator should continue to measure the char around the rest of the area. The resulting measurements can be charted and compared to find the area with heaviest damage (**FIGURE 11-13**).

FIGURE 11-13 Diagram showing the use of a depth char gauge. To use this gauge, hold the face of the gauge against the char, push the plunger in with an even pressure, and allow the blade to sink into soft char. Take the measurement; this is the depth of the char. Next, take an unburned 2 × 4 stud and line it up where the front of the burned stud would have been. Take a measurement to identify how much of the stud burned away and is missing. Add the two measurements together to obtain the total impact on the stud. Do the same in other areas of the room where there is char for comparison.

By comparing the measurements around the room, while taking into account any excessive fuel load and ventilation, investigators can determine variations in the depth of char and a direction of fire travel. This may point back to the area of origin.

> **TIP**
>
> Taking into account excessive fuel load and ventilation, variations in the depth of char may show the direction of fire travel. This may point back to the area of origin.

Floors

Patterns on the floor can be of great value depending on the circumstances. Before using these patterns in the fire scene analysis, the investigator must consider certain conditions. If the room experienced flashover, any previous pattern may be destroyed as the result of the massive heat when combustibles in the room reached their ignition temperature. Patterns left after flashover can be misinterpreted as something that they are not. Notably, studies have shown that post-flashover floor patterns can resemble ignitable liquid burn patterns.

> **TIP**
>
> Patterns left after flashover can be misinterpreted as something that they are not. Notably, studies have shown that post-flashover floor patterns can resemble ignitable liquid burn patterns.

Other cautions include situations involving worn wood floors in older structures such as historic buildings or old country stores. Foot traffic in the common areas, such as hallways or between aisles, can wear down those floors over the years. This may not be noticeable on a normal day, but these wear patterns expose wood fibers and create a rough surface that can char differently from areas that have not worn away. In some instances, the charring of these wear patterns may look similar to a pour pattern of an ignitable liquid.

Floor patterns are great indicators in many circumstances. In small simple fires starting near or at floor level, the burn pattern may clearly indicate the area of origin because it is the lowest burn. In other circumstances, more than one low burn point may exist. This can be the result of multiple points of origin, or the presence of one area of origin combined with other low burns from burning materials falling down, creating subsequent low burn points. By taking into account fuel load and other evidentiary clues, the investigator should be able to discern which low burn was the area of origin.

Of course, not all fires are small and simple. The vast majority leave the floor area covered in debris. When this happens, fire fighters should take steps to ensure no more damage to the structure occurs. Barricades should be established to stop anyone from walking across the floors in and around the area of origin unless absolutely necessary. If the area of origin has not been identified, then all potential areas of origin must be protected.

After the scene is documented, the investigator can then start removing debris in layers, similar to an archaeology dig, looking for evidence in each of the layers. Once the debris has been carefully removed, the investigator can document and examine any patterns on the floor. One mistake that an investigator may make is not to remove all the debris in and around the area of origin. Leaving any area unexamined will most likely lead to faulty conclusions.

> **TIP**
>
> After the scene is documented, the investigator can then start removing debris in layers, similar to an archaeology dig, looking for evidence.

Holes in the floor are always a safety concern. When debris drops onto the floor, burning embers may continue to burn as even more debris falls, covering the burning embers and protecting them. As a result of the continued combustion process of the coals, the coals may potentially burn all the way through the floor.

Holes can also burn down through the floor as the result of flame impingement from ventilation or from the underside of contents burning such as a couch or bed, creating radiation or flame impingement on the floor. Holes burned in floors do not happen often, but require explanation when they are found. Only then can the investigator make an accurate determination of the fire origin.

Holes can be created by heat and flame impingement on the underside of the floor as well. This may give an indication that the fire started on the level below the floor where the hole is located. Such a natural

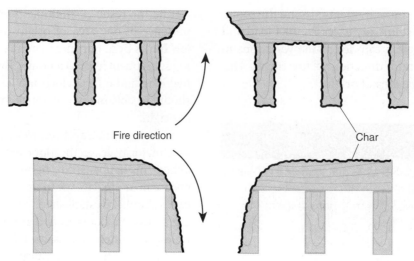

FIGURE 11-14 Drawings showing the fire damage from the fire breaching the floor from above and from below.

progression of the fire would be expected if the fuel load below the floor had enough energy to create the damage.

The direction of fire travel can usually be determined by looking at the burn pattern around the hole itself. Provided the hole was protected and was not damaged by suppression operations, indicators will signal that more fire was present on one side versus the other. When more damage occurs around the top of the hole than on the bottom, the investigator will observe a slope on the top side, like the beginning of a crater (**FIGURE 11-14**). The edges will look just like someone scooped out the upper layers of the floor, with the same slope noted on the edges. When the fire burns from below, it has the reverse effect: More damage will be visible on the underside. In this case, the slope of the sides of the hole will be toward the center of the top layer, with the underside surface being wider than the top side, just as if someone used a scoop from below.

TIP

The direction of fire travel can usually be determined by looking at the burn pattern around the hole itself.

Different types of floor finishes can create some unique situations. The following subsections discuss some of these floor finishes and what might be seen during the investigation.

Carpets

Even though a carpet may have a fire-resistive rating, it will eventually burn if exposed to enough heat

FIGURE 11-15 Burn pattern on a carpet resulting from the presence of a poured ignitable liquid.
Courtesy of Russ Chandler.

and flames. More importantly, if an ignitable liquid is present, the carpet will likely hold that liquid by absorbing it into the carpet backing or the foam padding. The carpet fibers may then act as a wick, aiding in the burning of the liquid fuel. A common indicator of the presence of an ignitable liquid is the sharp line of demarcation present on the carpet between the burned and unburned areas (**FIGURE 11-15**). These liquids may have been poured on the floor intentionally to create a trailer, allowing the fire to travel from one point in the structure to another.

However, ignitable liquids are not the only thing that can cause a sharp line of demarcation on carpets. In particular, ordinary combustibles can cause a line of demarcation on carpet and other floor finishings. This line of demarcation usually outlines a concentrated area, rather than forming a pour pattern leading out of a room or area. If there is any question about how the line of demarcation formed, the investigator should take samples and submit them to a laboratory for analysis. A positive result for ignitable liquids still requires the investigator to determine whether the fire was caused by an accident or was intentionally set. Further, a negative result for ignitable liquids from the laboratory could mean either that there was not enough ignitable liquid in the sample or there was no ignitable liquid in the first place.

Wood Flooring

As with all floor surfaces, charring or burning of wood flooring indicates a low burn, which can be associated with the area of origin or drop-down fire. The duration of the burn and the condition of the flooring and its finish can influence the type of pattern seen. A small amount of an ignitable liquid with a relatively short burn time on a sealed hardwood floor may show wisping and sooting. By contrast, a similar floor without the tight finish, where minute cracks actually occur between the boards, will allow the liquid to seep into these cracks and then burn as the vapors are released. This, in time, could create rounded charred edges on the individual boards.

Concrete

Concrete flooring is commonly found in basements and garages. However, in warmer climates where frost is not a problem, it is not unusual to have homes built on concrete pads. Concrete that has any foot traffic usually is sealed: The sealant prevents chemicals and other liquids from seeping into the concrete. Even if a sealant is applied to the concrete, it can eventually be worn away by high traffic.

The sealant does not play a role in the investigation other than assuring the investigator that any sample taken for laboratory examination has not been contaminated. Under most circumstances, the investigator will want to take a comparison sample of the concrete from an area that is known to have not been contaminated.

Spalling of the concrete floor is not an indication of anything other than the fact that there was moisture in the floor and the moisture was heated to 212°F (**FIGURE 11-16**). The investigator might question why the concrete reached that temperature in one place and not another. Such variations can result from the burning of the underside of a vehicle located in a garage where the heat may be reflected back down to the floor. The burning of the sides or underside of furniture can raise the temperature at floor level, as can the

Rushing Out of the House

The engine company arrived at the residence of a reported fire to find no smoke but a burn pattern, quite narrow, running the length of the carpet that led from the center of the house to the back door. The doors were open and it was winter. Checking with a neighbor, they were told that another neighbor took the homeowner to the hospital for burns.

The investigator was dispatched automatically and arrived just minutes after the engine company lieutenant had talked to the neighbor. Looking at the pattern together, they found it went from the door to the kitchen to the back door, on the carpet only. In the backyard, they found a clue to the mystery: a large 3-in.-deep cast-iron frying pan sitting in the yard with a burned potholder. They also found one burned slipper.

The investigator interviewed the homeowner at the hospital. She had gone to the kitchen that morning and planned to do some deep-fat frying with the iron skillet. She got distracted by a phone call until she heard the smoke alarm. Upon hanging up, she found that the grease in the pan, about 2 in. deep, had ignited. She grabbed the pan, and holding it behind her with just one hand (it was a large pan), she ran out of the house. When she got outside, the fire in the pan was out—but the carpet in the hall was burning where she left a trail of grease from the kitchen to the back door. She ran to a neighbor's house and reported the fire. After calling the fire department, the neighbor took the homeowner to the hospital for burns to her left foot and right hand. When she went to the neighbor's house, the grease in the carpet was still burning. Fortunately, by the time the engine company arrived, the fire had burned out.

Although there was a sharp line of demarcation and an ignitable liquid, this fire was accidental in nature.

FIGURE 11-16 Typical spalling on concrete as the result of moisture in the concrete expanding as it turns to steam, forcing the surface of the concrete to separate from the rest of the floor.
Courtesy of Russ Chandler.

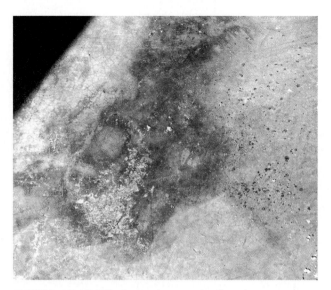

FIGURE 11-17 Wisping pattern on the floor from an ignitable liquid.
Courtesy of Russ Chandler.

presence of ignitable solids and liquids on the concrete floor itself. Unburned items on top of the concrete protect the floor, but if they ignite, they may be the source of heat that causes the spalling to occur.

> **TIP**
>
> Spalling of the concrete floor is not an indication of anything other than the fact that there was moisture in the floor and the moisture was heated to 212°F.

If an ignitable liquid was poured on the floor as a trailer, that pattern may be visible (**FIGURE 11-17**). In the process of removing the debris to examine the floor, the investigator may have to wash the floor. In such a case, concern might arise that any residue of an ignitable liquid on the floor will be washed away as well. As a precaution, while removing debris and after the debris is removed, the investigator should use a hydrocarbon detector to see whether there is any measurable sample in the area. Use of a K-9 accelerant dog is ideal in this scenario because the dog has the ability to cover a large area in a shorter period of time and can detect residue in smaller amounts than can an electronic detector. Using detectors can ensure that the investigator will get the best sample possible to send to the laboratory for analysis of the presence of a liquid accelerant.

Sometimes, in such cases as in flooded basements, it may not be an easy task to remove murky standing

FIGURE 11-18 Pouring clean water onto shallow, murky standing water on concrete can sometimes allow the investigator to see the pattern on the concrete surface.
Courtesy of Russ Chandler.

water. Depending on the depth of the standing water, clean water may be poured on the area to temporarily force away the murky water long enough for the quick snap of a photograph (**FIGURE 11-18**). This allows the investigator to see any patterns present on the concrete surface.

Protected Areas

When the fire burns and patterns from fire damage are left on the floor, investigators should look for some other things as well—such as a lack of fire damage. As the floor is cleaned off during the examination of debris, the investigator may notice areas that have been protected from the fire and that are relatively or completely undamaged. **FIGURE 11-19** shows that books strewn on the floor protected that

FIGURE 11-19 Protected areas where school books had been left on the floor.
Courtesy of Russ Chandler.

part of the floor from fire damage. These protected areas will show the proper placement of furnishings and may help the investigator tell the story of the fire. In the case shown in **FIGURE 11-19**, the investigator might wonder why several books were on the living room floor. In fact, a high school student was doing her homework on the floor when the fire occurred, and she left her books where they were as she escaped from the structure.

> **TIP**
>
> Protected areas will show the proper placement of furnishings.

Anything sitting on the floor will most likely leave an indication of its placement as the fire damages the floor surface. During firefighting and the overhaul phase of the event, some of these furnishings may be moved or even removed from the structure. The protected areas can help the investigator place the contents back where they were at the time of the fire. This will provide additional information and additional patterns to be observed.

Protected areas may also occur in places other than the floor. As the heat moves through a structure, furnishings can block and sometimes limit the amount of heat absorbed in various areas. The back of a solid chair placed near a wall can somewhat protect the wall behind the chair, creating a shadow. This imprint on the wall, should it survive the fire, shows the placement of the chair. Even photos hanging on walls can create a protected area on the wall. When the hanging picture falls from heat damage, the shadow it created on the wall may survive. The shadow can be the result of lesser heat damage or limited collection of soot in that specific area.

Patterns on Furnishings

With the floor cleaned off, the investigator will see the protected areas. By placing the furnishings back where they belong, or if they are already in place, the investigator can verify their location based on these indicators. With the surviving contents of the structure in place, the overall patterns from the ceiling, walls, floors, and contents can be compared. Sometimes the patterns on the furnishings can provide a much better picture of the direction of fire travel. If nothing else, they may confirm or refute the patterns documented on ceilings, walls, and floors.

The most damaged corners or sides of the furnishings will be noted on the side from where the fire came. For example, if the right side of a chair is heavily damaged but the left side of the chair has only moderate damage, then the fire came from the right and traveled to the left. The furniture may also be in the area of origin and can even be located within the V pattern.

At other times, the role of the furnishings in the fire may not be so obvious. Perhaps only minor differences may be found in the pattern, such as on a wood chair. The chair may look blackened by the fire but is still intact. Looking at the squared edges, the investigator may find the edges on the right are slightly rounded, whereas the edges are still sharp and pristine on the left side of the same piece of wood (**FIGURE 11-20**). In such cases, the pattern shows the fire moving from right to left (**FIGURE 11-21**).

Melted objects also give an indication of the amount of heat present in a particular area. Aluminum objects will melt at temperatures of 1100°F to 1400°F, and some other pot metals melt at even lower temperatures. If the drywall has failed, the wiring in the walls can even give an indication of the temperature. Where both aluminum and copper wires

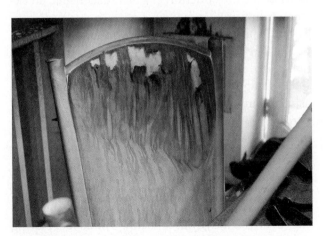

FIGURE 11-20 Light damage on a kitchen chair showing the direction of fire travel is from the right to the left.
Courtesy of Russ Chandler.

FIGURE 11-21 Diagram showing the rounded edges of the arms of a stuffed chair along with damage to the back of the chair indicating the direction of fire travel.

travel in walls, the investigator may find the aluminum wire melted but the copper wire intact. This indicates that the temperature was at least 1100°F but less than 1981°F, the average melting temperature of copper. By knowing these melting points, the investigator can identify the potential highest temperatures at any given point. This can also show the direction of fire travel. The temperature in an average house fire does not exceed 1600°F. When copper is found melted, it could signal a localized electrical short or the introduction of a fuel that created an extremely hot fire.

Steel springs today are most commonly found in bed mattresses, although they are not present in all mattresses. In homes with older furnishings, steel coil springs might be found in large stuffed chairs as well as in mattresses. These steel springs have been tempered to give them added strength, allowing them to spring back to their original shape. When heated excessively, coil springs lose their temper and collapse on themselves, flattening out—a process called **annealing**. However, steel springs also flatten out from constant use, such as the sagging in the center that occurs in an old heavily used mattress or couch. During

a fire situation, the springs will flatten out more prominently if weight is present on the springs, such as a basket of clothes on the bed or even a body.

Occasionally, annealing can give an indication of the direction of fire travel. If only one portion of the bedsprings is annealed, that may indicate how the fire moved through the room. For example, if the foot of the bed on the east side has collapsed, the fire may have come from the east and moved toward the west: The heat may have had a longer impact on the east corner of the bed than on the rest of the bed. Of course, as the heat builds up in the room, the entire mattress may collapse as a result of annealing. When this happens, it may eliminate the ability to determine fire direction from this one piece of evidence.

The loss of the mattress coverings can help the investigator as well. By the time the entire mattress anneals, the coverings most likely will have been consumed. Just one corner of a mattress being consumed constitutes a pattern on its own and may indicate the direction of fire travel. Annealing by itself should not be considered the defining indicator, but rather serves as one more fact to add to the culmination of evidence that leads to the final analysis.

Wrap-Up

SUMMARY

- Finding and recognizing burn patterns can greatly help the individual doing the cause and origin investigation. These tasks are best accomplished by working from the area with the least amount of burn patterns to the area with the most burn patterns. The char on the walls and ceilings will point back to where there is more char, eventually leading the investigator to the place where the fire started. This chapter has provided a great deal of information on burn patterns at the floor level, but the investigator must remember that the fire can just as easily start on or in walls, on countertops, and in ceiling areas.

- Along the path to the area of origin, the investigator may encounter heavy char and conflicting patterns. This area may be where the fire burned more intensely as a result of a heavy fuel load or a fuel that had a higher heat-release rate than the fuel in the area of origin.

- Ventilation created by the fire, such as the breaking of a window, may have allowed the fire to burn more intensely in that area, creating a deeper char at the ventilation source than in the area of origin. Both natural ventilation and ventilation created by fire fighters may provide air currents that could potentially change the direction of the fire travel. Working past these other patterns, and recognizing them as secondary issues, the investigator should be able to continue working back toward the area of origin.

- If the fire went to flashover, the investigator should exercise caution when interpreting burn patterns, especially those on the floor. In these situations, the final floor patterns often resemble ignitable liquid patterns, even when no ignitable liquids were present.

- Patterns that look like they indicate an intentionally set fire must be thoroughly examined to ensure that this is the case. The presence of an ignitable liquid, taken by itself, does not absolutely mean the fire was intentionally set. A sharp line of demarcation may be observed, but some ordinary combustibles can create similar patterns.

- Other indicators that can help the investigator work back toward the area of origin include the patterns left on the furnishings. Protected areas on the floor can help ensure that the furnishings are properly placed where they were at the time of the fire. The char on furnishings, such as a chair, may show more damage on one side, indicating the direction from which the fire traveled. The melting of certain metals, such as aluminum, will show the temperature that a specific area reached during the fire. Comparing this temperature with that experienced by other areas may help lead back to the area of origin.

- This chapter reinforces a most important tenet of fire investigation: No single finding can conclusively solve the riddle of how the fire started. The culmination of all the facts gathered as the result of a systematic analysis of the entire scene will eventually lead to the successful final determination or hypothesis of the fire origin and cause.

KEY TERMS

Annealing The collapse of coil springs, loss of temper, as the result of heat.

Arrow patterns As described in the NFPA 921, the patterns on wooden structural members that show a direction or path of fire travel.

Awning window A window designed with a hinge at the top that opens outward at the bottom with the use of a hand crank; some designs use a motor to open and close the window. Units usually still have the hand crank in case the motor fails.

Calcination The process of driving off the moisture in the gypsum (including the moisture chemically bonded within the gypsum), discoloring and softening the gypsum in the process.

Casement window A window hinged in the frame from one side, similar to the awning window; it operates with the use of a crank.

Craze (crazing) Multiple small, close fractures in glass created from rapid cooling; the opposite of fractures caused by rapid heating. (Contrast with tempered glass, which is designed to break into small fragments to limit injuries should it be accidently broken.)

Demarcation A distinct line between a fire-damaged area and a nondamaged area; not a gradual charring but an abrupt edge.

Double-hung window Two separate sashes installed in a window frame, with both the top and bottom sashes able to independently slide vertically up and down.

Drywall Also called sheetrock; a wallboard of gypsum with a coating of paper on each side.

Jalousie window A window in which glass, plastic, acrylic, or wood louvred panes slightly overlap and join, with a track operating each of the slats in unison; the panes tilt outward from the bottom edge with the use of a crank handle.

Oxidize The result of heating a metal surface; the heat consumes any covering, resulting in the rusting (oxidation) of the metal surface.

Single-hung window A window with two separate sashes installed in a window frame, with the top sash mounted as a stationary fixture and the bottom sash installed to slide vertically up and down.

Spalling A condition in which the surface of concrete pops off as a result of water in the concrete reaching boiling temperature and turning to steam. When this happens, the steam expands 1700 times in volume, creating the energy required to pop the concrete off from the surface.

Trailer The use of flammable liquids or easily ignitable combustibles (e.g., paper, cardboard, ropes) to enable a fire to travel from one area to another or to reach a specific target. Combustibles can be used in conjunction with the flammable liquids. Trailers can be both outside as well as in a structure.

REVIEW QUESTIONS

1. Define *burn pattern* as it relates to the investigation of fire origin and cause.

2. Describe two types of events that can affect the fire pattern and possibly give false leads as to the direction of fire travel.

3. What is a cone pattern?

4. Describe drywall and calcination.

5. What is the natural flow of fire and associated hot air and gases?

6. Define *spalling*.

7. Define *annealing*.

8. What is an arrow pattern?

9. What type of evidence might be found in relation to glass both inside and outside a fire-damaged structure?

10. What simple tool can be used to measure the depth of char?

DISCUSSION QUESTIONS

1. Which new household products on the market today could affect the growth and development of a fire, possibly skewing burn patterns?

2. At the scene of a recent fire, burn patterns indicate that an ignitable liquid may have been involved. Lab results will not be available for 2 weeks. Should the investigator wait until the results are back from the lab before interviewing the occupants?

CHAPTER 12

Examining the Scene and Finding the Origin

LEARNING OBJECTIVES

Upon completion of this chapter, you should be able to:

- Apply what has been learned in previous chapters to see how it all comes together to complete a fire scene examination to search for the fire origin.
- Define the role of the fire department and fire officer in the investigative assignment.
- Define the scientific method and its importance in the investigation.
- Describe safety issues the investigator may encounter in the investigation.
- Describe how to secure the scene during the investigation.
- Describe what the investigator can do during the operations phase of the fire scene.
- Describe what to look for and what may be found during the exterior examination.
- Describe the process and what is to be examined during the interior examination to include examining building systems and contents of the structure.
- Describe making the final step to determine the area of origin.
- Describe what to look for on the exterior of the structure that would indicate the area of origin of a fire.
- Describe what to look for when examining the interior of a structure that will lead you to find the area of origin and point of origin.

Case Study

One Saturday morning, Engine 1 responded to a report of a structure fire. Upon arrival, they found a mother and two small children in the front yard. The mom immediately advised the crew that there was no one in the structure. The officer asked again to make sure that no individuals or pets were inside. Moderate smoke was coming from the open front door. While the crew set up their attack line, the lieutenant did a quick walk around the outside of the structure. He noticed the window on side B (the left side of the rectangular-shaped structure) had smoke coming around the frame of the window. When he arrived back at the front, the crew had the attack line at the front door charged and ready.

Making an entry into the living area and working clockwise, they entered the first door to the left, which was the room where the lieutenant has seen smoke coming from around the window during his size up. Heavy smoke was coming from around a closed door in that room to what appeared to be a closet. Opening that door, the crew knocked down the flames, and after a quick tour around the rest of the house they found that there was no other fire and no other occupants.

While his crew conducted ventilation and overhaul, the lieutenant switched from his suppression role to the task of being the first responder investigator. Working from the least damaged area to the most damaged area brought him back to the closet in the first bedroom. All of the contents had been burnt, and all of the clothing that was hanging on the rod was now on the floor of the closet.

Searching the closet and sifting through the debris, the investigator could not find a heat source. Clearly, this was the area of origin. This was the area with the most damage; it was the area with the lowest burn, and the fire had moved upward and outward from this point. Looking on the floor about 2 ft. from the closet door, he found a disposable cigarette lighter. This was a child's room, with toys and bunkbeds, not a place where one was likely to find smoking materials.

Approaching the mom in the front yard, he asked her about what had happened that morning. The mom explained that she could hear her two sons playing in the other rooms while she was in the kitchen. She smelled smoke, the smoke alarm activated, and when she stepped into the living room, she saw smoke coming from the bedroom door. She grabbed the kids and left the house yelling for her neighbor to call the fire department. The lieutenant praised her for her appropriate actions and asked if he could talk to her two boys, and she gave her consent.

Taking a lesson he had learned from watching their fire marshal at previous fires, the lieutenant crouched so he was eye level with the kids, and talked to them in a calm and friendly voice. The oldest was 9 years old and was quite animated; he explained that he was playing with his matchbox cars on the living room floor when his brother came running out of the bedroom and sat on the couch, just moments before the smoke alarm went off. The 8-year-old started inching behind his mother while this conversation was happening.

The lieutenant asked the 8-year-old his name and asked him if he wanted to be a fire fighter one day. Coming out from behind his mom's legs, the young man said "yes." Showing this boy his helmet, the lieutenant told him if he wanted to be a fire fighter one day, he could never lie to his mom or to a fellow fire fighter. The boy said he knew that. Then, the lieutenant asked the younger boy if he knew what happened in the bedroom.

Hesitantly at first, the boy started to explain that he was using a lighter he found on the coffee table as a toy car. He said that as he pushed it across the table he thought it was cool that it made sparks. Finally, the boy said that he went into the closet and closed the door so he could see the sparks in a dark room. He said there were even more sparks but then the plastic on some of the clothes caught on fire. When asked what he did next, he hung his head and said he ran out of the closet, closed the door, and ran to the couch. He admitted that he was too scared to tell his mom. The lieutenant praised the boy for his truthfulness and took the opportunity to make this a teachable moment for both boys and the mom.

The mom confirmed that she had picked up her dry cleaning the previous day and hung them, still enclosed in the thin plastic, in the boy's closet. She also confirmed that she was a smoker and there were several lighters in the house, but they were usually left up high, out of the boys' reach. The mom promised that she would certainly be more cautious in the future.

Introduction

In this chapter, you will take all of the knowledge you have obtained from previous chapters and apply it to the physical examination of the scene in search of the area of origin. Although the terms *origin* and *cause* are frequently used together, one cannot find the cause of the fire until they have located the area of origin. To find the area of origin, investigators must work their way through the fire scene interpreting the burn patterns. By doing this properly, these patterns usually lead back to the area where the fire originated. There is also the term point of origin. This is a more precise description of where the heat source and first material ignited. It points to a specific, precise location. If the investigator can identify the point of origin, then they need to identify this in their report. For the sake of clarity, in this text, when mentioning area of origin, it should be understood that that there is a point of origin as well.

TIP

Although we frequently use the terms *origin* and *cause* together, the investigator cannot find the cause of the fire until they have located the area of origin.

Once again, neither one text nor one class could ever make someone a fire investigator. To attain the level of competence necessary to make accurate determinations of the area of origin, the investigator must have a strong knowledge of building construction. The investigator must also have experience with the natural progression of a fire and know how that process can be altered by fuel load, heat release rate of materials, ventilation, and suppression activities.

This chapter concentrates on the activity of the assigned investigator to show the complexity of the scene and to take into account all issues dealing with making the final determination of the fire area of origin. This information will aid the first responder investigator by helping them understand all roles in the fire investigation and will aid them with any scene they may work in the future.

The Assignment

As discussed in Chapter 3, "Role of First Responders," there must be a comprehensive policy and procedure on how fires are to be investigated by the locality.

National standards indicate that the preliminary fire origin and cause determination is the responsibility of the fire department. The National Fire Incident Reporting System (NFIRS) has an entire section on the origin and cause of the fire to be filled out by the fire officer in charge (OIC) or the designated person doing the report. Although not all fire departments use NFIRS, it is hoped that someday every fire department in the United States will participate in this process. A common phrase is that "we fight fire with facts"; the facts come from the NFIRS data. The more participants, the better the data. The better the data, the more we can accomplish.

TIP

National standards indicate that the preliminary fire origin and cause determination is the responsibility of the fire department.

TIP

A common phrase is that "we fight fire with facts"; the facts come from the NFIRS data. The more participants, the better the data. The better the data, the more we can accomplish.

Even if the local department does not use the NFIRS, it should create some type of report. When a citizen loses a home or even a vehicle as a result of fire, the insurance company usually requires a fire report from the locality or the responding fire department. Without this report, there could be a delay in handling the claim and in turn a delay in the victims getting their lives back together.

To ensure consistency and accuracy, the department must decide what role it will take in the determination of the fire origin and cause. Under most circumstances, for the average single-alarm fire, the OIC should identify the area of origin and write the report. Even for larger events where an assigned investigator may be automatically dispatched, the OIC of the first engine arriving may do the preliminary report and leave the details on the area of origin to be updated in the NFIRS by the assigned investigator.

Remember, the term *fire investigator* refers to the person who examines the scene to determine the origin and the cause of the fire. This may be a fire fighter, fire officer, or the assigned fire investigator

Statistics from the NFIRS are a valuable tool that can be used by the locality in multiple ways. When focusing on the need for personnel or apparatus, statistics on one's own department comparing with other similar department can help support an increase in budget when requesting. Providing this data when requesting increased funds from the AHJ may be beneficial. When deciding on next apparatus purchase, statistics can provide insight as to what type of apparatus may be needed such as a 30 percent increase in medical calls would justify more ambulances.

This data is critical for research and other fire studies. In the past, the annual statistics on how many fires occurred in the United States was based on a calculation on what was believed to be happening. These statistics were based on taking a known cross section of the country where fires were accurately reported and make an extrapolation to come up with the number of fires, fire deaths, injuries, etc. The NFIRS is a tool to improve the accuracy of those numbers. As more jurisdictions submit their reports, we will be able to see a better picture of what is actually happening.

FIGURE 12-1 The OIC with an accountability board, tracking the location of all personnel working at the emergency scene.
Courtesy of Bill Larkin.

the OIC's knowledge and consent. (See **FIGURE 12-1**.) This is predominantly a safety issue and therefore must come before all else.

such as a fire marshal or police detective. The assigned investigator usually does the more detailed investigation. An investigator may be assigned to an incident as the result of an automatic dispatch because of the size of the incident or because the engine company officer needs assistance finding the area or origin or cause.

> **TIP**
>
> The term *fire investigator* refers to the person who examines the scene to determine the origin and the cause of the fire. This may be a fire fighter, fire officer, or the assigned fire investigator such as a fire marshal.

When the assigned investigator arrives on the scene, regardless of rank or responsibility, the OIC of the fire is still the leading official until the fire is placed under control. The investigator must report to the OIC to advise of his or her arrival and planned activities. This is all explained in the National Fire Protection Association standards for proper incident command. If the investigator enters the structure before the fire is placed under control, it must be with

The Scientific Method: A Process

Whether the primary investigator is a fire officer looking at the scene after it is brought under control or an assigned investigator, the approach in finding the origin of the fire must be systematic and consistent. It must follow a logical path that is conducive to identifying the fire area of origin. By using one methodical approach on most, if not all, fire scenes, the investigator can prevent overlooking key indicators and evidence. Above all else, ***the investigator must be a seeker of truth***. The best way to accomplish this is using scientific methodology to develop a hypothesis as to the area of origin.

> **TIP**
>
> By using one methodical approach on most, if not all, fire scenes, the investigator can prevent overlooking key indicators and evidence.

> **TIP**
>
> Above all else, the investigator must be a seeker of truth.

A common approach is working from the least damaged area to the most damaged area. By doing so, the investigator should, under most circumstances, be working from the extension of the fire to the area where the fire originated. Smoke patterns and burn patterns can be the path to the area of origin and are indicators that assist the investigator. The process of searching the scene should be systematic and concise. This pattern of search should be used in each and every scene. This provides a methodology that becomes consistent.

Each fire is unique; in other words, no two fires will ever be the same. This is because of outside influences such as fuel load, ventilation from suppression activities, or from the fire burning through to the exterior of the structure. Regardless, there should be consistency in the burn patterns that will allow the investigator to follow the path back to the area or origin. But before they get there, they must consider what must be done first.

TIP

Each fire is unique; in other words, no two fires will ever be the same.

Safety Issues

As the fire is being knocked down, even before the last ember is extinguished, the investigator should be thinking about the first and primary concern: safety. Everything from the condition of the structural elements to the presence of unsafe electrical lines or even exposure to chemicals must be considered. After the OIC leaves the scene, it is the fire investigator's responsibility to ensure that the scene is safe not only for those who may be investigating but also for the curious public as well.

Structural Stability

As an element of assuring safety, the structural stability is a major concern. Knowledge of proper building construction is essential. If there are areas that appear unstable, then actions must be taken to render those areas safe before entry. If there is any question as to the safety of the structure, the local building official can be contacted to ask for assistance in identifying structural integrity concerns.

Elements That Pose a Hazard

Other resources can possibly assist as well. Larger departments may have *technical rescue teams* who have the necessary tools and supplies to shore up walls and overheads. If the insurance company representative has arrived, he or she too has an interest in determining the area of origin and possibly even doing an inventory. The insurance representative may be willing to provide resources in the form of contractors and supplies. Most insurance companies recognize that leaving an unstable structure is a liability and will want to assist.

TIP

A key concern is to make sure the investigator stays in control of the scene and guards it against contamination from equipment and supplies. The lead investigator should also have the responsibility for the safety of everyone on the scene while the investigation is conducted.

Sometimes damage may be so severe that the only recourse is to demolish those structural elements that pose a hazard. If this is the case, a crane or other heavy equipment can be called in to lift damaged items away from the structure. However, efforts must be taken to make sure that the heavy equipment does not contaminate the scene should it be necessary to take debris samples for ignitable liquids. Steps can be taken such as scrubbing all components of the heavy equipment that will be reaching into the scene, and checking it with a hydrocarbon detector to assure there are no contaminates. The investigator should also check all hydraulic and any fluid connections to assure no leakage.

Every step must be taken to prevent the use of any gasoline-powered equipment such as saws, fans, or generators on fire scene itself. This is essential to assure there will be no contamination of the scene putting any recovered evidence as risk of being denied as evidence in a trial. During the investigation, no gasoline-powered saws should be used, the investigator should have electric or battery-powered units. Even these electrically powered tools should be cleaned and checked to assure no hydrocarbons are on the blades.

If generators must be used, every effort must be taken to use them away from the physical scene, using longer electrical cords for power supply. There are other tools such as fans that may be gasoline-powered. If such units must be used, they should be as remote from the area to be investigated. Any gasoline-powered equipment, if used, must be free of any leaks. If they are to be refueled, this must be taken to an area remote from the fire investigative area. Once refueled, they should be cleaned to be sure that any spilled fuel is removed before being taken back on to the scene.

Utility Hazards

The first utility that is of most concern is the presence of electricity. The department or the electrical utility company may have already disconnected the meter. This is no assurance that all electrical sources have been eliminated. Some investigators have found extension cords running from neighbors' homes for one reason or another, legally or illegally. With today's technology, battery systems and solar units can still provide quite a jolt if encountered without protection. In addition, emergency generators from a nearby house or commercial property might inadvertently energize power lines that otherwise would be thought to be de-energized.

Every investigator should know how to use an electrical multi-tester properly. They are quite inexpensive and easy to use. These devices can determine whether there is energy on any given electrical circuit; if there is not, then the wires should be safe to examine. Testers generally have two probes, one going to the wire or post and the other to a ground. They then indicate the amount of voltage present. There are some devices on the market labeled as noncontact voltage detectors, as shown in **FIGURE 12-2**, that activate when the plastic tip is placed near an energized line. Of course, the investigator should always test all equipment before relying on the readings.

FIGURE 12-2 A noncontact detector that will activate in the presence of an electrical current.
Courtesy of Russ Chandler.

On occasion, there are overhead power lines that could create a hazard, even if they are undamaged. The electricity stopped at the meter base may still be supplied by the power company through an overhead line from the power pole to the structure. Every effort should be made to ensure that it is safe and that there will not be a need to get anywhere near live lines. If investigators may need to get near this line or if the line suffered any heat damage, the power company should be contacted to have the line de-energized.

The investigator should look to see whether the structure had gas lines. Merely checking records or asking neighbors if public gas lines are available is not enough. Looking in the back or side yard is also insufficient because some liquefied petroleum gas tanks are buried; if a tank is buried, there is a control head cover that could be visible if not covered in rubble. A close examination of the scene should help to ensure that the hazard is nonexistent or, if gas lines are found, that the supply is shut off at the source. The atmosphere should be checked with a multi-gas detector for combustible gases, oxygen, and carbon monoxide.

Even water lines can be a hazard. If the structure is a multistory building, leaking water lines in the upper areas could add additional weight to lower floors, weakening the structure. Water continually leaking into a basement can create problems by hiding other hazards under the water or potentially drowning someone if that person was to fall into the water-filled basement. Depending on the weather, water can create additional safety problems with icing, adding weight to the structure and creating slipping hazards.

Water weighs approximately 8.34 lbs/gallon. During fire suppression, hundreds to thousands of gallons of water can be introduced into a structure within a few minutes of the initial fire attack. Furniture, drapes, clothing, carpet and carpet pad serve as a sponge collectively adding additional weight to a structure thus compromising the structural integrity of the building. If the structure was supplied with well water, then cutting the electricity should solve the problem, unless there is a secondary source of electricity for the well pump. If the water comes from a public water source, the flow can be cut off at the street by the fire department or local water authority.

> **TIP**
>
> Water can create additional safety problems with icing, adding weight to the structure and creating slipping hazards.

If water is collecting on the floor, steps should be taken to allow it to drain off the floor and away from the structure. Water that collects in the basement may need to be pumped out should the basement need to be examined by the investigator. Some fire departments have portable pumps that can assist in this process.

Secure the Scene

During the suppression operations, efforts should have been made to cordon off the area for the safety of the general public. When the assigned investigator arrives on the scene, one of his or her first duties is to check the fire line and expand it, if necessary, to cover the entire fire scene. Depending on the situation, the investigator may want to find personnel from the fire department or police department to man the lines to keep everyone out of the fire area. If the fire officer is to be the investigator, then as soon as the fire is under control the fire officer should check the fire line or assign this duty to one of the fire fighters.

Security must be maintained throughout the investigation. Arrangements must be made for someone to monitor the exterior while the investigator is inside the structure. Depending on the laws of the locality, not even the building owners or tenants should be allowed to enter the scene unescorted until at least the preliminary scene examination is complete. Efforts

should be taken to see if law enforcement personnel may be available to provide scene security. Many times, an engine company may stay on the scene to render assistance and provide scene security.

> **TIP**
>
> Depending on the laws of the locality, not even the building owners or tenants should be allowed to enter the scene unescorted until at least the preliminary scene examination is complete.

Examining the Scene During Suppression Operations

For smaller events where the fire officer takes the role of investigator, not much will be done upon arrival on the scene. Depending on the situation, the fire officer, having set plans in motion such as giving the crew instructions, may then have a chance to look around and consider the area of origin. If this is an active structure fire, conducting the investigation is not a primary function at that time. The most important task at hand is safety, not just of the crew working at the scene, but of the officer as well. Wandering off alone is not a wise course of action. The fire scene examination as a whole can take place (during and after the fire is extinguished); the fire investigator should work with the incident commander to determine when fire suppression operations will transition to the fire investigation).

During suppression operations, there will be observations by both fire fighters and the fire officer that should be noted and passed on as soon as the fire is under control. Items to consider that may lead to suspicion of a possible arson would include: forced doors found upon arrival, windows open on a cold night, or flames coming from two or more unconnected parts of the structure suggesting that there are multiple points of origin. In addition, personal items or family heirlooms that seem to be missing may be an indicator of suspicious activity.

Photographs

Under most circumstances, the engine company officer will not have the equipment or the opportunity to take photos of the scene until the exigencies of the situation are over. Even then, most engine companies do not have sophisticated camera equipment.

However, today's technology does offer the ability to get photographs and videos.

Helmet cameras are recording the initial attack as well as all other activities at the scene. Most fire fighters also have their personal cells which they can use to capture photos of the scene. Law enforcement personnel arriving and working the scene also may have body cameras that may provide videos that could be of use. Both fire and law enforcement vehicles may have installed dash cameras that can also capture additional images.

The assigned fire investigator will most likely have digital 35-mm cameras, which have a full range of photographic features such as adjustable flash, wide angle, extreme zoom, and many other capabilities. They can also easily download their photos onto their computer, sometimes even on the scene. Other technology to be used in additional to the 35-mm digital cameras are the 360-degree cameras. These units are a phenomenal tool. Once the scan is downloaded, stitched, and processed, it can allow anyone to scroll through the fire scene with their computer being able to scroll from wall to wall and from ceiling to floor.

The vast majority of the population will have photo capability on their phones. At larger fires, it would be unusual for nobody to take pictures of the incident, except in the late night or early morning hours or in a rural setting. Fire service personnel may well have their phones with them. If they have an assignment, such as an engineer, they may have the opportunity to take some photos that may be of value to the fire investigator.

For larger events when an assigned investigator is dispatched, upon arrival the investigator will report to the OIC and provide their **accountability tag**, letting OIC know they will work the exterior first. After checking the security of the scene, the investigator can start making observations about the fire and suppression activities. What is the status of the fire? Where are there signs that the fire was knocked down? Tracing the fire hose may be a way of to see where the attack on the fire began, which can come in handy when interviewing the firefighters about the scene later in the investigation. During this observation phase, the investigator should take as many photographs of the overall scene as possible. Photographs should include the structure, surrounding area, and the crowd.

TIP

The investigator should take as many photographs of the overall scene as possible.

The investigator may find it beneficial to go into the structure even before the fire has been completely extinguished. This should be done only if the investigator is completely qualified to work in an **imminent danger to life and health (IDLH)** atmosphere with full turnout gear and self-contained breathing apparatus (SCBA). Once all equipment is donned and a second check is made that all is ready, the investigator should then return to the OIC to ascertain where he or she can or cannot work at that time. If there are areas that can be examined with relative safety, then the OIC will place the accountability tag appropriately and assign the investigator to that sector of the structure. If personnel are available, it is always best to work in pairs, but this is not always possible or practical at this stage of the emergency event. One important rule that applies during the entire time while on the scene is that the investigator must know at least two paths of escape at any given time. Everyone on the scene must be aware of their surroundings so that in case one avenue of escape is blocked, there is always the second path to get out safely.

TIP

If personnel are available, it is always best to work in pairs.

When Not to Take Photos

A word of caution: Many departments have banned their fire fighters from using cell phones on the scene. If the department does not have this policy, it is extremely unwise for fire fighters to take photos of victims for their own use. Even more inappropriate would be for them to share or post such photos. Unfortunately, fire personnel have posted photos of crash victims on the Internet; it is unacceptable for a victim's loved ones see them in such devastating peril or (worse yet) if a victim did not survive. Each department should have a photo policy to prevent this from ever happening again. Many photos taken on the scene can be of value to the investigator, or to prevention personnel, but only generic photos or those used with victim permission should be shared or distributed, even for educational purposes.

It must be understood that if the OIC feels that it is not safe for the investigator to enter, it is prudent for the investigator to comply. In many jurisdictions, the OIC and the investigator in charge have equal and concurrent jurisdiction at any emergency scene. Regardless, safety must come first, and until the fire is under control, the OIC should be the lead in all safety rulings.

If there are sufficient personnel, and with the approval of the OIC, the investigator and a partner may be able to go anywhere in the structure, provided they do not hamper suppression operations. With this ability, the investigator may see firsthand the fire's progress and its extinguishment. Observations may be limited in areas because of smoke and steam, but as the atmosphere clears, various patterns may be visible. This may enable the investigator to identify sections or areas of the structure to protect and where overhaul should be held off until the area can be properly documented.

As with any accountability system, should the investigator change from one sector of the fire scene to another, the OIC should be notified, usually in person, so that all personnel on the scene can be accounted for at any given time. The investigator should be equipped with a radio for personal safety and to stay appraised of any changing condition.

Exterior Examination

Once the fire is under control, the overall scene can be examined in earnest and in its entirety. If it is still dark, the investigator may choose to wait until sunrise to begin the investigation. This may be for the sake of safety or to ensure that no clues or burn patterns are overlooked. Regardless, the Supreme Court has decided (*Michigan v. Tyler*) that under most circumstances this may be acceptable. If there is a delay in the scene examination, the scene must be protected to ensure there is no alteration or damage to potential evidence. To learn more about *Michigan v. Tyler*, see Chapter 3.

> **TIP**
>
> If there is a delay in the scene examination, the scene must be protected to ensure there is no alteration or damage to potential evidence.

The patterns on the outside of the structure can tell a story in themselves. Documenting the patterns is essential. Photographs can show burn patterns that indicate wind direction or more damage in one location than another. Later in the investigation, it will be necessary

FIGURE 12-3 Conflicting burn patterns on the exterior of the structure showing wind direction different for each opening.
Courtesy of Russ Chandler.

to explain why there was more burning in one location than there was in another. Even if it cannot be explained now, it will need to be explained later in court.

Sometimes there may be conflicting patterns that will also need explaining. In **FIGURE 12-3**, the patterns from the door show the wind going from left to right, but the pattern in the adjacent window shows a pattern indicating a wind direction from right to left. In this case, statements from suppression forces advised that when they showed up on the scene, fire was coming from the door and the smoke was going away from the street into the backyard, to the right in the photo. As they knocked down the fire in that room, the fire continued burning in the adjacent room and the wind changed direction, sending the smoke toward the street and engulfing the engine and the engineer. Thus, the pattern in the door was first, and when the fire finally breached the window, the wind had changed direction.

One beneficial tool to a fire scene examination is the truck company apparatus. Whether in a ladder truck or an aerial scope, it is a great tool to be able to see the overall scene from above. Photographs from this area will be of great value.

The new technology of using drones for fire scene assessment and investigations is becoming a popular and more economical method of "getting the bigger picture" on fire scenes. Drones offer a "bird's-eye" look at a scene and possible uncover evidence that otherwise would have been missed or overlooked. They can provide both photos and videos for later use. It may be beneficial to check with law enforcement counterparts to share resources and staff in the use of drones. Note: Certain licenses (FAA) may be required to launch drones in such areas (see https://www.faa.gov/news/fact_sheets/news_story.cfm?newsId=22615). Also research to make sure there are no limitations from state and local laws.

As the investigator works toward the structure, they need to examine the debris to see whether there is any evidence of an explosion. Examine any natural ventilation areas or areas where there was forced ventilation by suppression forces, and again, document and take photographs as the investigation progresses.

Building Systems

The topic of building systems could fill volumes of books. The investigator may not need to know everything, but they will need to know their resources, that is, where they can obtain help to assist with the various building systems' questions as they arise. However, every investigator must know the basics of systems such as construction type, utilities, fire safety, and suppression systems.

Compartmentation is a component of the building system. The construction of the walls and their ability to resist fire travel, concealed spaces where fire can travel, and the control of openings into these spaces are components that can affect fire growth or can be key elements in resisting fire growth. For more information, see "Building Construction," Chapter 5.

Walls should have a fire rating in the form of sheetrock. The room should be completely enclosed, with no holes leading to concealed spaces to prevent extension of fire. The doors to the room should have automatically closing doors or fire doors that will seal the room when activated by a fire. The fire investigator must have a good working knowledge of building construction and safety systems such as fire doors. The investigator will also need to know about the balloon construction, wood paneling, open dumbwaiters, and decorative features that are far from fire resistive. However, this is in a perfect world.

Fire investigators must be familiar with fire suppression systems or have known resources to assist with the examination of such systems. These can include all forms of automatic sprinkler systems, deluge systems, as well as fire-extinguishing foam or dry-chemical extinguishing agent delivery systems, or total flooding systems such as carbon dioxide systems. These types of systems must be examined in both damaged and undamaged areas to ascertain their condition as well as determine if they reacted properly and performed as they were designed. Detection systems must also be examined to ascertain whether they activated properly and performed as required for that type of occupancy.

Of course, any examination of building systems will include the examination of utilities such as the heating, ventilation, and air conditioning (HVAC) system as well as the type of finish on the walls.

The key issue is the impact building systems have on fire growth and the extension of fire to other areas of the building. A detailed exam and documentation of the structure are instrumental to the future outcome of any investigation. This is an area where computer fire modeling has been instrumental by providing a visual demonstration of different impacts that building design can have on fire growth and travel.

The examination of all heat-producing building systems and appliances will be critical if there is to be a litigation. In order to point a heat source to state that is what caused the fire, all other potential heat sources must be eliminated.

Interior Examination

As the investigator completes the exterior exam there is a lot to remember; it would be wise for the investigator to write down any notes before starting the interior investigation. When the investigator enters the structure, they should begin with the least damaged area first and work their way to the most damaged area.

The investigator will examine the burn patterns as they are encountered within the structure to deduce the relative length of time the fire burned in each area. Each pattern must be considered in its entirety, with the surrounding area. The volume of the fuel load or the heat release rate of the available fuel in each area must be examined to determine whether that may have caused a heavy char rather than the length of time the material burned, which leads back to the area of origin. The potential for ventilation-generated charring must be examined in areas where either the fire self-ventilated or ventilation was created by suppression forces.

> **TIP**
>
> Each pattern must be considered in its entirety, with the surrounding area.

The lowest burn patterns are of importance because fires naturally burn upward and outward. However, some low burns may be caused by dropping curtains that ignited at the top and fell down when they burned away from the curtain rod. Some fires can burn downward depending on how the fuel is positioned, and the movement of air, heat, and hot gases. One should never forget that fires can start in a multitude of areas, so the investigator must keep an open mind, applying logic to all patterns observed (leaning on the scientific method).

> **TIP**
>
> The lowest burn patterns are of importance because fires naturally burn upward and outward.

The investigator must never forget that building construction can have a dramatic impact on how the fire grows. In a balloon-construction home, the fire could start in the basement only to extend through the interior of the walls all the way to the attic. The entire upper level of the structure could burn away, leaving a floor relatively undamaged between the basement and attic.

> **TIP**
>
> The investigator must never forget that the building construction can have a dramatic impact on how the fire grows.

A fire in a structure with concrete floors and walls with a metal roof could start in an area that contains relatively little fuel load. After smoldering and burning for hours, it may reach an area with massive amounts of fuel with high-heat release rates. Then, the fire grows in intensity, breaching the structure and alerting those passing by of the fire. The intensity of the fire where the fuel was located may throw off the investigator at first. However, examining the burn patterns should lead the investigator to question why the fire damage occurred in the remote location and whether the fire had burned longer in that area.

In following the burn patterns back to the area of origin, the investigator should use tools such as the depth of char gauge on wood char and check calcination to

help to determine the relative burn times. Along the way, a 60-watt lightbulb may have been distorted, pointing back to where the heat originated. Melting can assist as well. Look for similar items that may or may not have melted along the fire's path to give an indication of where the hottest temperature occurred.

> **TIP**
>
> The investigator should use tools such as the depth of char gauge on wood char and check calcination to help to determine the relative burn times.

Debris Removal

Fire suppression personnel must be taught to not hamper the accurate determination of the fire origin and cause; the debris must not be disturbed any more than absolutely necessary while extinguishing the fire. An example of why debris removal could be a problem to the investigator is shown in **FIGURE 12-4**, where debris has been thrown out the window, collecting on the porch roof and on the ground with the debris from the first floor. There is a fine balance in that the fire suppression personnel must conduct overhaul operations to extinguish each and every last ember, but at the same time disturb the evidence as little as possible. Failure to suppress the fire entirely could result in a rekindle, which could destroy any evidence remaining from the first fire.

> **TIP**
>
> Fire suppression personnel must be taught to not hamper the accurate determination of the fire origin and cause; the debris must not be disturbed any more than absolutely necessary.

FIGURE 12-4 Debris thrown out of the second-floor window, hampering a proper investigation of the fire scene.
Courtesy of Russ Chandler.

To truly make the final determination of the area of origin, the debris must be carefully removed from in and around all suspected areas. Any area where debris is not examined may lead to incorrect conclusions. Just as bad, if the case ends up in court, the jury may doubt the investigators conclusions because they failed to look at all the evidence. Either way, it may end up being a betrayal or injustice for all concerned.

The fire debris should be removed in layers and carefully examined for any potential evidence. After the debris is completely removed, all patterns can be more easily seen and evaluated. The debris itself may provide an indication as to what may have been in the area. A pile of newspapers will burn, but usually only on the surface and with a relatively low heat release rate. However, a bean-bag chair filled with polystyrene beads will have a tremendous heat release rate in comparison and, in turn, may create more char in the area. Residue will usually be left from each, giving the investigator clues as to what happened.

Spoliation of evidence is always a concern when there is unnecessary removal of debris. If debris removal is not necessary to prevent rekindle or lessen the weight on a weakened structure, it may be considered spoliation if it indeed hampers the scene examination and eventual determination of the fire origin or cause.

> **TIP**
>
> Spoliation is always a concern when there is unnecessary removal of debris.

Examining Contents

The burn patterns on contents can be vital to the discovery of the area of origin. They can support burn patterns on walls, ceilings, and floors and can help resolve any confusing burn patterns within the structure. The key issue is to make sure that any furnishings that were removed are placed back in the correct position. With the careful removal of the debris in the area, the investigator will be able to see any protected areas on the floor that will match up with the furniture. Shadows on walls can indicate the placement of the furniture as well.

> **TIP**
>
> With the careful removal of the debris in the area, the investigator will be able to see any protected areas on the floor that will match up with the furniture.

The burn patterns on furnishings can show the direction of fire travel. More damage on one side than on the other may indicate direction of fire travel. However, it could also mean that there was a sufficient fuel load on that side of the chair to cause damage and have nothing to do with the travel. Searching through the debris will identify any potential of a secondary fuel affecting the patterns in that area of the fire scene.

> **TIP**
>
> The burn patterns on furnishings can show the direction of fire travel.

The Area of Origin

With the thoughtful and careful examination of all burn patterns throughout the entire structure and with the examination of the debris and patterns on all furnishings, the area of origin can be identified. It is only after a systematic search of the entire fire scene, using a scientific methodology, that a successful hypothesis can be tested to identify the area where the fire started. Any conflicting patterns as the result of ventilation, fuel load, heat release rate, or fire suppression activities must be identified and taken into account before making the final hypothesis.

> **TIP**
>
> It is only after a systematic search of the entire fire scene, using a scientific methodology that a successful hypothesis can be tested to identify the area where the fire started.

With the area of origin identified only part of the investigation has been completed. The next step is to identify the first material ignited, the heat source, and the act that brought them all together, which is the cause of the fire.

Wrap-Up

SUMMARY

- As with everything dealing with a fire investigation, safety has to be the first concern of everyone on the scene. This would include all public safety personnel as well as the public. One way to accomplish this is to secure the scene with barricade tape and personnel, fire or police, to protect unauthorized personnel from entering. Cordoning off the scene will also protect all evidence, both known and yet to be discovered.

- There has to be a clear understanding of the role of everyone in the investigation of a fire. The assignment will make that clear for who investigates and the depth of their involvement in the investigation.

- Upon arrival at the scene, both the first responder investigator and the full-time investigator will start their investigation. It will be items observed by the fire fighters and fire officer who will take note but continue on their primary task of extinguishing the fire. Observing tool marks on doors or windows can be observed, noted, and then reported as soon as practical. The assigned full-time investigator upon arrival may want to suit up in full turnout gear and enter the structure. As with all personnel on the scene, they will report to the command post, and follow all accountability procedures. The observations during suppression may be extremely valuable in the process.

- Once the fire is under control, the investigators will start their process of conducting an exterior examination, working from the least damaged are to the most damaged area. This allows for the careful observations of burn patterns that may point back to the area of origin.

- The examination of the interior will follow the same process as in the exterior, working from the least damage are to most damaged area. Searching the interior should include looking at all building systems to determine if they were or were not involved in the ignition sequence or if they assisted in the spread of the fire. The patterns on interior surfaces and on the contents may allow the investigator to follow a path back to the area of origin.

- With the application of a systematic search along with scientific methodology allows the investigator to make an accurate determination of the area of origin of the fire. This is the first step in the final determination of the fire cause and identification of how to prevent future fires.

KEY TERMS

Accountability tag An identification tag with the holder's name and sometimes unit number or assigned company. More sophisticated tags also include the individual's medical history, including blood type and medical allergies. Used in the accountability system to track the location of all personnel at an emergency scene.

Area of origin The location of where the fire began, where the heat source and first fuel ignited came together. In some circumstances can also be the "point of origin."

Technical rescue teams Assigned teams who have the application of special knowledge, skills, and equipment to safely resolve unique and/or complex rescue situations. Also referred to as heavy technical rescue teams, structural collapse teams.

Imminent danger to life and health (IDLH) An atmosphere in which anyone entering endangers themselves unless proper protection is taken, which could include full turnout gear with self-contained breathing apparatus.

Officer in charge (OIC) The fire officer who has ultimate charge and control of the overall fire or emergency scene. Usually but not always the senior officer on the scene.

Point of origin The exact physical location within the area of origin where a heat source and a fuel first interact, resulting in a fire or explosion.

REVIEW QUESTIONS

1. Why does the investigator need to know two routes out of a structure at all times?

2. What is NFIRS and how can it help to prevent fires?

3. Why should investigators have to report to the OIC and let him or her know where they will be working?

4. Name three outside influences that can affect or change the fire-burn pattern.

5. How could a building official be of assistance at a fire scene investigation?

6. Why would the investigator photograph a gas line or electrical panel that was not involved in the fire and not even damaged?

7. How could water be a hazard to the fire investigator?

8. What tool could be used to determine how long the wood had burned in one area of a room in contrast to another area of the room?

9. Why would it be essential to not have suppression firefighters clean out a room, shoveling out the debris so that the investigator can look at the floor? Why would the investigator not want the fire fighters to completely clean out a room, shoveling debris and all contents to the outside?

10. How can burn patterns on furniture help point back to the area of origin?

DISCUSSION QUESTIONS

1. The National Fire Incident Reporting System collects fire investigative information from fire reports filled out by the fire companies. What type of statistical information on fires do you feel would be beneficial to help the community or the fire department?

2. Once the lead investigator arrives on the scene, she or he reports to the OIC. They are both of the same rank within their department and have been in the department for the same length of time. Who do you feel should be in charge from that point on? Why? What if the circumstances were that the fire was obviously an arson fire and part of a possible string of arsons that had recently occurred? Would that change who you think should be in charge?

CHAPTER 13

Fire Cause

LEARNING OBJECTIVES

Upon completion of this chapter, you should be able to:

- Describe what needs to be done in conjunction with the search for the area of origin to help determine the fire cause.
- Describe the safety and legal concerns that apply when on the fire scene.
- Describe evidence found on the exterior of a structure.
- Describe the importance of identifying the first material ignited and the act that brought the source of ignition together with the first material ignited.
- Describe why it is important to eliminate all other fire causes.
- Describe the four classifications of fire causes.
- Describe the process of identifying the product or person responsible for the ignition sequence.

Case Study

The oil-fueled furnace had just received its semi-annual servicing by a local oil and heating contractor. This process had taken place in the spring and fall each year for the preceding 10 years. The next day was the first cold day in the fall; the family was in the basement in a small den located adjacent to the furnace room. The thermostat was turned up, starting the furnace for the first time after the summer months.

It was not long before the family discovered smoke coming from the furnace room. Fortunately, they thought fast and cut off the furnace at the thermostat—but that did not stop the black smoke from permeating the structure, coating expensive original paintings and antique furniture with black, oily soot. The insurance company was faced with damages in excess of $150,000 from sooting alone.

The initial investigation started with the lieutenant of the first responding engine. She walked around the structure doing her assessment, reported her findings over the radio to other responding units. Arriving back at the front of the structure, the lieutenant entered it with her crew, heading to the basement. She noted that this thick black smoke was unusual and identified that the fire was confined to the furnace room. Fire on the exposed stud walls of the furnace room was extinguished and the remaining smoke came from the furnace itself. One of the fire fighters cut off the fuel to the furnace and noticed that the flames were coming from the air intake port on the furnace, which was located just below the burner assembly.

After the fire was knocked down, the lieutenant reported to the officer in charge (OIC) at the front of the structure that the fire was limited to the furnace room and that the area of origin was the furnace, due to a malfunction in the equipment. This information was passed on to the assigned investigator, who confirmed the lieutenant's findings.

The next day the insurance company called the fire department asking for the fire report and any information about the cause of the fire. The assigned investigator related the lieutenant's findings. The insurance company also confirmed that the damage was in excess of $200,000, consisting mostly of smoke damage to the home's contents. The insurance company hired an independent (private) fire investigator to document the exact cause so legal action could be taken in the future.

The investigation started like any other systematic search of the property, working from the least damaged area to the most damaged area. The smoke patterns clearly led to the basement area. In the basement, the patterns led back to the furnace area. The wall adjacent to the furnace was burned clean away, all the way to the floor (**FIGURE 13-1**).

Burn patterns indicated that a flame came out of the furnace and impacted the wall, spreading heat and fire upward and outward. The only opening where the flame could have escaped from the furnace was the small 1 in.^{2} hole designed for air intake (**FIGURE 13-2**).

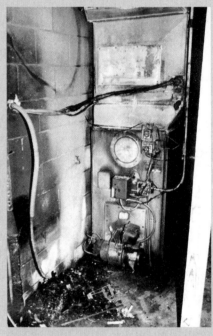

FIGURE 13-1 The furnace in the area of origin. The wall adjacent to the furnace is burned clean.
Courtesy of Russ Chandler.

FIGURE 13-2 The burner assembly removed from the furnace by the investigator. In the center of the photo is a 1 in.^{2} hole at the base of the furnace, the air intake. By all accounts, it appeared that this was where the flames escaped the furnace.
Courtesy of Russ Chandler.

Under normal circumstances, this opening has a cover with one screw. The cover can be adjusted to allow just enough air into the burn chamber for a proper efficient flame. For flames to come out of this small hole, flames in the burn chamber would have had to go up and over the burn chamber and then down to the hole to escape; this was not likely under normal circumstances.

In pulling out the **burner assembly**, the private fire investigator proved that the oil **jet nozzle** had not been replaced as it should have been during the recent service call. The nozzle showed considerable age and wear. During the removal of the assembly, an excessive amount of soot and **scale** came out with the unit. If the furnace had been properly serviced only two days before, this area would have shown signs that it had been vacuumed clean, as would have the area where the heat was exchanged just above the burn box area.

Pulling out the burn chamber provided the largest shock of all. The firebox was a cylindrical steel chamber with an open top that allowed the heat to escape upward. In this case, the chamber had melted away on both sides of the box (**FIGURE 13-3**), which allowed flames to escape the chamber and come out of the air intake. In the chamber was a bed of soot and scale (**FIGURE 13-4**); the amount of time it would take to accumulate that much damage and debris defies the imagination.

Above the burn chamber was a round opening to access the heat exchange area. As part of every maintenance service, this opening would be accessed to vacuum out this area. To get into this area, the cover plate had to be removed. However, when the investigator tried to remove this plate, one screw broke off and a second screw would not come out; he used the latter screw to pivot the plate to gain access. The plate itself had to be pried away from the face of the furnace. This was not fire damage, but rather damage from age. The first look showed scale and soot piled up at the plate (**FIGURE 13-5**). By using an external,

FIGURE 13-4 In the area where the burn chamber sits, the designed vacant space was filled with soot and scaling from the destruction of the burn chamber itself.
Courtesy of Russ Chandler.

FIGURE 13-3 Steel burn chamber showing where fire burned through the steel as a result of debris clogging the oil jet and skewing the flame to the side. Over time, more debris in the jet skewed the flame to the other side so that holes were burned on both sides of the chamber.
Courtesy of Russ Chandler.

FIGURE 13-5 The access area to the heat exchange with the plate turned up out of the way. This area had allegedly been cleaned the day before, but soot and scale were readily evident, piled up at the face of the opening.
Courtesy of Russ Chandler.

remote flash, the investigator could see the soot inside the exchange area, which obviously had not been cleaned in several heating seasons (**FIGURE 13-6**).

The area of origin of this fire was at floor level, just outside of the furnace, in the area of the air intake. The first material ignited was a cardboard box placed adjacent to the furnace and up against the cinderblock wall.

The homeowner was able to provide receipts for 10 years' worth of service calls, twice a year, at a cost of a little less than $200 for each call. Each document had marks on the form showing the jet was replaced, the electrodes were aligned, and the burn chamber

FIGURE 13-6 Using a remote flash allowed the investigator to see into the heat exchange area, which showed heavy soot buildup.
Courtesy of Russ Chandler.

was inspected and vacuumed, as was the heat exchange area. The furnace was then allegedly adjusted to give an efficient flame; this would have entailed moving the plate over the 1 in.^{2} hole. Interestingly, this small metal plate was never found in the debris.

In this situation, the failure of the furnace was the cause of the fire. However, because the service company accepted money for a job it never did, it committed fraud. When the civil suit (subrogation) was initiated, the insurance company obtained a subpoena for the service records from the oil company. The average service call takes between 1 and 2 hours, depending on the experience of the technician; at this company, however, the average technician made anywhere from 12 to 20 service calls in an 8-hour day. The case was settled: A condition of the settlement was that the oil company shut down its service department and only deliver oil. The second part of the settlement was that the oil company had to pay for each home it had purportedly serviced that fall to be revisited by another company to properly service those furnaces and ensure that a similar incident would not occur in those homes.

This investigation started with the engine company officer acting as the first responder investigator. The assigned investigator took the next step. The investigator contracted by the insurance company was able to provide a detailed technical report that not only allowed the insurance company to recover its cost to the homeowner in this case, but also prevented future fires in other residences.

Access Navigate for more resources.

Introduction

As described in Chapter 12, *Examining the Scene and Finding the Origin*, the investigator can use all the knowledge gained from training, past experiences, and the current investigation to seek out the facts that will lead to an accurate hypothesis as to how the fire started. The investigator must first consider issues of safety, followed by legal issues such as access to the scene and spoliation. Knowledge of chemistry and science principles is invaluable when examining the various burn patterns found at the scene of the fire and can even assist the investigator in identifying an accurate determination of the cause of the fire. If the evidence is not collected and preserved properly, all of the work and effort will be for nothing. Not only must investigators identify how a fire started, but

they must also prove their hypothesis to their peers, and they may potentially have to testify about it in a court of law.

> **TIP**
>
> First, consider issues of safety.

The first necessity in the search for the cause of the fire is locating the fire origin, as discussed in Chapter 12, *Examining the Scene and Finding the Origin*. During the search for the origin, the investigator might find many pieces of evidence that will assist in determining the cause of the fire. Burn and smoke patterns can give some indication of the fire cause, especially if an unusual amount of charring occurred that would not be expected with the normal type and amount of contents in the structure. Documentation is critical along the path to the fire origin. Once the investigator determines the area of origin, a detailed examination must be undertaken to discover which material first ignited, along with the heat source that brought this material to its ignition temperature.

As with the search for the origin, the search for the cause must be done systematically, using scientific methodology to come up with the hypothesis that explains the investigator's conclusions about how the fire started. This process may sometimes result in a hypothesis that does not withstand testing, requiring the investigator to reexamine his or her notes or the scene itself and create a new hypothesis. Thus, some fires may require a more in-depth investigation to come up with an accurate determination of the cause.

> **TIP**
>
> The search for the cause must be done systematically using scientific methodology.

Legal Issues

The fire service is allowed to stay on the fire scene both to handle the emergency and, for the good of the public, to identify the area of origin and the cause of the fire. All portions of the structure should be examined, including those areas without fire damage. However, investigators must focus on identifying the extension of the fire or eliminating potential heat or fuel sources. They have no right to look in furniture, such as drawers in undamaged areas. If fire investigators' actions have no connection to looking for extension of the fire, then they would constitute an illegal search. In addition to being ethically wrong, if any evidence is found in this process, it may not be admissible in a court of law.

If at any point an investigator uncovers evidence that the fire may not be accidental in nature and may be an intentionally set fire, the focus of the investigation changes. If the investigator is a fire officer, this is the time to call for an assigned fire investigator. The legal issues may differ from one jurisdiction to another, but it would be prudent for the investigator to place the scene under guard, keeping all parties out of the scene, and to approach the proper authority, such as a magistrate, for a search warrant to continue the examination. Based on the evidence collected, the investigator would take an oath stating the facts in the case that confirm probable cause exists that a crime has

been committed. In addition, the investigator would identify both the remaining areas to be searched and what is being sought in the process. Only then can a valid search warrant be issued. This process protects any further evidence collected from being thrown out of court because it was illegally obtained.

> **TIP**
>
> If at any point an investigator uncovers evidence that the fire may not be accidental in nature and may be an intentionally set fire, the focus of the investigation changes. If the investigator is a fire officer, this is the time to call for an assigned fire investigator.

One major caution: If evidence is found out in the open and may potentially disappear or be damaged, then it is acceptable to document, collect, and preserve that evidence. The most important consideration is that the investigator must consult legal counsel in advance to ensure that investigative procedures are up to date and appropriate to the laws of the jurisdiction.

> **TIP**
>
> If evidence is found out in the open and may potentially disappear or be damaged, then it is acceptable to document, collect, and preserve that evidence.

Protecting the Scene

Investigators can protect the scene and any potential evidence by using barrier tape and guards. Investigators are just as obligated to protect evidence from an accidental fire as they are in a criminal fire case. The evidence in an accidental fire can be just as crucial in a civil case. For example, proving a product failure may well prevent the failure of similar devices or products in the future; this, in turn, can save lives and property.

Exterior Examination

When investigators start their actual examination for the fire origin, they are also searching for evidence of the fire cause. Although the cause of the fire can be determined only after the area of origin has been located, plenty of supporting evidence may be uncovered along the path to the fire origin.

The investigator starts the examination at the farthest point away from the structure where there is no damage at all. Working systematically, the investigator then examines the surrounding area for any indications of debris or evidence that may indicate a cause of the fire. Although many items identified outside the structure may be evidence of an incendiary fire, the cause of the fire is still undetermined during this phase of the investigation. The investigator then continues the examination, working from the least damaged area to the most damaged area.

Once in an area where the burn is consistently the same, it does not matter whether the investigator moves in a clockwise or counterclockwise direction, or sets up a grid system. The important fact is that the examination is conducted in a consistent manner from one fire to the next, as much as is practical.

The investigator should photograph all evidence in place as it is discovered and before it is moved. It should be documented in place; then, if the evidence will be collected, the investigator should complete further detailed documentation, such as cataloging and preserving the evidence for further examination by a laboratory. Items found in the outside area could include, for example, glass from the windows. Broken glass that is clean with no soot found some distance from the structure may indicate an explosion before the fire. Sooted glass may indicate the fire occurred first, with the explosion occurring secondary to or being caused by the fire.

Other items may be indicators of an incendiary fire. For example, smoking materials discarded in an unusual area, especially if they do not appear weathered and no one in the family smokes, can be suspicious. The presence of refreshment bottles or cans, both soda and alcoholic, can be suspicious, especially if they are not weathered and there is no history of anyone drinking anything in that area or discarding such containers in that area. A container that may have contained an ignitable liquid could be in the area and arose suspicion if it does not belong to the occupants of the structure. A word of caution: Never place your nose over any container in an effort to detect its contents; this could lead to an unnecessary potential exposure to hydrocarbons, other hazardous vapors, and carcinogens. Using a simple hydrocarbon detector is the best way to test for residue of an ignitable liquid.

Other incendiary indicators may include ignitable liquids used as a **trailer** leading from the structure to a hiding place outside the structure. This type of trailer does not always work as intended. Its chances of success depend on the type of fuel, the amount of fuel, the type of soil (some soils may absorb the liquid quickly), and the temperature (which may dictate the amount of vapors given off by the fuel). Regardless, the trailer often leaves a telltale residue that will make it visible and obvious (**FIGURE 13-7**). The specific trailer of diesel fuel shown in Figure 13-7 was a failure; however, the perpetrator reentered the structure and used gasoline inside to get ignition.

Fresh footprints in and around the area may be of interest, as would unexpected tire tracks found at the

FIGURE 13-7 A trailer of poured diesel fuel that leads from the kitchen, out the door, down the steps, across the backyard, and behind an aboveground pool.

Courtesy of Russ Chandler.

scene. Even though they may not be of any known value, it is best to protect these items until the entire scene can be examined. This can be accomplished by placing items such as boxes or buckets over the print or track and then marking it with a tag as evidence (**FIGURE 13-8**). For example, if someone finds a tire track, he or she can place two traffic cones at each end of the track, and then place an unfolded attic ladder over each cone. The horizontal ladder should be hovering just over the print. For added protection, cover the site with a salvage cover. If available, an evidence marker can be placed on the ladder or salvage cover. If the examination of the scene indicates an incendiary fire, the tire track may potentially have evidentiary value. The full-time investigator will have supplies to make a plaster cast of the track.

It is essential to photograph all sides of a structure regardless of whether patterns are visible there. These photographs of the structure provide a reference for recall at a later time. Photos are good for showing areas that suffered damage and areas where the fire did not extend. It is also important to photograph all fuel and heat sources entering a structure. In particular, the investigator should photograph the electrical meter base, the overhead wiring, or the conduit from underground lines and document their condition. Likewise, the investigator should examine, document, and photography any gas lines entering the structure. In addition, the investigator should photograph any chimneys or flues coming through the roof, at least from the ground and potentially from above by perching on an aerial apparatus, ladder truck, or use of a drone.

> **TIP**
>
> The investigator should photograph all fuel and heat sources entering a structure.

FIGURE 13-8 An evidence tag sitting on top of a bucket that protects a footprint in the soft earth.

Courtesy of Russ Chandler.

Interior Examination

The path to the area of origin inside the structure may also yield potential evidence. The investigator must keep an open mind and analyze what is found along the way. More often than not, evidence located outside the area of origin is a potential incendiary indicator. Of course, the investigator is not just looking for evidence of arson, nor is that the only evidence that will be found. Even so, evidence of an accidental fire is more often found at or near the area of origin.

Clues that may assist in the investigation can be found in many areas. For example, an abundance of spliced extension cords is not evidence by itself, but suggests that the investigator should take a close look at electrical problems near or at the area of origin.

> **TIP**
>
> An abundance of spliced extension cords is not evidence by itself, but suggests that the investigator should take a close look at electrical problems near or at the area of origin.

What to Search for and Document

The investigator must search all rooms and areas of the structure. This activity does not involve a detailed search of any furnishings, but rather a search of the structure for extension of the fire and to eliminate any potential fuel or ignition sources, as well as unsuccessful fire starting points. As heat or fuel sources are located, the investigator must photograph and document them to show either potential involvement or no involvement in the fire. All too often, an investigator goes to court in the case of a product liability or incendiary fire and the opposing counsel brings up the problem with a furnace or faulty electric heater in the bedroom and insists that it had to be the cause of the fire. With the proper documentation, the investigator can easily refute this assertion with a photograph of the heater or furnace in question, clearly showing that it did not start the fire and, in fact, suffered no fire damage. This documentation could certainly help the jury or judge render an opinion on the case.

Accidental Indicators

When thinking about indicators of an accidental fire, look for clues that may indicate what to search for around the area of origin. For example, when ashtrays around the house are full of cigarette butts or when burn marks from smoking materials are found in areas not damaged by the fire, these findings indicate carelessness with smoking materials. This is not evidence itself, but does offer clues about what needs either to be confirmed or eliminated. Of course, as noted throughout this text, the process of elimination is just as important as confirming a potential correlation between the clues and actual proof of the fire cause.

Electrical wires and connections in the undamaged areas of the structure require the investigator to confirm or eliminate structural electricity as being involved in the incident. When the investigator finds open containers of paint thinner or other chemicals in areas of the structure unaffected by fire, the investigator must confirm or eliminate their involvement in the ignition sequence in the area of origin. As with all indicators, the investigator must keep an open mind and not become blind to other sources of ignition. However, it is essential that these potential hazards be eliminated in the overall report should the actual cause of the fire turn out to be something different. Specifically, the scientific method requires that the investigator include all hypotheses that were eliminated in the process of making the final determination of the cause of the fire.

TIP

The investigator must keep an open mind and not become blind to other sources of ignition.

Incendiary Indicators

If the fire was potentially incendiary, the investigator may consider having the local police department on the scene to help with security. Depending on local policy, the police may already be on the scene if it is their responsibility to examine any fire scene where a potentially criminal act took place.

TIP

If the fire was potentially incendiary, the investigator may consider having the local police department on the scene to help with security.

Burn patterns, such as **pooling patterns** and sharp lines of **demarcation** on the carpet, may indicate that ignitable liquids were used to get the fire to start burning. These liquids can also be employed as trailers to help the fire travel from one area to another.

Many materials can be used as trailers. Sometimes it is not so much materials' composition as their form that suggests they played this role. Stacking magazines on the floor from one room to the next may not result in fire travel. However, tearing out the pages of those magazines and crumpling them up can result in a faster-moving fire that will travel wherever the trailer leads.

Common household items such as fabric softener sheets or wax paper streaming from one piece of furniture to another can act as trailers. These items will almost completely disappear, except in areas where they were protected from burning, such as under cushions. In contrast, crumpled paper from magazines will leave behind carbon residue that, if not destroyed by suppression forces, will still appear as a trailer; some pages may even hold their crumpled shape after burning.

The mere presence of an ignitable liquid pattern or even a positive sample is not an indication by itself of an incendiary fire. Unusual as it may seem, there may be a legitimate reason for that flammable liquid to be in the structure at that location. Of further note, post-flashover burn patterns can be very similar to ignitable liquid pour patterns.

TIP

The mere presence of an ignitable liquid pattern or even a positive sample is not an indication by itself of an incendiary nature of the fire.

When taking samples, the investigator is well advised to use an accelerant detector. For example, an electronic hydrocarbon detector can identify the best location to obtain a sample.

K-9 resources are a great tool in the collection of evidence of a hydrocarbon accelerant. They can narrow down the search area and help to locate the best location to collect the samples for laboratory testing. Accelerant dogs were pioneered by the Bureau of Alcohol, Tobacco, and Firearms (ATF) through its Accelerant Detection Canine Program (ADCP). The ADCP's first operational canine was Mattie, who began her training in conjunction with the Connecticut State Police. She proved enormously valuable not only to ATF and Connecticut, but also to hundreds of fire and police departments across the country. Today, the ADCP is still recognized as the leader in the use and training of accelerant detection dogs.

If possible, it is beneficial to collect comparison samples. A comparison sample is taken from a remote area that was not involved in the fire.

Source and Form of Heat of Ignition

Chapter 9, *Evidence*, discussed the many sources of heat substantial enough to bring a material to its ignition temperature. Everything from smoking materials to faulty electrical wiring and escaping sparks from a fireplace or chimney can provide sufficient energy to start a fire. The investigator's role is to identify the specific source of heat that came in contact with the first material ignited within the area of origin.

The ability to recognize residue is a major skill of the investigator. Many items, once burned, are difficult to identify, and the investigator needs to know what to look for in the debris. Some of this experience may come from being a fire fighter and looking at scenes after a fire. By picking up debris during overhaul and examining it, fire fighters can discover what the object was—for example, the base of a coffee maker. Other investigators get appropriate experience during their training program, which may require them to investigate a room after a fire. They can examine the debris and then watch a video or view photos of what was in the room before the fire to gain "before and after" perspectives on burned items. Even better training is when the team that put the room and content training fire together sees firsthand what a hairdryer or curling iron looks like after being subjected to 1000-degree heat.

> **TIP**
>
> Many items, once burned, are difficult to identify, and the investigator needs to know what to look for in the debris.

Many times, just examining an item in its burned and misshapen form can reveal its original form and use. The item shown in **FIGURE 13-9** was found adjacent to a furnace. In fact, 10 of these items were found surrounding the furnace. By looking at the bottom, it became evident that the object was a 1-gal milk jug. Small pockets of liquid were preserved as the sides melted down on themselves. Laboratory confirmed the liquid was gasoline. The homeowner attested to having a serious problem with the furnace. However, the oil tank valve was in the off position and weeds had grown into the valve handle, making it impossible to open or close without killing the weeds. When confronted, the homeowner admitted that he set the fire using the milk jugs, hoping to make the cause look like a malfunctioning furnace.

FIGURE 13-9 Careful examination of the scene revealed melted plastic that was the residue of a 1-gal milk jug that contained gasoline at the time of the fire.
Courtesy of Russ Chandler.

Sometimes the evidence of the source of ignition may not be found in the area of origin. Perhaps an arsonist carried a lighter away from the scene. A fire alongside a road could occur when a spark from the muffler or the heat from a catalytic converter set dried leaves on fire as a vehicle drove away. A brush fire may lead back to a burn barrel, but the burn patterns may begin about 15 ft from the barrel. In this case, perhaps a spark from the barrel started the fire, but the likelihood of finding that burning ember may be slim. The point is that in some circumstances, the heat source may not be present at the site, but through eyewitness accounts and other evidence, the investigator may still be able to label the heat source with a reasonable degree of certainty.

> **TIP**
>
> Sometimes the evidence of the source of ignition may not be found in the area of origin.

First Material Ignited

Following a few ground rules can ensure consistency in reporting data on fires. During a training session for filling out the forms used by the National Fire Incident Reporting System (NFIRS), a student was given the scenario of a child playing with matches while in a closet. The flame from the matches came in contact with the plastic film covering clothing that had just returned from the dry cleaners. The student entered the first material ignited as wood. He was correct to a certain extent—the match-head chemicals were attached to a wood stick, the matchstick. In his mind, the match chemicals ignited the wood match and everything else occurred secondary to that. In contrast, another student argued that the plastic film was the first material ignited in this scenario. Like the child that used a lighter in a closet as discussed in the previous chapter, a match, a lighter, hobby torch, etc. is the heat source. The first material ignited is the dry cleaning bag.

Although technically correct, the fuel from the heat source does not meet the NFIRS's intent of gathering information for a report or an investigation. To say that the first material ignited was the wood from the match has little value. The investigator must be more specific. If it was a wood log, 12 in. in diameter, and the heat source was a pilot light, the wood would take some time to ignite and that fact is of importance. Likewise, if the first material to ignite consisted of wood shavings, they would ignite rapidly.

The first material ignited tells the story of what happened. The heat source could be a gas burner on the stove, with the first material ignited being a dish towel left near the burner. This is the beginning of the explanation of how the fire starts. This information tells us that more training on keeping combustibles away from gas range burners is a preventive solution. Proving that the pilot ignited the gas serves no value other than to prove the appliance operated as it was designed.

Some items need a little more clarification. Plastics, for instance, take many shapes and sizes and have varying characteristics. Plastic includes the soft sponge foam in a couch cushion as well as a hard plastic toy. The plastic in toys can be hard and brittle, or it can be soft and bendable. Each type of plastic has its own burning characteristics that need to be examined. Of particular concern are the plastics that can turn to liquid upon exposure to heat, and then ignite and flow. Sometimes the telltale signs of flowing plastics resemble those of flammable liquid pours, but not always. If plastic is a potential first material ignited, then it would be prudent to obtain a similar material and examine and test it to see whether it just chars or is capable of spreading the fire.

> **TIP**
>
> Sometimes the telltale signs of flowing plastics resemble those of flammable liquid pours.

Ignition Sequence

If investigators are ever in a position to become public safety investigators, they may realize that it is truly amazing that even more fires don't happen on a routine basis. Many times an appliance or device malfunctions, but circumstances are just not right for ignition. Often, this failure happens by design, such as when manufacturers recognize a potential heat source and insulate or isolate it from ignitable items. At other times, manufacturers ignore this risk—which can lead to devastating losses, recalls, and civil suits.

When a heat source and a fuel do come together, ignition is usually a possibility. It is the role of the investigator to place these two items at the proper location and time and to develop a hypothesis that explains what happened to result in a fire. When considering the heat source and the fuel coming together, there is no implied movement at that particular moment in time. Perhaps the two were together over a long period of time before flaming combustion took place. The two could have been in contact with no action until an external force interceded and caused ignition, such as a strong wind current fanning a pilot flame.

During this part of the investigation, the fire investigator must apply scientific principles to show that the available heat source was sufficient to ignite the specific first material ignited. Sometimes the ignition sequence is so obvious that verification is readily documented—such as a wooden kitchen match dropped onto crumpled writing paper. At other times, matters are not so simple. When doubt arises about whether the heat source was capable of causing ignition, the investigator can consult various NFPA texts and documents to verify the ignition temperature of various materials. The final hypothesis should identify and explain the heat source in detail. It should also identify the first material to be ignited and explain why the two were in close proximity. In addition, the fire investigator might need to explain why that material was in that area. In some instances, such as a gas leak, it may be necessary to explain the location of the leak that led to the event. It is a matter of common sense when writing the report.

> **TIP**
>
> The fire investigator must apply scientific principles to show that the available heat source was sufficient to ignite the specific first material ignited.

> **TIP**
>
> The final hypothesis should identify and explain the heat source in detail. It should also identify the first material to be ignited and explain why the two were in close proximity.

Elimination of All Other Causes

Before solidifying the final hypothesis, the investigator should seek to answer questions before they are asked. In particular, the investigator should check each and every heat source and potential fuel to ensure that these items could not have been involved. The investigator might have already done this in the scene examination process, but should verify and clearly document these findings in the notes and in the report, including the reason why that heat source or fuel could not have been involved in the fire. The photos taken during the scene examination of the various heat sources can support the investigator's elimination determinations. In some situations, a fuel source has been removed. For example, in **FIGURE 13-10**, the liquid propane tanks had been removed from the property prior to the fire.

Sometimes it may be difficult to eliminate some heat sources as potential fire causes, such as smoking materials. If no one in the house smoked and there were no guests in the preceding 24 hours, it may be logical to eliminate smoking materials as being involved. To be sure, the investigator's interviews should include questions about whether someone used to smoke and is in the process of quitting. If teens reside in the house, the investigator will want to believe they were not smoking, but can decide if this belief is warranted based on the results of the interview process.

Classification of Fire Cause

The terms used to classify the types of fire causes may vary, but are essentially equivalent in most jurisdictions. For consistency, it is recommended that fire investigators follow the fire cause classification provided by the National Fire Protection Association (NFPA) in NFPA 921, Guide for Fire and Explosion Investigations.

The first tenet of fire cause determination is that all fires are considered accidental until proven otherwise. Some states mandate this assumption through case law. However, it is good practice that investigators adopt the attitude that they are true seekers of the truth. Going into a fire scene with a clear mind and no preconceived notions of the cause keeps investigators' minds open to new ideas, thoughts, and concepts as they uncover evidence about the cause of the fire.

> **TIP**
>
> Going into a fire scene with a clear mind and no preconceived notions of the cause keeps investigators' minds open to new ideas, thoughts, and concepts.

Natural

Some texts or articles may say a fire was an "act of God." Depending on one's beliefs, this could be accurate. To clarify, the term *natural* implies that the fire was not started by the actions of a human. This category includes fires caused by lightning, lava flows, or flooding that creates electrical shorts. As long as no act by a person or man-made product created the cause of the fire, the fire should be labeled as natural.

Accidental

The label "accidental" means just what it implies: The fire was started because of an accidental act. In other words, someone made a mistake and some action of a human resulted in the heat source coming in contact with the first material ignited. This includes product liability cases where the fire was the result of a design error, a manufacturing error, or an error in how the device was used.

For example, a steam iron is a good tool, but if left face down on a fabric while the person doing the ironing runs to answer a phone, it can result in a fire—an accidental fire. Electrical wiring installed using the wrong components can cause an electrical short, and the resulting fire is considered an accident. It could be argued that the electrician consciously used the wrong connector, knowing that it could start a fire.

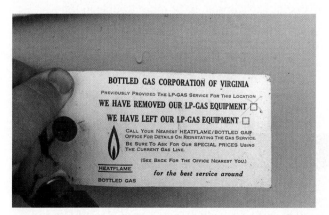

FIGURE 13-10 Tag left on the gas supply line outside the structure showing that the liquid propane gas bottles had been removed prior to the fire. If there is no date on the tag, the gas provider can give exact dates of removal and reasons why.

Courtesy of Russ Chandler.

But unless there is proof that the electrician intended for the fire to start, the fire should be labeled accidental. However, this classification does not remove blame for the electrician's actions. Civil suits are available for those seeking recourse in a loss resulting from an accidental fire.

When a child plays with matches and starts a fire, the fire is considered by most to be accidental in nature. Because the child is too young to know the consequences of his or her actions, the fire was accidental. In many jurisdictions, the courts have ruled that children younger than the age of 7 years could not have understood their actions. However, some 5-year-olds clearly demonstrate that they know what they are doing is wrong. Even so, the courts are unlikely to prosecute except in a compelling case where the goal is to get assistance for the child.

TIP

When a child plays with matches and starts a fire, the fire is considered by most to be accidental in nature.

Incendiary

Many fires are intentionally set, but that does not make them incendiary. Setting fire to the logs in a fireplace and igniting the trash in a legal burn barrel both produce intentionally set fires. The key difference between these types of fires and incendiary fires is the intent—that is, the frame of mind of the individual who starts the fire. A key component of this frame of mind is malice[2]: If someone starts a fire with malice, that act goes against the common good of the public. The motivation need not be hate, as malice includes being reckless toward the law and the rights of others. Malice can also imply an evil intent, such as committing insurance fraud by burning a car for the insurance money. For more information on arson fires, see Chapter 15, *Arson*.

The term arson means essentially the same thing as *incendiary*. Nevertheless, case law in many jurisdictions has dictated that it is up to the courts—not the investigator—to decide whether a fire is arson. The definitions of arson and incendiary are similar in *Black's Law Dictionary*, but how those words are used in the investigators' own court system makes the difference as to which term they should use in their report.

Pint-Size Problem

The engine company pulled up on the scene of a small brush fire. As the crew exited the engine, the officer spotted a small child running into the house next door from the open lot with the fire. The door slammed loudly. As the fire fighters pulled off the booster line to extinguish the small fire, three other boys in the open lot pointed to the house next door and said, "Jimmy started the fire." Jimmy, of course, was the boy whom the officer saw slamming the door.

An assigned investigator was called because it was a potential juvenile case. The investigator knocked on the house's door, and little Jimmy opened the door a crack. The investigator identified himself and asked Jimmy if his parents were home. The young boy looked up at the investigator and shouted that he was 5 years old and that there was nothing the investigator could do about it. Then Jimmy slammed the door shut. As tragic as this may seem, this young man was likely destined for more trouble in the future. But the engine company personnel got quite a chuckle when the 6-ft-plus investigator had a door slammed in his face by a 5-year-old.

Suspicious

Investigators may recognize suspicious circumstances, issues, items, or actions that lead them to believe a fire was intentionally set. However, there is no such classification as suspicious. If investigators cannot find the exact area/point of origin and the precise cause of the fire, then the fire must be classified as *undetermined* on the fire incident report. "Undetermined" is a legal definition and should not to be confused with the process of data collection.

Undetermined

No matter how good an investigator's skills, the cause and origin of some fires will remain undetermined. The investigator may find the origin, but the cause may elude him or her because of conflicting patterns or because the evidence sustained too much damage. For example, this can be the case when the fire burns too long or when the fuel load was so large that it caused extensive damage.

There should be no shame in reporting an undetermined fire. Unfortunately, senior officers in municipal departments may not always understand investigations and may insist on having a cause for each and every fire. This practice is dangerous because it can result in documentation of false data and can create problems later if the true cause of the fire is discovered. Worse yet, making a guess as to a cause may result in the implied guilt of an innocent party.

Determining Responsibility

With the cause identified, the investigator may have sufficient evidence, after interviews, to identify *who* or *what* caused the fire. This determination must meet the requirements of a systematic investigation using scientific methodology. The *what* can be a product failure, which may allow the victim or the victim's insurance company to recover their funds through subrogation. The *who* may be an individual who intentionally set the fire either with or without malice. If malice was present, criminal charges may be filed. A no-malice situation might be something like a child playing with matches or the homeowner accidentally using gasoline in a kerosene heater.

Product

Product failure is not unusual in this age of technology. A number of kitchen appliances have been recalled because of their propensity to overheat, creating a fire hazard. The investigation into the failure of an appliance or any other device is intended to identify whether the failure was a result of a design flow or improper manufacturing.

If You Are Going to Do It, Do It Right!

The fire almost totally consumed the two-story, multiple-room house. The entire structure was reduced to a 3-ft-high pile of debris. The local investigator labeled the fire as electrical in nature because the homeowner mentioned that he had been having problems with a light fixture. One week later, a fire investigator for the insurance company looked at the scene. Only one small corner had been disturbed by the local investigator. To conduct a thorough examination, the insurance investigator brought in labor in the form of off-duty fire fighters, who dug through the fire scene. When they got to the bathroom, which by design was 3 ft lower than the rest of the house, they found a 55-gal drum on its side, with holes poked in the drum in areas that were against the floor. Further investigation gave clear indication of an incendiary fire.

Although additional evidence of an incendiary fire was found and documented, the homeowner maintained his innocence. The homeowner's only defense during depositions was the report created by the local investigator, which said that the fire was electrical in nature. Charges were never brought, but the insurance company successfully denied the insurance claim.

Temporary Repairs

The responsibility for any fire could belong to a third party. In one incident, a homeowner found flames shooting out of her gas water heater and had the presence of mind to immediately shut off the fuel supply. She called the gas company, which sent out a service-repair person to look at the water heater. He advised the homeowner that the valve assembly leading to her pilot light was faulty. The repairman told her that he would put a temporary fix on the valve and get her an estimate on the repair before ordering the part.

Less than an hour after the repairman left, the homeowner smelled something burning. She found that the flame from the pilot light was 3 ft tall and had ignited some nearby combustibles. The homeowner called the fire department, which extinguished the fire. The first responder investigator (engine officer) who did the preliminary investigation labeled the cause as faulty equipment. The insurance company sent an investigator. In the presence of the gas company representative, the homeowner, and the repairman who worked on the heater, the investigator disassembled the valve for the pilot light. He found a wad of aluminum foil, what appeared to be a gum wrapper (**FIGURE 13-11**).

The homeowner said she had noticed the repairman take out a stick of gum and wad up the wrapper when he was doing her repair. The repairman confessed it was a common practice for him to push foil in the valve, limiting the gas flow to save money for his customers. Regretfully, this is not a good or acceptable practice and was the direct cause of the subsequent fire.

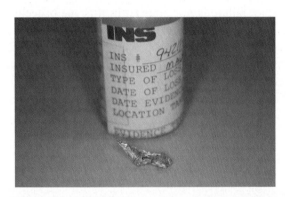

FIGURE 13-11 Small foil from a gum wrapper improperly used to limit the flow of gas in a water heater pilot light. Notice the use of a film canister as an evidence container.
Courtesy of Russ Chandler.

Another product issue arises when the product is functioning correctly, but the instructions were lacking or the individual did not follow the instructions. Many instructions and warnings may appear ridiculously obvious, such as "coffee is hot" or "do not use clothes iron on clothing being worn." In fact, the reason these instructions exist is because someone hurt himself or herself this way, and a jury awarded the individual a settlement.

When fire reports are filled out properly, they may eventually indicate a noticeable number of fires started by one specific product. That could result in further research and, in turn, in a recall of the product. The end result is the reduction or elimination of fires caused by faulty products, which will definitely avoid future damage and possibly save lives.

Person or Persons

Identifying the person who maliciously set a fire, making an arrest, prosecuting the offender, and having him or her incarcerated prevents that person from setting another fire, at least while the individual is behind bars. The typical arsonist is not usually a dirt bag, low-life scum, or deviant bum; instead, most arsonists are people not much different from anyone you might see on any street. They are businesspeople, homeowners, and people with families; they may even live next door. The arrest made in an arson case will not only put away the arsonist, but also act as a deterrent for those who may be thinking of turning an insurance policy into a fiscal profit. It just might make a homeowner with cracked walls caused by water in the basement, which is not covered by insurance, think hard before burning the house down to collect on the fire insurance policy.

TIP

The typical arsonist is not usually a dirt bag, low-life scum, or deviant bum; instead, most arsonists are people not much different from anyone you might see on any street.

Deterrence is a tool in the fire prevention toolbox. People will think twice before setting a malicious fire if they risk going to jail. Deterrence relies on not just the fact that the fire department made the arrest, but on the fact that the department is actively looking at fires and making determinations as to the cause of fires. This fire department's vigilance in conducting investigations may provide the deterrent necessary to keep someone from committing a crime.

Wrap-Up

SUMMARY

- It is essential to determine the area of origin for a fire. Without the origin, the investigator cannot possibly determine the exact cause of the fire. In conjunction with the search for the fire origin, all other potential heat sources must be eliminated as candidates for the cause of the fire. To do so, the investigator must conduct a systematic search using scientific methodology. By eliminating all other possible causes of the fire, the investigator helps to establish the credibility of the final findings of the investigation.

- In the final hypothesis of the determination of the area of origin and the cause of the fire, the investigator must identify the first material ignited along with the source of ignition energy. The final step needed is to determine the action that brought the fuel and the ignition energy together.

- The cause of a fire can be natural, accidental, incendiary, or undetermined. It is difficult to leave any fire scene with an unknown cause of the fire—but as a result of destruction of evidence and other indicators, it may be the only solution.

- Once the area of origin and cause have been confirmed, the investigator must identify the persons or things responsible for the fire. Most homes contain a multitude of heat-producing appliances, and the improper placement of these items or misuse could have led to the failure. Other potential causes include poor design or improper manufacturing of the appliance. Depending on the outcome of the investigation, the investigator may be subpoenaed to testify in civil court about his or her findings.

- If the investigation reveals that a person was responsible for the fire, the investigator must determine whether malice was present. If someone inadvertently stacked wood in a fireplace so that a burning log rolled out onto the floor and ignited the rug, that incident involves human error but not an incendiary act, because the fire was an accident and no malice was present. The same is true when someone places gasoline in the kitchen sink to clean auto parts (not something that an average person would do, but still a real incident); the vapors could reach an ignition source and ignite. In such a fire, no malice is present, though there is plenty of stupidity (people are rarely arrested for stupidity).

- Everyone has a job to do on the fire scene. The overall focus is to save lives and property by conducting searches to find trapped victims and by extinguishing the fire as soon as possible to limit the fire damage. Yet the responsibility of the fire service does not stop when crews are packing up the hose to leave the scene. Tools are needed that can prevent fires, and providing complete and accurate fire incident reports is part of that process. The fire officer as the first responder investigator or the assigned investigator must make a determined effort to find the area of origin and the cause of the fire. Transferring this information to the NFIRS ensures that others can benefit from these data. The combined data from these reports will eventually provide guidance for future fire prevention efforts.

KEY TERMS

Arson A criminal act of maliciously burning of property belonging to another or one's self if done with fraudulent intent. Arson statues vary from state to state, with several states having a classification of arson of first, second and third degree. There area also classification of aiding and abetting of arson. The term of arson is a noun defining a crime. Incendiary as a general rule is the adjective meaning to actual cause the fire.

Burner assembly A unit that houses the electrodes in a furnace that ignite the oil pumped out of the jet nozzle; it also contains the fan, which blows air into the chamber, allowing for a bright and efficient flame.

Demarcation A distinctive line between a burned area and an unburned or lesser-burned area; can also be used to describe other delineations such as smoke patterns.

Jet nozzle In an oil furnace, a nozzle with an extremely small opening. The oil is pumped up to this nozzle, where it squirts out and is ignited by the electrodes, providing a proper flame for the burn chamber. Nozzles come in varying sizes.

Malice A wrongful act that includes the intentional desire to cause harm.

Pooling patterns An accumulation of a flammable liquid, a puddle. Identified by staining of the floor surface or if the flammable liquid burned, it will show a sharp line of demarcation between the burned and unburned area.

Scale A flaking of metal as the result of oxidation.

Suspicious Upon examination of the scene there are items, issues, or indications that are questionable,

either the facts surrounding the fire scene or the actions of an individual that may indicate that the fire cause to be other than accidental.

Trailer A deliberate use of a flammable liquid or other combustible items to intentionally move the fire from one point to another.

REVIEW QUESTIONS

1. What is the primary reason why fire crews are allowed to stay on a fire scene and search for the origin and cause of the fire they just extinguished?

2. If the investigator discovers evidence that a fire is incendiary in nature, what is being risked by not obtaining a search warrant at that point?

3. What is the best device to use to obtain a good sample of a potential accelerant to send to the laboratory?

4. All fires are considered _____ until proven otherwise.

5. What is the advantage of doing a detailed determination of the fire cause if you know that the fire was accidental?

6. Give an example of a fire that could be classified as having a natural cause.

7. How should a fire resulting from a 3-year-old child playing with a lighter be classified?

8. What key component must be present to accuse a person of a criminal intent in setting a fire?

9. Why must you have an area of origin before you can find the cause of the fire?

10. Under what circumstance, mentioned in this chapter, would it be necessary to obtain a search warrant in the process of the investigation?

DISCUSSION QUESTIONS

1. In the case study at the beginning of this chapter, the area of origin was labeled as just outside the furnace. If the furnace failed on the inside, should that be the area of origin instead? Why?

2. In the case study, the cause of the fire was the malfunction of the furnace. The person responsible was the technician. Was he the only problem?

3. In the case study, the oil company settled the case by agreeing to shut down its service section. It also had to pay a third party, another service company, to visit all of the clients who had a service call to ensure that their furnaces were properly serviced. Do you think this was fair, or should the case be continued in the courts in hopes of a stronger punishment?

REFERENCES

1. National Fire Protection Association. (2017). *Firefighters and cancer*. Retrieved from https://www.nfpa.org/News -and-Research/Resources/Emergency-Responders/Health -and-Wellness/Firefighters-and-cancer

2. National Fire Protection Association. (2008). *NFPA 921: Guide for fire and explosion investigations* (Section 11.5.6). Quincy, MA: National Fire Protection Association.

CHAPTER **14**

Vehicle Fires

LEARNING OBJECTIVES

Upon completion of this chapter, you should be able to:

- Describe the safety precautions that must be taken before starting the investigation.
- Describe the fuels and fluids used in vehicles that could be ignitable.
- Identify the available hot services in vehicles.
- Discuss the electrical systems found in most vehicles.
- Describe hybrid electrical systems and their associated hazards.
- Discuss the potential heat sources and fuels available in most vehicles today.
- Explain the examination of a burned vehicle, including both the exterior and the interior.
- Identify specific factors related to fires involving motor homes.
- Identify specific factors related to fires in boats.

Case Study

School was quite a commute for one student, more than 50 miles each way. But because of the student's physical disability, staying in a dormitory was out of the question. The student had suffered a major injury, losing all sense of feeling below the arms. The family was able to find a used car that had hand controls that enabled him to at least be mobile.

On this particular day, the student got in his vehicle after class. Even before he reached the interstate, he could smell something burning. It was an unusually cold day, so he had to keep the windows closed. The smell did seem to dissipate, so he kept driving home. When he arrived home, he got in his wheelchair and went into the house. Almost immediately, his mother asked him what was burning. The smell had not dissipated so much as his body had continually ignored it and compensated and adjusted, not sending the message to his brain that something was burning.

He leaned forward and his mother noticed a blackened area on his shirt. The student moved to his bed and laid on his stomach. Pulling the shirt up, she found blisters and a very reddened area. He had experienced first- and second-degree burns. She called 911.

The student was hospitalized to treat the burns and as a precaution to prevent infection. His mother called local authorities, only to discover that the town where they lived did not have an official fire investigator. Going through the phone book, she finally found a fire investigator in the private sector who would at least come and look at the vehicle. After hearing the story and visiting the scene, the investigator offered to work the fire **pro bono** (free of charge).

At first glance, it was obvious that the seats were heated. The investigator was amazed when he looked at the upright portion of the seat (**FIGURE 14-1**): The fabric was not even burned. However, he noted some char in the area where the back of the seat met the bottom of the seat; it was slightly smaller than the size of a dime.

The design of the seat included a zipper on the back, which allowed the cover to be pulled off. Inside was the heating element—that the investigator easily pulled out of the seat. It had shorted out and continued to burn until the supply wire melted off, which was where the small hole burned through the fabric (**FIGURE 14-2**).

Not all stories have a happy conclusion. Because the vehicle was purchased secondhand, the manufacturer would not consider doing any repairs or even discussing the issue. The family could not find an attorney to take on the case—at least one willing to work within their limited budget. Because the fire

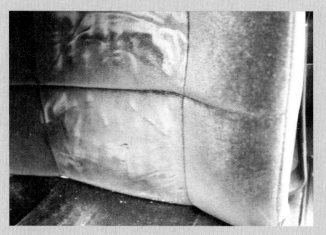

FIGURE 14-1 Front seat driver's side showing heat damage of seat fabric.
Courtesy of Russ Chandler.

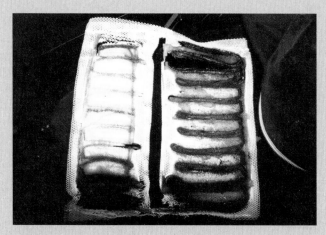

FIGURE 14-2 The heating element that was located directly behind the fabric on the seat's upright portion.
Courtesy of Russ Chandler.

department did not respond, there no fire report to document the potential danger. However, the family sent photos of the seat and copies of the private investigator's report to the manufacturer and to the Consumer Product Safety Commission, the National Highway Traffic Safety Administration, and the National Transportation Safety Board.

The student suffered first- and second-degree burns to 10 percent of his back. Because of his condition, he did not feel the pain so much as the extreme inconvenience. He could no longer drive to school and had to stay home until his burns healed. It was hoped that he would one day return to college. In the meantime, his mother had to change his bandages daily, had to carry the guilt that she bought the vehicle, and worried that her son might not go back to school.

Introduction

Vehicle fires are undoubtedly some of the most confusing and difficult fires to work. Some people believe they are the easiest to investigate—probably because they are not exactly sure what they are looking at. For most people, a vehicle is a mode of transportation; for some, it is a status symbol. Vehicles can also include farm equipment, heavy machinery, off-road vehicles, and the like. The key to investigating a vehicle fire is not just determining the origin and cause, but also discovering how the vehicle looked before the fire.

> **TIP**
>
> Vehicle fires are undoubtedly some of the most confusing and difficult fires to work.

We teach investigators throughout their career that in investigations they need to do fire scene reconstruction—to remove all the debris and place the contents back to their positions prior to the fire. This can help them determine a more accurate origin and cause. This rule applies to both structure and vehicle fires. As with structure fires, the investigator needs to work the vehicle fire scene from the area of least damage to the area of most damage.

> **TIP**
>
> As with structure fires, the investigator needs to work the vehicle fire scene from the area of least damage to the area of most damage.

Investigators must not become complacent when investigating vehicle fires. They must take each step to make sure they have covered all the basics; they should not treat these fires as minor incidents. Vehicle fires are just as important as any structure fire. The investigator needs to ask for assistance in the areas with which he or she is not familiar or comfortable. The Internet is a great resource; you can find dealership and manufacturer sites for specific vehicles, from which you can obtain information on components or even the particulars of the vehicle's construction features.

The passenger vehicle today is an extremely complex device, although it is very simple to use. It is full of hundreds of feet of wires, tubing, and combustible material, and for the most part uses ignitable liquids for combustion. These characteristics add to the fuel load of most vehicles, enhancing the burning process. This point is important to remember when it comes to all heavy equipment, as they have even more ignitable fluids and heat sources available. This chapter seeks to give the first responder investigator the tools to investigate the vehicle fire initially and to recognize when an assigned investigator needs to respond.

National Fire Protection Association (NFPA) statistics show that 174,000 fires occurred in highway-type vehicles in 2015. This was a slight decrease from previous years, but because of the rising costs of vehicles, the dollar loss was higher. These statistics also show that fires occur in highway vehicles at the rate of one every 3 minutes 1 second. According to the NFPA, highway vehicle fires resulted in $1.2 billion in property losses in 2015. An estimated 23,000 vehicle fires were intentionally set in that year, excluding fires whose causes were unknown—a highly significant increase of 21.1 percent from the number in 2014.[1]

Safety and Initial Examination

To reinforce a key message in this text, the issue of utmost importance in every investigation is safety. Investigators should take all necessary precautions to protect themselves and others from injury. Unfortunately, fire investigators tend to not be as cautious with vehicle fires as they are with structure fires, probably because they are usually working outside of a structure in the open air. The mindset is that open air equals clean air—which is far from the case.

The materials in modern vehicles contain numerous chemicals that can lead to the release of toxic fumes from burning components. The foams and plastics in vehicles give off numerous toxins when subjected to heat. These toxic chemicals range from hydrogen cyanide to carbon monoxide. **Off-gassing** is a term used in the fire service to describe the process by which materials—fabrics, in particular—absorb the by-products of combustion during the fire and release the gases after the fire. Carbon monoxide is a colorless, odorless gas, so you cannot see it. When absorbed in a large quantity in the lungs, however, it can be detrimental to health. Fire gear may also off-gas the smoke and particulate matter produced by fires. It is for this reason that all personnel who investigate fires must use self-contained breathing apparatus (SCBA) or respirators, even when dealing with cold scenes. Debris that may be cold on top may be holding heat within or underneath. Following the investigation, all gear should be properly decontaminated.

> **TIP**
>
> Off-gassing is the process in which materials—fabrics, in particular—absorb the by-products of combustion during the fire and release the gases after the fire.

The first responder investigator should not assume that the initial fire crews have secured the vehicle. Is the vehicle in a position such that it may roll or move during the investigation? One interesting note about electric hybrid vehicles is that the vehicle may engage and start up if the accelerator is pressed, even though the motor is not running. If the vehicle's electrical system has been damaged, shorting may occur, causing the vehicle to move or, even worse, to begin to burn again. Chocking the wheels and making sure that the vehicle's battery has been disabled by fire crews are musts.

> **TIP**
>
> Electric hybrid vehicles may engage and start up if the accelerator is pressed, even though the motor is not running.

Air bag deployment is a huge area of concern in almost all vehicles today. There are side air bags, front-passenger air bags, driver's-side air bags, and a multitude of configurations to protect the passengers from impact. All of these bags contain a chemical propellant—for example, sodium azide. When an impact occurs, a rapid chemical reaction takes place within the inflator that creates mostly nitrogen gas, which inflates the air bag. Bags can inflate in 1/20th of a second,[2] leaving little time for the investigator to move out of the way. Typically, the air bag will deploy during the fire if the canister containing the pressurized gas reaches a temperature greater than 300°F.[3]

As a fire investigator, you should not crawl under the vehicle during your investigation. Not only are you possibly contaminating your protective clothing by lying in the unidentified liquids that have leaked from the vehicle onto the ground surface, but you are also putting yourself in danger from the vehicle. Shifting weight within the vehicle could cause it to move or collapse, causing crushing injuries, or jagged steel items cut by fire crews or ripped and torn by the vehicle accident impact may protrude from the underside of the vehicle, causing harm. One way to avoid these risks is to have the towing service raise the vehicle

with all safety precautions in place. This will give you a chance to briefly look at the underside of the vehicle. At no time should you crawl under the vehicle while it is lifted. Instead, make all observations from a safe area off to the side. By having the vehicle raised, you can observe smoke and soot stains or flame impingement from a safe distance. Any items of evidence that may be recovered from the physical site of the accident will remain after the vehicle is removed. In some case, the impound lot and garage may have lifts that could be used to examine and photograph the underside of the vehicle.

> **TIP**
>
> Do not crawl under the vehicle during a fire investigation.

Always be aware of sharp edges and broken glass, and wear protective equipment at all times. Latex gloves do not afford any protection from sharp edges and glass. Heavy equipment has many moving parts, so pay particular attention to pinch points or areas where the parts are barely attached due to the heat. These parts are extremely heavy and may dislodge at any time. If conducting the scene examination when it is dark, it is essential to have adequate lighting to enhance safety and allow for the efficient investigation of the fire.

> **TIP**
>
> Always be aware of sharp edges and broken glass, and wear protective equipment at all times. Latex gloves do not afford any protection from sharp edges and glass.

Type of Vehicle and Ignitable Fuels

The manufacturing divisions of the automotive industry are constantly making changes to their products. Competition, popular demand, safety innovations, and new technology keep the vehicles we use daily ever changing in appearance and functionality. Currently available models include **diesel, biodiesel, gasoline, electric hybrids, ethanol, methanol, compressed natural gas (CNG)**, and **liquid propane-driven vehicles**, both passenger and commercial. All have their own distinctive characteristics, yet contain very similar components. The basic structures of various vehicles are the same—a framework, wheels, interior, fuel and electrical systems, and so on.

The engine area is where major differences arise between various types of vehicles, with the fuel storage being the second area of concern. One item that stays the same on most vehicles, with the exception of a fully electric vehicle, is the cylinder in the engine that serves as the combustion chamber. This is where fuel and air come together to form an ignitable mixture, which is then ignited by a spark. This explosive mixture drives the pistons, which then turn the crankshaft. Vehicles have a transmission that attaches to the engine and, when engaged, transfers the turning of the crankshaft to the gears and makes the vehicle move.

Gasoline

Gasoline is the most commonly used vehicle fuel today, with alternative fuels largely being used as supplements to gasoline. Gasoline has a flash point of −43°F to −45°F. It is not soluble in water, and its ignition temperature is 536°F.[4] In a vehicle, gasoline is stored in a sealed tank, attached to the vehicle frame. Fuel is pumped from the tank through a fuel filter and to the injectors. The fuel is sprayed into the cylinder, with a spark then igniting the mixed air and fuel, thereby pushing the piston downward on the power stroke. This process is called electronic fuel injection (EFI). The EFI system is under great pressure, with some operating at more than 100 psi.[5] Lack of maintenance or worn and loose parts can cause the fuel to be sprayed outside the cylinder, atomizing it. This could be a possible cause for the spread of the fire in the event this fuel comes into contact with an ignition source.

Diesel Engines

Diesel engines operate on a compressed combustion chamber (cylinder). Fuel is sprayed into the cylinder prior to the pressure stroke. The fuel–air mixture is compressed in the cylinder and an explosion occurs, pushing the piston back down to the exhaust stroke. This action results in the revolution of the crankshaft, driving the piston down and back up the cylinder to the intake stroke of the motor. As air is drawn into the cylinder, the process starts over.

The diesel engine needs diesel fuel to run. Diesel fuel is a hydrocarbon (petroleum distillate) combustible liquid with a flash point exceeding 100°F. By contrast, biodiesel is not a hydrocarbon fuel; it is made from vegetation and animal fat along with other items such as alcohol and lye, and will burn in a diesel

engine without modification. Its flash point varies, but starts around 266°F and can be higher depending on its components. Biodiesel will combust in the engine cylinder.[6] Diesel fuel can also contain a certain percentage of biodiesel.

In diesel-fueled vehicle, the fuel is stored in a sealed tank and is pumped from a fuel pump through lines and filters, which run along the frame rails of the body of the vehicle. The fuel is then pumped under pressure—upward of 78 psi—into the engine's cylinder to begin the combustion process.[7] Failure of the system at any given point may be a possible point of fuel release in some vehicle fires.

Maintenance issues come into play here in relation to the upkeep of the vehicle. Were the fuel lines tightened correctly? Did leaks supply the fuel to an ignition source? Was the right fuel used in the vehicle? One common consumer mistake is pumping gasoline into a diesel fuel tank. The nozzle of most gas pump dispensers will fit into the neck of the diesel tank. In contrast, the nozzle on most diesel dispensers will not fit into the neck of a gasoline-fed vehicle.

FIGURE 14-3 shows a picture of a logging skidder that had a fuel line break from debris, ripping it away from the fuel pump. Constant daily maintenance of heavy equipment, especially, can be a major issue for the operator. If debris is not cleaned away from moving mechanical parts or heat sources, or if it dislodges hoses, fires can and will occur. In this case, by using a like or similar machine and obtaining the assistance of heavy equipment specialists, the investigator can make comparisons to judge how the fire started and the reason why. **FIGURE 14-4** shows how comparing the fire-involved vehicle with a similar machine can help the investigator determine the area of origin and identify which parts were where.

FIGURE 14-3 Logging skidder burned as a result of a fuel leak.
Courtesy of Bobby Bailey.

FIGURE 14-4 A comparison skidder to show missing parts.
Courtesy of Bobby Bailey.

Gasoline Engines

Gasoline, a volatile fuel with a relatively low ignition temperature, raises concerns because it is the first fuel ignited. With the fuel tank usually located in the rear of the vehicle, the fuel lines run up the frame of the vehicle to the engine. The force required to get the fuel from the tank to the engine comes from the **fuel pump**. The majority of fuel pumps in today's vehicles are located within the fuel tank. However, older vehicles and vehicle kits may use a fuel pump that can be located anywhere between the tank and the engine.

Gasoline engines operate on flammable liquids that have a flash point of less than 100°F. A multitude of additives may be mixed with the gasoline, including ethanol and methanol.

Ethanol—also known as ethyl alcohol, drinking alcohol, grain alcohol, or simply alcohol—is a flammable, colorless, slightly toxic chemical compound with a distinctive perfume-like odor; it is the same alcohol consumed in alcoholic beverages. According to a 2018 report from the U.S. Department of Agriculture,[8] most ethanol is produced from corn, with as much as 40 percent of the corn grown in the United States being transformed into ethanol. In 2018, according to the U.S. Energy Information Administration, the United States consumed approximately 142 billion gallons of gasoline.[9]

Ethanol fuel is a biofuel alternative to gasoline. It can be combined with gasoline in any concentration up to pure ethanol E-100. As a fuel, ethanol is primarily used in two forms. E-10 is a blend of 10 percent ethanol with 90 percent unleaded gasoline; this form can be used in any vehicle. E-85 is 85 percent ethanol blended with 15 percent unleaded gasoline; it can be used only in specially built vehicles. On a practical level, ethanol can be found in mixtures containing 10 to 85 percent ethanol in gas pumps and as 95 percent pure ethanol with 5 percent gasoline added in rail cars, tank trucks, and barges.

> **TIP**
>
> Ethanol can be found in mixtures containing 10 to 85 percent ethanol in gas pumps and as 95 percent pure ethanol with 5 percent gasoline added in rail cars, tank trucks, and barges.

Both methanol and ethanol are water-soluble. Both also have characteristics that enhance the burning process, creating less exhaust emission. This is the trade-off for a cleaner environment: a vehicle that will burn hotter and faster, with less emission gases.

Other Liquids

Other liquids in a vehicle include transmission fluid, windshield washer solvents, ethylene glycol, motor oil, brake fluids, and other hydraulic fluids. All of these liquids have different flash points but may become ignitable under extreme conditions, either contributing to or causing a fire.

Commercial vehicles in particular carry large amounts of these other liquids, as well as fuel. Some carry as much hydraulic fluid to run different equipment attachments as they carry fuel. For example, trucks with snowplows, tailgate lifts, and farm equipment for implements carry large amounts of hydraulic fluid. Care should used when conducting fire investigations of such equipment because pressure may have built up in the hydraulic lines for the attachments. If you attempt to push or close a valve or knob, without knowledge of the vehicle, the implement could be lowered or dropped, causing injury.

> **Physical Data on Ethanol[7]**
>
> - Boiling point: 173°F (75°C)
> - Melting point: −179°F (−117°C)
> - Specific gravity: 0.7893
> - Volatility: 100 percent
> - Vapor pressure: 40 mm Hg at 13°C
> - Evaporation rate: 1.4 (carbon tetrachloride = 1)
> - Solubility in water: Complete
> - Appearance: Clear, liquid
> - Odor: Pleasant odor
> - Flash point: 55°F

Personal vehicles such as cars, pickup trucks, or motorcycles also have fluids that can be ignited when not properly contained. Like hydraulic fluid, most liquids found in vehicles are ignitable in varying degrees. When the vehicle is operating properly and the liquids remain in their containment areas, they pose no problem. It is only when the liquid leaks or is sprayed, or the container ruptures, that a problem may occur.

Hot Surfaces

The manifold, catalytic converter, and exhaust pipe operate at sufficient temperatures to ignite a wide range of fuels. For example, windshield washer solvent, if sprayed onto a hot surface, may ignite. This could occur if the rubber line between the windshield reservoir and the wiper becomes detached. The liquid may then spray onto the hot engine manifold instead of the windshield, possibly igniting.

> **TIP**
>
> Windshield washer solvent, if sprayed onto a hot surface, may ignite.

Leaking brake fluid is most often the first fuel ignited when it comes into contact with a hot surface. Transmission fluid, especially if heated from the vehicle being overloaded, leaking onto the catalytic converter of associated exhaust pipe is also capable of ignition.

Gasoline, however, usually will not ignite when sprayed onto a hot surface. Many variables such as ventilation or humidity can influence whether the gasoline ignites in a particular scenario. The investigator should not dismiss this outcome as not probable, but rather should continue the investigation, seeking any possible heat source and fuel that may have been involved in the ignition sequence.

In **FIGURE 14-5**, the heat from the vehicle's many moving parts ignited the combustible material of dried grass or straw. This hay baler suffered major damage as a result of many people becoming too complacent in their duties of cleaning or servicing machinery. Fatigue is a major contributor of many farm-related fires. Improper or lack of lubricating of parts is another cause.

Electrical Systems

Most vehicles—but not all—operate on a 12-V direct current (DC) system. Notably, some commercial

FIGURE 14-5 Hay baler fire as a result of heat being too close to combustibles.
Courtesy of Bobby Bailey.

vehicles, such as tractor-trailers, farm equipment, and boats, operate on a 24-V system. Hybrids—relatively new vehicles that have an electric motor to supplement the gas power engine—have opened up a whole new realm in the investigation of vehicle fires. The voltage in hybrid vehicles can reach more than 300 V of alternating current (AC).

> **TIP**
>
> Some commercial vehicles, such as tractor-trailers, farm equipment, and boats, operate on a 24-V system.

> **TIP**
>
> The voltage in hybrid vehicles can reach more than 300 V of alternating current.

Direct-current (DC) 12-V vehicles operate with a lead acid battery that feeds circuits and fuses in the vehicle. The voltage is supplied to the circuits through wires that run from the battery to the fuses and switches. The energy is then distributed throughout the vehicle to various applications. The wires coming from the battery to the main fuses (located in the engine compartment) and switches are typically 2 to 6 American Wire Gauge (AWG) in size, often called the primary wiring. The wires from the fuses to the distribution areas can run in size from 8 to 18 AWG; these wires are referred to as secondary wiring. All of these wires are insulated with a coating to protect them.

FIGURE 14-6 Cracking of wires from age and from continuous heating and cooling of their insulation.
Courtesy of Bobby Bailey.

FIGURE 14-7 Battery pack and components of the Honda Insight located in the rear passenger area under the seat.
Courtesy of Bobby Bailey.

The secondary wiring (located in the passenger compartment) includes the wires that run from the fuse panel to other switches, outlets, and lights, which carry the load of the vehicle's applications that the driver of the vehicle uses every day. These wires operate the lights, radio, windshield wipers, heated seats, and anything the vehicle uses besides ignition. The secondary wiring can come into play in fires stemming from wear and tear and misuse. Overloading these wires can occur with the addition of cell phones, computers, DC/AC converters, and so forth. Wiring to these accessories typically consists of 16 to 18 AWG wires, which are designed to carry no more than 100 to 200 W. Vehicle operators sometimes disregard the maximum capacity of these outlets and strain the wire. This strain, which is called resistance heating, can cause the insulation around the wire to break down or melt. The cracking or melting of the insulation may expose the wire to a short or an arc, an illuminous discharge across a gap that can cause a fire. In **FIGURE 14-6**, the wires show signs of cracking as a result of age and heating and cooling of their insulation.

Hybrid Vehicles

Hybrid vehicles combine a smaller internal combustion gasoline engine with an electric, battery-powered motor. There are basically two types of hybrid systems: series and parallel. A series hybrid system switches between gasoline and the electric power source, whereas a parallel system supplements the engine.

Hybrid vehicles contain a battery pack within the interior of the car. The batteries are made of high-voltage nickel–metal hydride (Ni-MH). The packs consist of individual battery cells that contain potassium hydroxide (KOH). KOH has a pH of 13.5 and is highly alkaline. KOH liquid is absorbed in each cell by a thin membrane paper; the result is a gel, or a dry cell battery. The voltage of the battery packs may range from 100 V to upward of 300 V, considered a high voltage. Care should be taken when working in or around these vehicles after a fire.

The Honda Insight has approximately 54 C-size cell batteries in each battery pack, with four to six packs found in each vehicle battery system (**FIGURE 14-7**). The battery system is located in the rear of the Honda under the passenger seat. The battery packs are tied together in series, in which the main feed wires run to the DC/AC converter located in the engine compartment area of the vehicle.

In a series hybrid, the gasoline engine will automatically stop and start as the vehicle is running. The gasoline engine stops when the vehicle comes to a halt, such as at a traffic light. Pressing the accelerator pedal restarts the gasoline engine. The gasoline engine may run to recharge the battery pack. This mechanism may cause an issue for the investigator when looking at the scene. If the vehicle is not shut off and the gas pedal is accidently touched, it may engage the engine. In a parallel system, the electric motor runs in conjunction with the gasoline engine; that is, the vehicle does not operate on the electric motor alone. The gasoline engine shuts down only when the vehicle comes to a complete stop.

> **TIP**
>
> If a series hybrid vehicle is not shut off and the gas pedal is accidently touched, it may engage the engine.

Both Honda and Toyota battery packs are well insulated and protected from impact damage during a collision because of their location above the rear axle. The KOH in these batteries will react violently if it comes in contact with metals such as aluminum, zinc, and tin. The battery packs themselves have several shut-off switches located on the vehicle. Just because the investigator shuts off one of the switches, however, that does not mean the system is secure. In **FIGURE 14-8**, the shut-off switch for this vehicle is located under the hood in the engine compartment. It shuts the power off to the electronic box to the engine but does not shut the power off from the batteries; a second switch that needs to be turned off and secured to shut off the batteries.

If a hybrid vehicle is partially or completely submerged, responders and investigators should not touch any high-voltage components because of the electrocution hazard. As mentioned earlier, components of the high-voltage system may have in excess of 300 V. Most manufacturers use a bright orange cable to identify the voltage system. The wires, which are anywhere from 0 to 4 AWG in size, carry high voltage from the battery pack to the DC converter of the vehicle. These cables can run either inside (**FIGURE 14-9**) or under the vehicle (**FIGURE 14-10**) depending on the manufacturer. In addition, a 12-V DC battery in the vehicle operates the accessories.

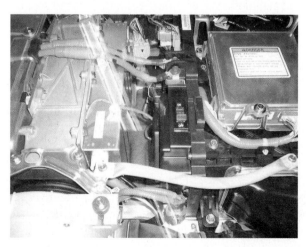

FIGURE 14-8 Shut-off switch under the hood of a Honda Insight.
Courtesy of Bobby Bailey.

FIGURE 14-9 Cable (orange) running inside the Honda Insight; the cable runs under carpeting beneath the driver's seat.
Courtesy of Bobby Bailey.

FIGURE 14-10 Underside of a Toyota hybrid, showing cable (orange) from battery pack.
Courtesy of Bobby Bailey.

Other Forms of Heat

The investigator assigned to a vehicle fire needs to examine the vehicle's heater and fans, as well as all wires attached to these components. These items need to be examined by someone who is familiar with the vehicle's components—looking at all these wires can be a specialty in itself. Just because you may find a bead on some wires, it is not definite proof that a short there caused the fire; instead, the bead could have been created as a result of the fire. Basically, the only thing that such a bead shows initially is that the wire carried current at the time of the fire. The investigator needs to eliminate heated seats, power mirrors and seats, and other electrical applications built into the vehicle as sources of heat of ignition. The investigator must also look for items, or their residue, indicating that aftermarket items were installed. If so, they should be examined to ensure proper installation and proper operation.

A properly operating catalytic converter, located on the underside of the vehicle, can reach temperatures of more than 700°F externally.[10] The converter is positioned in close proximity to the floorboard. The interior covering on the floorboard includes an insulation barrier, with the carpeting then placed over that insulation. Heat can build up on the underside of the vehicle from a clogged system that has not been maintained, possibly causing heat transfer through the metal floor to the combustibles in the interior. Obtaining owner or driver feedback on the maintenance and how the vehicle was running prior to the fire is important to verify the damage found at this point.

In one fire incident, a driver, weary from travel, parked his passenger car next to the curb at the rear of a shopping mall to nap. The vehicle was backed into a parking space where leaves and debris had accumulated. The vehicle was running for approximately 6 hours to keep the occupant warm on the cold night. The vehicle's catalytic converter and the exhaust system ignited the dry leaves under the car and ignited the vehicle. Leaking fluids from lack of maintenance fed the flames as the fire consumed the vehicle—unfortunately with the passenger inside.

Turbo chargers are located within the engine compartment and add as much as 40 percent more horsepower to the vehicle's engine. The turbo charger collects air from the exhaust and turns a turbine at a speed of more than 100,000 rpm.[11] The turbine then forces collected clean air into the engine, in addition to the fuel, to increase combustion and power. These appliances create tremendous amounts of heat and may be a source of ignition. The air is pulled in from the exhaust and pushes the turbine, which is positioned on a shaft. The shaft turns an opposite turbine that forces air into the combustion chamber.

Smoking-related causes of vehicle fires can be difficult to find. However, looking at the ashtray of the vehicle and talking with the occupants of the vehicle may lead to discussions of how the fire began. Potentially, a slow smoldering fire may have begun low inside the vehicle or in the seats. The smoking-related items must have combustibles to ignite or the fire will most likely burn out. Information from the occupants about smelling smoke for a while may indicate that a lit cigarette or match was dropped in the floor or seat area.

Vehicle Burns

Most vehicles are extremely safe and reliable. But like any other mechanical device, they must be maintained on a regular basis. If the vehicle was kept in clean and running condition and a fire still occurs, the incident may have involved arson. If the first responder investigator looks at the vehicle and cannot come to a conclusion by using deductive reasoning and a systematic search of the fire, an assigned investigator needs to be called.

The owner may have burned the vehicle for the purpose of insurance fraud or to get out from under a lease agreement. In some cases, a vehicle may be burned by another person as a vendetta against the owner. Sometimes the vehicle may have been burned simply to escape a high auto-loan payment or even high fuel prices. Given the numerous sport-utility vehicles (SUVs) built and purchased over the last 10 to 15 years, this type of arson may become more prevalent over the next couple of years, if gas prices rise substantially. Either way, a more experienced investigator needs to be called because this determination exceeds the first responder's role in the investigation. First responders need to recognize that this is the time to ask for assistance.

At all times, the investigator has to work the vehicle fire scene from the area of least damage to the area of most damage. Investigators will likely miss a great deal when they jump into the scene before checking the outside perimeter. There is nothing worse than having potential evidence stepped on and crushed into the ground, never to be recovered, because the investigator went directly into the interior of the vehicle.

The investigator should first photograph the scene from all sides. Photos of the area from the vehicle looking outward need to be taken as well. This is particularly important when a stolen vehicle is found, as such photos will show the egress path and direction to the scene. Are there tire tracks leading into the scene? On your exterior walk-around, did you see evidence located nearby that may be associated with the fire? All of these items need to be photographed and documented before conducting the examination of the vehicle. In fact, all evidence needs to be documented and photographed before moving and collecting it.

Examination of the Exterior of the Vehicle

The extent of damage may show the direction of fire travel; burn marks or demarcation lines may direct the investigator to the path of travel; and the patterns left behind may show the direction from which the fire began and spread, typically up and out. Specifically, a fire that begins inside a vehicle will show lines of demarcation on the metal. As it burns upward, the fire will spread from side to side. The result will be a V pattern, much like that seen in a structure fire, which will show the possible direction of the fire.

Looking at the vehicle's damage will indicate whether the fire originated inside or outside of the vehicle. If no damage from flame impingement and/or smoke is observed on the underside of the vehicle, it is very possible that the fire originated inside. The investigator also needs to check the doorjambs for damage, possibly indicating that the door was left open or that someone tried to remove the **vehicle identification number (VIN)** plate.

The VIN plate is usually located in the driver's-side doorjamb or on the lower corner of the inside windshield on the driver's side. This plate is usually made of aluminum alloy and may melt at very low temperatures. The VIN contains information about the vehicle, such as its year, make, model, engine, and paint color. Although the VIN plate may burn partially or completely away on the dash, law enforcement officers may be able to retrieve this information from several other secret VINs stamped on the vehicle.

FIGURE 14-11 shows a VIN plate on a vehicle located on the driver's side of the dashboard. This VIN plate has been cleaned using water and cloth. If debris does not come off the plate with this treatment, you could use a small piece of steel wool to gently remove debris so that the number can be read.

Burn patterns may not be as obvious on total-burn vehicles. These cases will be harder to solve because of the extent of damage. The metal remaining on such vehicles will be distorted and buckled, and the first responder may not see any lines of demarcation. Nevertheless, any remaining paint left in areas of the vehicle

FIGURE 14-11 Vehicle identification number (VIN) plate after it has been cleaned.
Courtesy of Bobby Bailey.

and plastics that may not have melted may give clues to the fire's direction. If the inside and the outside of the vehicle are totally burned but the bottom shows no damage, it is probable that the fire originated inside the vehicle.

One interesting facet to examine is the vehicle's glass. Most vehicles use tempered glass for the side and rear window; this glass shatters into small pieces upon impact or fire. Tempered glass will fall into the vehicle or just outside it in the event of a fire. The position of the glass pieces enables the investigator to determine whether the glass was intact with the window in the up (closed) position prior to the fire. If the investigator finds a large percentage of glass inside the door, the window may have been down at the time of the fire. If still intact, the window mechanism will indicate the position of the window at the time of the fire. However, most newer vehicles use plastic parts for this assembly, whereas older vehicles may have metal construction. In any case, the glass will show soot or smoke stains that may have been left prior to breaking from the fire.

Unlike the other vehicle windows, the front windshield is made of laminated safety glass. Fire will cause long cracks on the front windshield where the laminate separates on the side of the area of higher heat. These long cracks can appear to point to the area where the laminate in the safety glass has separated as a result of the heat and flame. The glass, if found intact and still in the vehicle, may give a clue as to the side and location of the fire. For example, a fire that begins in the driver's-side floorboard will break out the lower driver's-side area of the windshield, and the fingers or cracks will run from this area all the way across to the passenger side. This shows that the fire occurred low in the vehicle, with flame impingement occurring through the dash and just over the dash.

FIGURE 14-12 Photo of laminated windshield with flame from top down, showing fire originated above the floor.
Courtesy of Bobby Bailey.

FIGURE 14-13 Vehicle that has burned, with indicators showing that the fire originated inside the passenger area and not in the engine compartment. Note the remaining materials on the front bumper and the paint on the bottom of the driver's-side door exterior.
Courtesy of Bobby Bailey.

In contrast, a fire that occurs in the rear of the vehicle will create cracks on the windshield running down toward the hood of the vehicle as well as from side to side. In the photos provided in this section, note that the glass has cracked and indicates where the flame has come through.

In **FIGURE 14-12**, the fire came through the windshield high near the roof line. The fire originated in the rear seat of the vehicle. Note that the laminate separates halfway down the windshield, with flame coming through on the top. This idea that the vehicle's windshield can provide information on the fire's direction and travel is valid only if this tool is used early enough after fire suppression activities, as part of the systematic approach to the investigation, and combined with deductive reasoning and a hypothesis.

Look at the tires remaining on the vehicle. Do they show consistent damage? Are the front tires destroyed and the rear tires intact, or vice versa? Some newer vehicles have a plastic fuel tank that may rupture and spill fuel contents below the vehicle, which enhances the burning process on the underside and the potential ignition of the tires. This can give the investigator a false reading on where the fire occurred because of additional V patterns on the vehicle.

In **FIGURE 14-13**, the vehicle was set on fire in the seats. Notice the burn marks left on the doors—they are up several inches from the bottom of the door. The front bumper is still intact, as are the plastic headlight covers. The roof area is buckled over the center of the vehicle. The area over the driver's-side door shows flame impingement because rust is already developing. This effect is not seen on the other doors. The investigation should also show whether the door windows were up or down prior to the fire. Note that the tires are flat on the driver's side from heat. However, they did not help fuel the fire, as the exterior burn

patterns show. The tires may also indicate the direction of flame, in particular on the inside side wall of the tire. The rubber in the tire will begin to melt in the area of flame impingement from the outside in.

Examination of the Vehicle Interior

Look at the vehicle's contents. Items may or may not be present in the glove box. Did the owner leave the keys in the vehicle? Can the keys be located? The metal components of the steering column, including the key assembly, will sometimes fall onto the floorboard on the driver's side. The keys, unless initially visible, may not be seen without the use of an X-ray machine. The carpeting and materials need to be collected to see if the keys are, in fact, there.

Most of the interior of a vehicle is combustible, composed of either plastics or fabrics. These items melt or burn very quickly, leaving melted globules or the metal frame in their stead. However, these remains may also provide information about the fire's direction. **FIGURE 14-14** and **FIGURE 14-15** present before and after photos of the inside of a Mack truck. The fire originated in the floorboard of the passenger side of the cab—an electrical fire from frayed wiring. Notice the low burn patterns on the interior from the floor area of the passenger side.

What do the hood and trunk areas look like? If the burn patterns appear on the vehicle's exterior and not on the inside, likely nothing in the engine compartment caused the fire. The trunk usually will not show any impingement inside unless the fire traveled

Questionable Manuscript

The contents should be examined to see both what is in the vehicle as well as what is not in the vehicle that might normally be expected to be found there. If there are no personal belongings, a line of questioning may clarify that this was not unusual. If there are reported contents but they are not found at the time of the investigation, this may be a concern. For example, an author was typing his manuscript with a typewriter instead of a computer. He had been given an advance on his book. He packed up his vehicle for the 200-mi trip to the publisher, including his written manuscript, which was placed on the floor or the front passenger seat. He left his home at 4 AM so as to arrive when the publisher's office opened in the morning.

The author reported to the engine company officer that he smelled gasoline just before smoke started pouring into the passenger compartment. He pulled over and saw flames coming from under the dashboard. The vehicle was burned from bumper to bumper. The vehicle owner repeatedly stated that his life's work was in the vehicle and now it was gone forever. He explained that his publisher was expecting the finished product that morning. He then asked where he could get a copy of the fire report.

The engine lieutenant, as the first investigator on the scene, looked on the front passenger floor. The dash there had melted and dropped onto the floor. When he asked the author how thick his manuscript would have been, the replay was 5 or 6 in. The lieutenant called for a full-time investigator.

When the investigator arrived, the pair cut away the area of the front passenger floor in hopes of providing what remained of the manuscript to the vehicle owner. Prying the melted plastic from the floor mat revealed that nothing was there—no carbon, no paper residue, no manuscript. The vehicle owner confessed that he had been experiencing writer's block, that the publisher was expecting a product that he had not produced, and that he was already well over his deadline. He set the fire to cover up his failure to complete the manuscript, in hopes that the publisher would not ask for the return of the advance money and would give him more time to write his book.

Instead, the author was charged with arson and providing false information to a public official, and taken away. When the day of his trial arrived, the author truly showed remorse, explained his dilemma to the courts, and pled guilty. He was sentenced to 2 years; his sentence was suspended and he had to pay restitution to the county for the engine company's response. In addition, the insurance company denied his claim to collect any monies on the burned vehicle.

FIGURE 14-14 The before picture of the interior of the cab of a Mack truck.
Courtesy of Russ Chandler.

FIGURE 14-15 The after photo of the same Mack truck interior cab after burning.
Courtesy of Russ Chandler.

through the rear passenger seat area into the interior of the trunk. Even so, the trunk may contain potential heat sources. For example, the vehicle owner may install items in the trunk such as large audio equipment. Some vehicles have factory-installed equipment in the trunk such as retractable antennas, and some even have the batteries located in the trunk.

TIP

The trunk may contain potential heat sources—for example, owner-installed audio equipment, factory-installed retractable antennas, and even batteries.

The hood will show burn patterns from the heat and flame impingement if the fire occurred inside the engine compartment. The paint will begin to melt and burn away, spreading outward across the hood. Then, the hood's metal will start to bubble or distort. Shortly thereafter, the hood will begin to rust very quickly owing to oxidation and the introduction of moisture from the hose streams during fire suppression, along with rapid cooling of the metal. Hood damage will possibly show the side of the engine area involved.

Motor Homes

Motorized homes on wheels (commonly called motor homes or recreational vehicles [RVs]) bring a structure to life on the road. The motor home contains the components of a vehicle combined with the combustibles of a structure—furniture, plastics, and electronic equipment, to name a few such items. Add a fuel tank and an engine, and if not properly maintained, the motor home could burn with massive damage. In **FIGURE 14-16**, the fire in the motor home began on the inside from food on the stove. The close proximity of cabinets and combustibles in such a vehicle does not allow for a lot of room for error with a heat source nearby.

Motor homes are unusual compared to a truck or car, and fires involving them must be investigated with care. They may contain liquid propane gas for heat and fuel for the cooking equipment. Some may have a built-in generator that uses a flammable or combustible liquid. Sometimes the tanks that contain the fuel are shared with the engine, so the tanks are very large. Electrical panel boxes located on the unit

tie into AC current with an electrical cord attached to a power source.

> **TIP**
>
> Motor homes are unusual compared to a truck or car, and fires involving them must be investigated with care. They may contain liquid propane gas for heat and fuel for the cooking equipment.

In motor home fire investigations, everything involved in a vehicle fire investigation applies, and these additional items of interest need to be addressed as well. Working from the least damaged to most damaged areas and from the exterior to the interior is of the utmost importance. You must find all sources of ignition and look at all of the burn patterns.

Boats

Some boats may not seem much different from a vehicle or a motor home, yet all boats have some inherent differences. When the boat uses gasoline, these fumes are heavier than air and can collect in the lowest part of the hull. Such boats usually have built-in ventilation to ensure that the hull is free of ignitable vapors, but this system can fail, and there could be a potential ignition source nearby.

Depending on the size of the boat, it can be very similar to a motor home, providing all the comforts of home, along with all the associated fuel load and ignition sources. A full kitchen is not uncommon, and even elaborate entertainment systems are available.

> **TIP**
>
> Depending on the size of the boat, it can be very similar to a motor home, providing all the comforts of home, along with all the associated fuel load and ignition sources.

FIGURE 14-16 This motor home fire occurred in the kitchen area.
Courtesy of Bobby Bailey.

Other unique situations may also arise with boat fires. Perhaps the boat sank (such as the boat shown in **FIGURE 14-17**) and was subsequently raised, and now much of the contents and some of the evidence are at the bottom of that body of water. When larger boats experience fires and have either been sunk by the fire department to put out the fire or sank as a result of the fire, the

FIGURE 14-17 This boat burned, sank, and was raised. Most of the debris remained within the hull of the boat.
Courtesy of Russ Chandler.

FIGURE 14-18 The engine compartment has little fire damage and much less damage than the crew quarters.
Courtesy of Russ Chandler.

FIGURE 14-19 Debris removed from galley area, revealing char patterns.
Courtesy of Russ Chandler.

FIGURE 14-20 Inspecting the area adjacent to the location where the boat was moored may provide additional information and evidence and is essentially identical to doing an exterior examination of the property surrounding a house fire.
Courtesy of Russ Chandler.

insurance company may even hire a specialized investigator who has scuba diving skills, enabling the first phase of the investigation to take place before the boat is raised.

As with all fires, the investigation should move from the least damaged area to the most damaged area. Check the overall condition of the boat just as you would a motor vehicle. Whereas you would check the tires on a vehicle, look at the hull of the boat for damage or lack of upkeep. Is the propeller (sometimes referred to as a wheel) pitted and worn, bent, or out of round? This could signal the owner was facing the prospect of an expensive repair soon.

On boats with inboard motors, you should look into the engine compartment just as you would with a vehicle fire. These areas are quite similar to those in an automobile. Fuel tanks and associated fuel lines will be present. **FIGURE 14-18** shows the motor mounts and the area surrounding where the two engines were located (motors removed for examination). This area clearly shows less fire damage than the living quarters do.

Working from the least damaged area to the worst damaged area, the investigator should systematically examine the living compartments in the same way as the investigator would examine a home. In the boat shown in **FIGURE 14-19**, this section was covered in debris and required searching and removal of debris, just as in a house.

If anything is out of the ordinary or if the area, origin, or cause of the fire is not immediately apparent, then the assigned investigator should be notified. If the assigned investigator arrives and is not comfortable with boat fires, he or she may want to ascertain whether the insurance company plans to send a specialist to investigate the fire scene. If so, it may be in everyone's best interest to protect the vessel while awaiting the arrival of that investigator, who has specialized knowledge in marine fires. However, the fire investigator should, at the very least, take a complete series of photographs to document the entire scene, as well as create a diagram of the boat and make notes on the scene. The area where the boat was moored should be closely examined, documenting all damage. **FIGURE 14-20** shows the burn pattern on a dock, which should match the patterns on the boat. Also of interest is the electrical shore connection on the dock. The investigator should determine whether it was energized and ascertain its condition both before and after the fire.

Although conducting interviews after the examination of the fire scene may have some benefits, in this scenario it may be best to at least get preliminary information from any witnesses and suppression forces. Working with an experienced marine fire investigator will also add to the sum of the assigned investigator's experience so that he or she will be able to conduct full marine fire investigations in the future.

Finishing Up the Investigation

Investigators do not need to be mechanics, but they do need to understand vehicles. When they encounter something unusual in a vehicle, they can usually call upon resources such as auto mechanics from vehicle dealerships. If they are not familiar with boats, many experts in this field may be of assistance—boat brokers, mechanics, and even specialty investigators in the private sector who deal with nothing but boat fires.

Has the vehicle been secured from movement or engaging itself? The investigator should ensure that battery has been disconnected, and all shut-off switches locked out. Wearing protective clothing when examining a fire-damaged vehicle is critical because sharp edges of metal and glass, along with contaminated liquids remaining in the vehicle, can cause injury to investigators and bystanders alike.

The investigator must look at all facets of the case. It is important to work the fire from the exterior to the interior and from the least damaged to the most damaged areas. Take into account how the vehicle arrived at the location where it burned. Document and photograph the scene on all sides, including the surrounding area as well. If the full-time fire investigator becomes involved in the investigation, he or she may decide to systematically sift through the scene debris for evidence as part of using scientific methodology, so as to produce an accurate and credible hypothesis.

If the investigator does not know exactly how the fire occurred or if the investigation of the fire brings more questions than answers, it is important to get a second opinion, or bring in another investigator to validate the initial hypothesis and give an honest second look at exactly what happened. Make sure this investigator is not unduly influenced by the initial opinion if it does not completely explain what happened. Deductive reasoning needs to be used.

Not all cases will be solved. Statistics from the U.S. Fire Administration show that 23 percent of vehicle fires have undetermined causes. However, before reaching such a conclusion, all possible scenarios must be exhausted. For every action, there is a reaction. Something, either accidental or malicious, caused the vehicle to burn. An individual's incompetence or lack of knowledge is not a crime. Likewise, failing to properly maintain a vehicle is not a crime. However, destroying one's personal property or the property of another with malice is a crime.

Wrap-Up

SUMMARY

- As with all investigations, safety must be the first and utmost concern with vehicle fires. In these fires, there is almost always an ignitable fuel at the scene that can create additional concerns. Make certain to wear full turnout gear, which will protect against both contaminants and sharp edges.

- Investigators must be familiar with all types of vehicles and the fuels they may use. They also must be familiar with the electrical systems that could be a source of heat for the ignition of the fire. In addition, other heat sources may exist, ranging from the engine manifold to the exhaust system, including the catalytic converter.

- Familiarity with hybrid vehicles will be important for fire investigators in the near future, if they have not already encountered such vehicles. Knowing the fundamentals of their construction and operation will lead to a successful investigation as well as assure the safety of the investigator and other personnel at the scene. As with all other vehicles, numerous resources are available that can provide both fundamental and advanced information on hybrid vehicles.

- The overall examination of the vehicle fire, its progression, and the burn patterns can help point the investigator toward the area of origin, which in turn will lead to identifying the cause. This determination also relies on examination of the exterior of the vehicle, including the underside, and examination of the entire interior of the vehicle from the engine compartment to the trunk.

- Through this process, the investigator will often be able to make an accurate determination of the area of origin and the cause of the fire. In some cases, however, a vehicle fire may have to be labeled as undetermined.

- Both motor homes and boats have characteristics similar to those other motorized vehicles. Likewise, they have characteristics similar to those of a structure; thus, fires involving these vehicles should be investigated no differently from any other structure fire. These similarities can aid the investigator in the search for the area of origin and the fire cause.

KEY TERMS

Biodiesel fuel Fuel derived from vegetable sources with added animal fat and alcohol that can be used in diesel engines with no modification of the engine.

Compressed natural gas (CNG) A natural lighter-than-air gas compressed for use as a fuel; it consists principally of methane in gaseous form plus naturally occurring mixtures of hydrocarbon gases.

Crankshaft A shaft that takes the power developed by the engine and transmits it through mechanical motion to other parts of the vehicle.

Diesel fuel A combustible petroleum distillate that is used in diesel engines.

Electric hybrid A vehicle that uses both an electric engine and a conventional engine (e.g., a gasoline gas). This type of vehicle provides an economical form of transportation compared to gasoline or diesel vehicles.

Electronic fuel injection (EFI) A mechanical system, controlled electronically, that reduces a fuel to a fine spray and injects it into the cylinders of an internal combustion engine.

Ethanol A type of alcohol produced by natural fermentation of sugars; usually associated with liquors but can also be used as an additive to vehicle fuels. It is a colorless volatile flammable liquid.

Fuel pump

Gasoline A hydrocarbon fuel refined from crude oil to be used primarily in internal combustion engines.

Liquid propane-driven vehicle Vehicles designed or converted to run on propane gas. These vehicles can be new vehicles manufactured to run off propane gas or converted vehicles that had gasoline engines and

are transformed to run off propane. There are two categories, dedicated or bi-fuel vehicle. Dedicated is designed to run only on propane where bi-fuel vehicles have two separate fueling systems allowing the vehicle to run off propane or gasoline

Manifold (1) The engine component that sends the fuel–air mixture to the cylinders (intake manifold). (2) An engine component that attaches to the cylinder head, with the exhaust pipes then attaching to the manifold (exhaust manifold).

Methanol A flammable liquid (alcohol) that is formed from the distillation of wood but can also be manufactured synthetically; used as antifreeze in vehicles and solvents, among other application.

Off-gassing The release of gas or vapors in the process of aging or decomposing. Also, vapors or gases that were absorbed by fabric, carpet, and so forth during a fire and then released after the fire is extinguished.

Pro bono Work done free of charge; at no cost.

Resistance heating heat that is produced when electrical current passes through a conductor.

Turbo charger A device that uses the energy of the vehicle exhaust gases to compress (increase) the air going into the engine, which in turn increases the power of the engine.

Vehicle identification number (VIN) A unique set of numbers and letters that serves as the serial number of a vehicle. These numbers, when decoded, can provide information about the vehicle, such as where the vehicle was built, the manufacturer, the vehicle brand, the engine size and type, the vehicle model year, and the plant where vehicle was assembled.

REVIEW QUESTIONS

1. Two different types of window glass are discussed in this chapter. What are they and how can they assist the investigator in determining the direction of fire travel and cause and origin?

2. Discuss the differences between parallel and series electric hybrid vehicles.

3. What temperature will the exterior surface of a catalytic converter reach when it is operating properly?

4. What is the flash point of ethanol? How does it affect the investigation when used in conjunction with gasoline in a vehicle fire?

5. Discuss the use of self-contained breathing apparatus (SCBA), and explain why it is important to use this apparatus when working vehicle fires.

6. When should a first responder investigator call for the assigned investigator, and why?

7. Describe how to properly secure a vehicle (electric, diesel, or gasoline) prior to investigating the incident for safety.

8. List flammable and combustible fluids used in the motor vehicles, and explain how they could be involved in the cause of a vehicle fire.

9. Discuss how an engine with no spark can operate on a combustible liquid.

10. List five potential and credible heat sources in vehicles, and explain how their failures may lead to an accidental fire.

DISCUSSION QUESTIONS

1. One larger fire department decided it needed to cut back on services because of budget cuts. As such, the department decided not to investigate vehicle fires. What problem might this create in that jurisdiction? What problem might it create in neighboring jurisdictions?

2. Hybrid vehicles are creating some unique problems for both the fire fighter and the fire investigator. What actions can be taken by manufacturers to alleviate safety problems related to these vehicles?

REFERENCES

1. Evarts, B. (2018, September). *Fire loss in the United States during 2015*. Quincy, MA: National Fire Protection Association, Fire Analysis and Research Division.

2. How Stuff Works. *How airbags work: Air bag inflation*. Retrieved from https://www.explainthatstuff.com/airbags.html

3. United States Department of Labor (1990, August 30). *OSHA Hazard information bulletins automobile air bag safety*. Retrieved from https://www.osha.gov/dts/hib/hib_data/hib 19900830.html.

4. National Fire Protection Association. (2008). *Fire protection handbook* (20th ed., Table 18.13.18, pp. 8–165). Quincy, MA: National Fire Protection Association.

5. Schwaller, A. (2005). *Total automotive technology* (4th ed., Table 14-4, p. 188). Clifton Park, NY: Thomson Delmar Learning.

6. Economy Watch. (2009, November 20). *How biodiesel functions: Information on complete biodiesel functions, renewable energy*. Retrieved from https://www.economywatch.com /renewable-energy/biodiesel-functions.html

7. Cole, L. S. (2001). *Investigation of motor vehicle fires* (4th ed., p. 53). San Anselmo, CA: Lee Books.

8. U.S. Department of Agriculture. (2019, August 20). *Feedgrains sector at a glance*. Retrieved from https://www .ers.usda.gov/topics/crops/corn-and-other-feedgrains /feedgrains-sector-at-a-glance/

9. U.S. Energy Information Administration. (2019, April 16). *Today in energy: In 2018, the United States consumed more energy than ever before*. https://www.eia.gov/todayinenergy /detail.php?id=39092.

10. National Fire Protection Association. (2008). *Guide for fire and explosion investigations* (25.4.3.1, pp. 921–212). Quincy, MA: National Fire Protection Association.

11. Schwaller, A. (2005). *Total automotive technology* (4th ed., Table 14-4, p. 458). Clifton Park, NY: Thomson Delmar Learning.

CHAPTER 15

Arson

LEARNING OBJECTIVES

Upon completion of this chapter, you should be able to:

- Distinguish between an incendiary fire and arson.
- Describe the impact that arson has on the community.
- Identify the three elements of arson.
- Describe the necessity of proving motive.
- Describe and give examples of the different types of motives.
- Describe the categories of youth fire-setters.
- Discuss various findings that could be indicators of arson.
- Describe fire patterns and characteristics that may indicate that a fire was intentionally set.

Case Study

A local homeowner left for vacation on Sunday night. He and his family had a 12-hour drive to get to the cabin they had rented on a lake in another state. He arranged for a neighbor to check in on his home, but gave specific instructions to go over only on Tuesday evening. A few hours after the homeowner left for vacation, a strong thunderstorm came through the area and knocked down several trees, taking out the power lines. It took almost 20 hours for the power to come back on. When the neighbor came home from work on Monday and found that the power had come back on, he felt he should not wait until Tuesday to check on the home next door. When the neighbor opened the back door, he could smell something burning. When checking the house, he finally made it down to the basement, where he found a 4-ft by 8-ft model railroad set on fire. It took three buckets of water, but the neighbor finally put out the fire.

Once the fire was out, he realized that an electric soldering iron had its trigger switch taped into the on position. The soldering iron was hanging from a hook in the ceiling by its power cord and was plugged into a 24-hour electric timer, which was attached to an extension cord plugged into an outlet. On the table were remnants of what was burning: The neighbor realized it was crumpled-up toilet paper stacked approximately 2 ft high, with the soldering iron stuffed in the middle of the paper. The neighbor immediately called the sheriff's office, and then called the homeowner at his vacation home.

When the homeowner arrived back at his house the following day, he was met by the sheriff's office personnel. The homeowner's prints were taken as part of the process of elimination for prints found in the home. The homeowner's prints matched the prints found on the sticky side of the electrical tape that had been used to set up the time delay device. The homeowner had set up the family trip to draw suspicion away from himself and to help him look innocent. He had arranged for his neighbor to check on the home, but only after he knew the home would have burned down.

The timing device (**FIGURE 15-1**) could be set for a little less than 24 hours, giving the homeowner plenty of time to get far from the house in hopes of escaping suspicion. The power failure certainly messed up his plans, as did the good neighbor.

The investigation revealed additional information about the homeowner's fiscal problems. Several creditors were threatening to take legal action against him, and he was close to facing foreclosure on his home. The homeowner confessed to the crime and was given 5 years in jail. The insurance company denied his claim, and the bank foreclosed on what remained of the house and land.

FIGURE 15-1 The electrical timer allowed for a time delay of a little less than 24 hours.
Courtesy of Russ Chandler.

Access Navigate for more resources.

Introduction

Under most circumstances, the first responder investigator is not the person who completes the investigation of an incendiary or intentionally set fire. Instead, a part-time or full-time investigator is assigned to handle such cases. Depending on the jurisdiction, the assigned investigator could be someone who works for the fire department, such as the fire marshal, or an investigator for the local police department. Depending on the state, the fire investigator may be assigned to the state fire marshal's office, the state police, or the state insurance bureau. However, the initial investigation may be started by the fire officer, so it is beneficial for the first responder investigator to know what happens in the next stage of an investigation of an intentionally set fire.

Previous chapters of this text have provided some information about arson or incendiary fires. As a general rule, first responder and assigned investigators use the term *incendiary* in their reports, because in many jurisdictions the courts decide whether a fire is considered arson. In those cases, the report must document the facts, such as that Mr. Jones set the fire using a match to ignite balled-up paper in the northeast corner of the garage.

This chapter discusses legal issues and the elements of the crime of arson. It discusses the motives for arson and gives special consideration to juveniles involved in such incidents. Various indicators are discussed, including sites where the arsonist may set the fire and the various tools of the trade he or she may use.

Arson Versus Incendiary

Although covered earlier in Chapter 13, *Fire Cause*, one point bears mentioning again in this chapter: *Black's Law Dictionary* includes an extensive definition of arson that states it involves the malicious burning of the house of another.[1] This is the earliest form of the charge of arson. Over the years, the law has changed and expanded what is covered under arson. States with model laws apply different penalties depending on the specific types of structures or items that are set on fire. For years, arson has been considered a crime against property, but now it is also recognized as a crime against people, which allows for greater penalties.

Black's Law Dictionary includes a lesser definition of *incendiary*, as what happens when a person sets a house or building on fire.[2] *Webster's Dictionary* goes into a little more detail, saying that *incendiary* means the willful destruction of property, and mentioning the possibility of incendiary devices.[3] As a general term, an incendiary fire is one that is intentionally and willfully set.

Each fire department should establish a policy regarding the proper terminology for arson and incendiary fires that is approved by that jurisdiction. The policy can be reviewed by legal counsel who are part of that jurisdiction's staff, but the final approval should be given by the local prosecutor, because this person will actually argue the department's cases in a court of law, after an arrest has been made.

Impact on the Community

The **criminal act** of arson has an impact on each and every person in the community. It can even have a far-reaching impact at the national level, when funding is taken from other worthy projects to support the investigation and suppression of arson. What people pay for vehicle, homeowner's, and renter's insurance is predicated on the anticipated losses to insurers—including the losses that result from arson or insurance fraud associated with a fire loss.

TIP

The criminal act of arson has an impact on each and every person in the community.

A more direct impact in the community is the potential for injury or loss of life as the result of an arson fire. This includes not only direct losses from the fire but also indirect losses, such as when emergency units are tied up with an arson fire and are not available to respond to other calls for assistance, such as for vehicle accidents, illnesses, or even other fires. Put simply, units handling an intentionally set fire are not available to assist other citizens. For example, when Engines 6 and 7 are at an arson fire, the citizens living near those stations will have to wait longer for the next closest units to respond to their needs. That response may take only an extra 5, 10, or 15 minutes. This may not sound like much, but if it is a cardiac situation, that extra time could reduce the victim's chance of survival.

Arson fires place fire fighters at unnecessary risk. They are not only endangered by the flames, but also sometimes at risk from the different tactics arsonists use to keep fire fighters from extinguishing the fire. For example, an arsonist may cut holes in floors at the door entrance or arrange for structure failure just as the fire fighters are making their attack.

TIP

Arson fires place fire fighters at unnecessary risk.

Often, arson is directed at older structures that some people may perceive as having less value, yet others see as irreplaceable pieces of a community's history. Even if fire does not destroy everything, additional damage will result from the suppression activities, such as water damage. Inside some of these structures are irreplaceable artifacts that may seem like junk, but they, too, could be pieces of the past that someone had cherished.

TIP

Often, arson is directed at older structures that some people may perceive as having less value, yet others see as irreplaceable pieces of a community's history.

One of Our Own

One of the facts we have to face is that sometimes a fire fighter may be the **perpetrator** of the crime we so detest in the fire service.

It had been quiet a week, with no fire runs at all. The older fire fighters tended to enjoy these types of duty tours, but the young ones were anxious to hone their skills and use their newly acquired fire suppression techniques.

One of the younger fire fighters came running into the station, shouting that there was a fire next door in a shed. The alarm was called in as the engine pulled out and set up in front of the small structure. The fire was knocked down quickly, but a road flare was readily visible in the window where the structure was set on fire.

Investigators spent the rest of the night interviewing each fire fighter who responded to that fire. The last to be interviewed broke down and told the truth: He had set the fire because he was bored. The tragedy was that the structure in question was an historic building, the last standing one-room schoolhouse in that jurisdiction. Inside were artifacts from the lessons taught in the previous century. What was a shack to one person was a treasure to the community, now lost.

What made this situation worse was that the young fire fighter tried to insinuate that one of the other volunteer members had set the fire. Fortunately, his lie was not believed. After a night in which all the fire fighters at the scene were interviewed, the perpetrator—the last to be interviewed—confessed to setting the fire. He was arrested and went to jail.

The young man made bail and got an attorney, and all awaited the trial. The full-time investigators who worked the investigation and the volunteers who had fought the fire were in court for the preliminary hearing. Before the case was called, the perpetrator's attorney and the prosecuting attorney approached the judge. After brief consultation, the next case was called. The investigator and fire fighters were surprised to see the defense attorney approach the perpetrator, with both then walking out of the court.

The investigator approached the prosecuting attorney and was advised that the attorneys discussed a plea deal and presented it to the judge, who accepted the suggested deal. The fire fighter was found guilty of illegal burning (not arson), was given a fine and a sentence of time already served, and was advised he could never be a fire fighter in that jurisdiction. However, it was later learned he became a volunteer fire fighter in a neighboring jurisdiction.

Arson: The Crime

To have a crime of arson, there must be a **corpus delicti**, or the body of the crime. If nothing was damaged, no crime of arson occurred. In most jurisdictions, the crime of arson is defined by three elements:

1. There must be clear evidence that something burned.
2. The burning had to have been an intentional act by the perpetrator.
3. There must have been **malice** in the setting of the fire.

TIP

To have a crime of arson, there must be a corpus delicti, or the body of the crime.

This text cannot answer the legal questions that may arise in any particular jurisdiction. Instead, each department must research the legal issues and establish a policy and procedures manual to address them.

In addition, the investigators and the prosecuting attorneys in the locality must recognize that they are a team and need to work together.

Youth Fire-Setters

Playing with fire can result from natural curiosity. In some cases, however, a child may use fire as a means to express frustration, to get revenge, or for some other reason. There are three recognized age groups of youth fire-setters.

Child Fire-Setters

Child fire-setters are usually 2 to 6 years old—an age group characterized by simple curiosity. They typically set fires around the home, which is the child's normal domain. Many states have established case law that states children younger than the age of 7 are incapable of committing crimes.

Juvenile Fire-Setters

Juveniles are defined as children from 7 to 13 years old. Fire-setters in the age group may set fires around

Solitary Confinement

An inmate was locked up in solitary confinement. While there, he tore open the mattress, pulled out the stuffing, and stacked it in a corner. Federal law requires the outside of an institution mattress is to be fire resistant; however, it does not regulate the interior of the mattress. Once the stuffing was stacked in the corner, the inmate pulled out a match and lit the stuffing. It burned poorly but put off a massive amount of smoke.

The prisoner's motive was to escape when the jailers came to rescue him. He knew there were only 4 jailers on duty, but more than 50 inmates housed in the prison that particular night. His plan did not work. The inmates were sent into the fenced-in compound, not out the front of the building, as the prisoner had anticipated. Both the inmate and the guard doing the rescue took in quite a lot of smoke getting to the other end of the structure, but they were given a clean bill of health by the attending physician at the scene.

The fire did not damage the structure. One blister of paint occurred on the wall, but the only other damage involved the mattress and massive smoke damage throughout the structure. The first responder investigator (engine lieutenant) and the full-time investigator both worked the scene. This was a very serious incident in which inmates, jailers, and other responders were put at risk. To charge the inmate with **felony arson**, there had to be damage to the structure. If just contents were burned, that would be a **misdemeanor**. The investigators consulted with the local prosecuting attorney, who decided to charge the inmate with destruction of government property and reckless endangerment of the staff and inmates in the jail. Those crimes carried a heavier punishment than misdemeanor arson and added 2 years to the inmate's time to serve. However, his fellow inmates had to clean the jail following the incident. Needless to say, he had a somewhat uncomfortable stay for the rest of his sentence—thankfully, there was no outright violence, but he lived to regret his actions.

the home and in their schools. These events could reflect responses to a broken home, physical abuse, or emotional abuse. Although juveniles can be charged for any crime they commit, they will be adjudicated under a special court, most often referred to as Juvenile and Domestic Relations Court. In this environment, the accused minors are protected and the courts will seek solutions that better fit this age group.

Adolescent Fire-Setters

Adolescents from ages 14 to 16 may set fires outside the home, targeting schools, churches, or abandoned buildings. The adolescents who engage in this behavior may have been raised in an unhealthy environment. They may be experiencing poor academic performance along with peer pressure. In some situations, they may work in pairs or in groups with one strong leader. Often, adolescent fire-setters have a history of delinquency. Their fire-setting behavior may represent an outburst due to stress, anxiety, or even anger.

Like juveniles, adolescents can be charged with a crime. and they, too, will be adjudicated in a juvenile court. Depending on the youth's age and the crime, an adolescent can sometimes, if approved by the courts, be charged as an adult. Some courts have defined this policy as "if they commit an adult crime, then they will face adult punishment." The U.S. Supreme Court has established that no juvenile can be given capital punishment (death penalty), but other maximum punishments for adolescent offenders vary from state to state.

Motives

In many states, it is not necessary for the prosecutor to establish a **motive** for a crime to get a conviction. However, with today's juries, a motive can be a vital part of the prosecution's case in court. Today, more than at any time in the past, the general public is well trained in legal issues. Over the years, an increasing number of TV shows have focused on crime and solving crimes. This trend started many years ago, especially with *Dragnet* and *Adam-12*, and progressed to *Hill Street Blues* and various murder mystery shows. Today, with the many variations of the *Law and Order* and *CSI* series, the public is highly educated on legal issues, and expects to understand the offender's motive and see evidence on why the perpetrator committed the crime. These educated members of the public, of course, also serve on juries.

When an arson case comes forward, a conviction is legally possible without ever knowing the motive or reason why the individual committed the crime. However, when the jury deliberates, the members may have a hard time coming up with a verdict without knowing why the individual committed the crime.

Regardless of the legal minimum requirements, the prosecution should always strive to answer all potential questions that may come up about each case. In the case of arson, the prosecution should seek to identify the motive of the individual who may have committed the crime with which he or she has been charged.

TIP

When a jury deliberates, the members may have a hard time coming up with a verdict without knowing why the individual committed the crime.

TIP

Research conducted by the National Center for the Analysis of Violent Crime has identified six motive categories: vandalism, excitement, revenge, crime concealment, profit, and extremist.

TIP

Pyromania is an impulse that can be ignored if the individual so desires.

The investigator must have supporting documentation—such as fiscal discovery, interview comments, or data collected from the fire—to present in court when giving testimony as to the motive of the individual. If not supported by tangible evidence, a motive should not be introduced.

A study by the Department of Justice (DOJ), the Federal Bureau of Investigation (FBI), and the Federal Emergency Management Agency (FEMA), and research conducted by the National Center for the Analysis of Violent Crime (NCAVC) identified six motive categories: vandalism, excitement, revenge, crime concealment, profit, and extremist.[4] The NCAVC, along with the American Psychological Association, has concluded that citing pyromania as a motive is an oversimplification of a more complex issue that may be indicative of other disorders or problems. As such,

the NCAVC has taken it off the list of motives for fire-setting.[4] Studies have indicated that pyromania is an impulse that can be ignored if the individual so desires.

Within each category of motives is a wide range of variations. Consider the elderly widow who is destitute because her husband died without any insurance; she has no income except her Social Security benefits. The widow has debts that will soon see her out of her home. She does not drive but has three vehicles. She tries to sell them, to no avail. One night, in a moment of weakness, she sets one car on fire. Her attempt is extremely poor: The vehicle burns, but overwhelming evidence shows that she committed the crime. Clearly, the motive was profit. But the investigator, now knowing the full set of facts, must decide whether he will arrest a grandmother just before the holiday season.

Good Girl

She was 13, and normally a straight A student, but the teacher had noticed that her grades had dropped off in recent weeks. It was test time. The classroom had 24 chair–desk combinations where the writing surface lifted up to store books or other items. The young lady was the first to finish her test, turned it in at the teacher's desk, and went back and sat down. She lifted the lid to her desk, balled up some paper, used a lighter to set the paper on fire, and closed the lid.

In no time, the students and teacher saw the smoke. All of the students were rushed out of the classroom as the teacher pulled the fire alarm. The principal ran to the room and extinguished

the fire with a fire extinguisher, and the fire department arrived shortly thereafter. The principal called for the fire marshal and the girls' parents.

Once everyone arrived, they went into the conference room and the fire marshal asked the girl what happened. She confessed that she set her desk on fire. When he asked her why, the student explained that her parents were getting a divorce. She said that both her mother and father continually kept telling her that she was a wonderful girl and the divorce was not her fault. She thought that if she was not a good girl, perhaps her parents would stay married. No charges were filed, and the parents promised counseling for all three of them.

Vandalism

Vandalism is a type of crime usually associated with juveniles and adolescents. It is hard to define the chronological age groups who are most likely to commit this type of senseless crime. Instead, vandalism is usually done by individuals who have not mentally matured.

> **TIP**
>
> Vandalism is usually done by individuals who have not mentally matured.

Vandalism fires can be associated with petty theft and sometimes with breaking and entering by the perpetrators. They are also associated with gangs, who may deface property with paint. These types of crimes are usually, but not always, done in groups. The targets may be limited to those with which the vandal is associated, such as a school. What may start out as a petty crime can change quickly.

> **TIP**
>
> Vandalism fires can be associated with petty theft and sometimes with breaking and entering by the perpetrators. They are also associated with gangs, who may deface property with paint.

In this age of computers, individuals can readily learn about more creative and more violent forms of vandalism. By accessing these Internet resources, vandals can learn to create devices using common household ingredients that not only set fires, but also cause small explosions.

Excitement

The fire fighter from the "One of Our Own" case earlier in this chapter is an example of someone with an excitement motive. This individual was bored and wanted the thrill of putting out the fire and the attention he would get in the process. It is an embarrassment to all in the fire service when one of their own commits arson. Such an event goes against the grain and gives the fire service a feeling of betrayal. Such an act is usually committed by just one person.

Other individuals with this motive may start with small fires and work up to dumpsters, vacant buildings, and possibly occupied structures after time. They usually stay in areas with which they are familiar, such as around their home or work. They have been known to watch the fire fighters' response to the fire, even offering to help, so as to live vicariously through the fire fighters' actions. They like the thrill of the lights and sirens, and they enjoy having a secret: They know how the fire started and no one else knows that information yet. They may even seek out recognition and attention by helping

Teen Prank

Three young teens were not happy with their shop teacher for the bad grades they got on a recent project. With nothing to do on a Sunday afternoon, they decided to go to their junior high school, which was just around the corner. With no plan of exactly what to do, they found a door that could be opened. The trio went to the shop and found three full 1-gal gasoline cans. The students picked up the gasoline cans and emptied them on the floor throughout the shop area, then grabbed a road flare from a toolbox and went out the rear door. The boys managed to strike the flare and threw it in the room, slamming the door behind them as they ran. Why their clothing did not ignite, we will never know. The room did not ignite because the atmosphere was too rich—too much ignitable vapor and too little oxygen.

The shop teacher arrived Monday morning and found gasoline still puddled on the floor and the residue from a burned road flare by the back door.

Outside the back door was a smooth concrete pad, where investigators found perfect outlines of three sets of sneakers; they could even read the sneaker manufacturer on two pairs of imprints. The students had walked through the gasoline, which had broken down the soles of their shoes, and while they stood still, it left the evidence behind.

When asked, the shop teacher said that only three students had failed their project this quarter. However, he was quick to add that they were good boys and would never have done anything like this. The investigators then visited the boys' parents; when asked, they all allowed the investigators to look in their sons' rooms. In each instance, investigators found sneakers that matched the imprints at the school. Faced with the evidence in hand, three confessions followed. What started out as a small rebellious vandalism event could have burned down the school and, worse yet, taken the boys' lives.

pull hose lines or tell the investigator what they saw before the fire department arrived.

> **TIP**
>
> Individuals with an excitement motive often start with small fires and work up to dumpsters, vacant buildings, and possibly occupied structures after time.

Most of the time, fire fighters are too busy to recognize familiar faces in the crowd. However, the engineer operating the apparatus may have the opportunity from time to time to check the crowd for a familiar face. Some departments issued disposable cameras for each engine so that the crew could take pictures of the fire and the crowd. Today, they are more likely to rely their engineers to use their personal cell phones.

Revenge

A fire motivated by revenge can be set to harm an individual, a group, or an institution. Sometimes the revenge is directed at none of these entities, but rather at society itself. For some people, things have gone so wrong in their lives that they decide to strike out at anything that represents society as a whole. They just want to get even for what life has dealt them and the life they lead. There usually is no pattern with these attacks; the strikes will be random.

> **TIP**
>
> A fire motivated by revenge can be set to harm an individual, a group, or an institution. Sometimes the revenge is directed at none of these entities, but rather at society itself.

Revenge Targeting a Person

When the revenge is aimed at a person, it tends to be more obvious. In addition to the fire set, there may be vandalism of personal property. Broken pictures, torn-up clothing, and holes cut in pants or dresses suggest a broken relationship as the motive. Anything that may have been a favorite item of the victim may be the target of the perpetrator's outrage. These types of fires are usually planned, but rarely involve time-delay devices. Instead, the fire-setter may lay in wait for the right opportunity to strike at the individual, with destruction and fire being the primary tools used.

In these types of fires, interviews are key to discovering the motive. During the discussion with all the parties involved, the investigator may get an indication of what is happening. Checking into relationships between couples, both heterosexual and homosexual, may be revealing. A betrayal could have caused someone to lash out, but the person may not necessarily attack the partner. Instead, the victim may be a third party who is presumed to be breaking up the relationship, with that third party's possessions being destroyed and burned.

Investigating office relationships may also reveal that the perpetrator suffered an actual or perceived hardship. Perhaps the individual was laid off or did not get a promotion he or she felt was deserved. In turn, the perpetrator may strike out against the person who got the job or the supervisor who failed to give the promotion or conveyed the news of the layoff. Even if perpetrators brought their situation upon themselves, they may still want to strike out at someone. The target could be the personal items in the office, the individual's vehicle, or even their home.

Revenge Targeting a Group or Institution

A perceived injustice by a group can manifest in the perpetrator lashing out at something that represents that group. Churches and synagogues are common targets for revenge crimes. The Bureau of Alcohol, Tobacco, Firearms, and Explosives heads up task forces to examine these types of crimes against places of worship.

> **TIP**
>
> Churches and synagogues are common targets for revenge crimes.

The target of this type of revenge is not really an individual, but rather an organization—in some cases, a loosely organized group. For example, gangs may get territorial and lash out at anything that they perceive threatens their life or existence. In other cases, an unorganized crowd may, as a whole, at one time, strike out against a target when they perceive they have a common cause. Although the crowd may not advocate committing a revenge crime, some members of the crowd may decide to take things into their own hands.

Commonly, those seeking revenge lash out at symbols of what they perceive as wrong—a cross, the Star of David, or corporate logos. These usually

Obvious Evidence

The English teacher at the local high school went out to get into her Volkswagen Rabbit in the morning and found the vehicle's fuel door open and a rag stuffed into the fill tube. The rag was burned, as was the plastic trim that came in contact with the burned rag. A 911 call resulted in an investigator being sent instead of an engine company.

After photographing the rag, the investigator pulled it out of the fuel tube. The part stuffed into the fuel fill had been tightly wound up and suffered no damage. The rag turned out to be a towel, and embroidered onto the towel was the name of the local restaurant that just happened to be next door to the high school. With no other incidents occurring overnight, the investigator asked the teacher if anything had happened at school the previous day. She explained that she had to fail one of her students; as

a result, that student would not be able to graduate as planned.

The investigator checked with this student, and she had an alibi for the previous night. When asked if she had a boyfriend, she said yes, but stated that he had graduated the year before. Sure enough, he worked as a dishwasher at the restaurant next door to the high school—the same restaurant identified on the towel. When interviewed, the boyfriend eventually confessed after failing a lie detector test. He was upset because he wanted to get married, but his girlfriend's parents would not consent until she graduated from high school.

One point of humor about the case: The temperature was below freezing, and the teacher's car used diesel fuel. There was no way this attempt would burn any vehicle.

well-planned events may involve a more sophisticated means of setting the fire—for example, using devices with ignitors to not only cause fires but also create explosions, causing considerable destruction.

Concealment of a Crime

Whether the crime was a well-planned act or a crime of passion, criminals usually seek to cover up their actions. This is especially true when perpetrators believe that they have left evidence behind that may indicate their guilt. For this reason, some criminals turn to arson in hopes of destroying the crime scene and allowing their other crimes to go undetected. The assigned fire investigator must be properly trained on the investigation of other crimes, if for no other reason than to be able to recognize pertinent evidence and process it appropriately.

TIP

Whether the crime was a well-planned act or a crime of passion, criminals usually seek to cover up their actions.

When fire investigators find evidence that suggests other crimes may have been committed at the scene, they usually call their counterpart in the local police or sheriff's office so that they can work as a team. Should they make their case and find the culprit, two experts will be testifying in court on the facts surrounding the incident. This teamwork atmosphere must be worked out in advance to ensure a smooth flow of information sharing and cooperation from the very beginning. If for some reason territorial issues or command questions might arise, the fire scene is not the place to work these disagreements out; instead, this groundwork must be laid in advance. Establishing a formal process and policy on jointly investigating and handling crimes can provide a positive atmosphere for all parties working together. Regular joint training and meetings can also provide a positive environment for future investigations.

Almost any serious crime can result in the perpetrator setting a fire for concealment. In the past, before DNA analysis became available, criminals actually did get away with murder in some cases—but this does not happen as often today. New forensic technology has brought us a long way, and there is new science just around the corner to help even more.

Serious crimes that may lead to setting a fire for cover-up include burglary, robbery that leads to murder when the victim recognizes the suspect, fraud, and embezzlement. For example, an employee who is selling parts out of the company's inventory and pocketing the money, then realizes that an annual company inventory will be taken, may be motivated to set fire to the stock. These types of crimes are only limited by the imagination.

Check Cashing

One company paid its staff every other Friday. The payroll clerk signed the checks using her employer's signature stamp. She would then take the checks to the staff in the workshop. Many preferred to be paid in cash, so she got them to endorse their checks, took the checks to the bank, and returned with the correct cash for each employee. This seemingly benevolent act was actually part of her ruse: Although the next Friday was not payday, the clerk created payroll checks for the cash-based employees, stamped the checks with the employer's signature, endorsed the checks, went to the bank, and kept that cash for herself.

One Friday, an auditor showed up and looked at the books for accounts receivable. He told the payroll clerk that he would look at the books for accounts payable on Monday morning. That Saturday night, the business suffered a major fire. The fire started in the accounting office. All the desks were in proper order, but the desk drawers of the payroll clerk were left wide open. The books, which were usually left locked in the desk drawer, were found in the center drawer. Gasoline was poured over the desk, books, and surrounding area. Interestingly, the books survived; they were charred, but legible.

A warrant was issued for the clerk's arrest on charges of embezzlement. After she was arrested, law enforcement intended to interview her and charge her with arson as well as the financial crimes. However, she was nowhere to be found.

Approximately 2 months later, a similar business located about 5 miles from the first business went to the police with an amazing story: The business owners thought the payroll clerk was creating unauthorized payroll checks and cashing them. By the time investigators arrived at her office, the clerk was gone. Going to her home, they found that they had just missed her again. Her last known location was somewhere in the Caribbean.

Ultimately, even if investigators find the area of origin and the cause of a fire, and even if they identify who is responsible, they may not always get that person to court.

Profit

Sometimes the reasons for setting a fire for profit are obvious, including such situations as when businesses facing hard times with outdated stock are set on fire for the insurance money or when someone with a gas-guzzling vehicle responds to a sharp rise in gas prices by burning the vehicle to get the insurance proceeds to buy a more economical vehicle. This motive can also sometimes cause one business to seek the demise of a competitor by burning it out of business. Further, the crime of extortion is committed when a business is threatened with a fire if it does not pay a certain amount to a crime boss or local gang each week.

Other profit motives are more obscure. One property owner intentionally burned his neighbor's barn because it blocked the view of the water from his house. He did not actually want to see the water; instead, he expected his house to sell for more money because it would have a scenic view. In the city, land is at a premium. It is not unusual for empty lots to be more valuable than the lots with existing structures because demolition costs are high. Thus, a developer must figure in demolition costs for structures sitting on the land. If the structure burns, however, the property owner realizes lower demolition costs and has a more valuable lot.

Insurance Fraud

For every type of insurance policy, there is a way to commit fraud. In regard to fire loss, the fraud constitutes the setting of the fire with the intent to defraud the insurance company. In a second scenario, the fire could be accidental, yet the property owner decides to take advantage of the situation and put in a claim for losses that did not occur, such as by overstating the value of items or claiming contents that were not consumed in the fire.

TIP

For every type of insurance policy, there is a way to commit fraud.

Insurance fraud is so costly for the insurance industry that the major companies have special investigative units to look into some fire losses. These units either have their own investigative staff or hire fire investigators in the private sector to look at fire scenes. The intent is to see whether the insurer has any reasons to deny the claim based on the policy holder being involved in setting the fire or defrauding the insurance company after the fire.

An insurance investigator, either from the insurance company or working as a private investigator on contract, has no legal right to enter the scene before the government investigator finishes his or her examination. However, depending on the situation and the individuals involved, nothing may prohibit the investigators from working together. Fire service investigators must be aware of the thin line between cooperation and **collusion**. Cooperation is essential, to the point that many states have arson immunity laws that allow the insurance company to share information without fear of civil or criminal prosecution. In contrast, caution must be exercised to ensure that no collusion occurs—namely, that the government investigator and the insurance investigator do nothing to abridge the rights of the insured in any way. The two should not set plans in motion to prove the person's guilt. The investigator is a seeker of truth and should be just as interested in evidence of innocence as evidence of guilt.

Insurance Policies

Every state has an insurance bureau or agency that regulates the insurance industry. This agency or the legislative body of the state may adopt standard language that is used for the sale of all policies within the state. Interestingly, most policies do not mention arson. Instead, they adopt language such as "concealment of fraud." Essentially, policies are null and void if the insured parties misrepresent any fact or circumstance or misrepresent themselves by committing fraud or falsely swearing to any material fact.

After a fire, the insurance company requires a written **proof of loss** form. This is essentially a signed statement that the fire occurred, along with a list of all contents lost in the fire and the value of each item. Under the requirements of most insurance policies, the insured must, if requested, present himself or herself for an interview with the insurance representative for an examination under oath (EUO). An EUO is essentially a **deposition**, where the insurance company, with its legal counsel, is able to ask a series of questions about the incident. A court reporter will be present and the insured parties may bring their own counsel, if they so desire. Keeping this in perspective, the EUO is an opportunity for the policy holder to explain any unclear circumstances surrounding the event. Should the policy holder refuse to participate, and after reasonable attempts are made to obtain cooperation, the insurance company can deny the claim and not pay the policy holder for the claim. To make sure that insurance companies do not abuse this ability, the courts have imposed penalties in case of bad faith; that is, penalties can be imposed if the insurance company does not act in good faith toward its policy holders.

The assigned investigator must have knowledge about the insurance industry because the investigator may be subject to a deposition should testimony be required in a civil case. The deposition is an opportunity for all parties to hear what will be in their testimony if they go to court. Sometimes the evidence presented in deposition is so compelling that the case is settled out of court. If the case does go to court, the deposition and the investigator's answers can be used to verify what is said while the investigator is on the stand.

All witnesses are subject to a deposition, including the municipal investigator, who may be compelled by the courts to testify. If subjected to a deposition, the deponent (in this case, the investigator) should never waive the right to review. Instead, the investigator should always get a copy of the deposition and read it carefully. If a mistake occurred in the recording, the investigator must insist on having it verified and corrected. If the investigator misspoke and the statement made was not correct, he or she must make the correction in writing and submit it to all sides in the case. This might be quite embarrassing at the time, but failure to do so could be more embarrassing when the investigator is confronted with the misspoken error should the case go to court.

Extremism

Just as the name implies, extremists include any person or group of people who possess an extremely different outlook on life in comparison to the world around them or those who take extreme measures to make a point (their point) or to get their way. Terrorism is the best example of the actions of this type of fire-setter. Today, examples of extremism can often be found on the nightly news—but it is not just a recent phenomenon. In fact, this type of attempt at social change has existed in one form or another since recorded history began. Conflict between Northern Ireland and the British government existed for decades. The Provisional Irish Republican Army (IRA) took extremism to new heights when they formed in 1969 to oppose British rule.[5] In recent decades, radical groups have been created around the world that oppose various forms of organized society. These extremists are willing to use whatever means necessary to force their ideals upon others. Sometimes their actions are not geared toward the ideals they espouse so much as the destruction and terror they can create. Today, extremists from the Taliban, al Qaida, and ISIS (Islamic State of Iraq in Syria) have taken terror to a new level, and, regrettably, the world has not seen the limits of the horrors that can be unleashed.

Extremist groups do not have to come from other lands or from other cultures. Indeed, the United States has its own share of extremist organizations. These groups speak for the rights of animals, the rights of the wilderness, and the right not to be taxed or to do as they please in or around their compounds. Some extremist ideals are based on religion, whereas others are based on a philosophical theory.

Some groups, after becoming upset about the damage to the environment, have lashed out and set SUVs on fire. In one state, a set of high-dollar homes under construction in a cul-de-sac were perceived as humanity's growth intruding on the wilderness—so extremists set the homes on fire. The list of such crimes goes on and on.

Opportunity

In addition to documenting the scene where the fire occurred, the investigator needs to collect evidence on any person of interest in the event to show whether that individual had the opportunity to commit the crime.

To do this, the investigator must establish a timeline for the fire event. This timeline can be only as precise as the data available. Reliable times that can be documented include the first call to the emergency communications center and the time when the first engine arrived. The next time points are to be documented are based on any eyewitness accounts, which can vary and may not be 100 percent reliable, but at least provide something to work with when placed on the timeline. The fire scene can provide some time indicators, but reliability will depend on the circumstances. For example, electrical analog clocks that stopped either as a result of fire damage or because the circuit was interrupted by overcurrent protection devices can provide a time, as shown in **FIGURE 15-2** and **FIGURE 15-3**. A word of caution: Ascertain from the homeowners whether the clocks were actually working and properly set prior to the fire. In this day and age,

FIGURE 15-2 A clock from the bedroom nightstand. Always check whether a clock was plugged in, and ask the homeowners if it was in working order.
Courtesy of Russ Chandler.

FIGURE 15-3 An analog clock located on the kitchen oven. Check with the occupants to ascertain whether such a clock was accurate and set to the correct time.
Courtesy of Russ Chandler.

analog clocks are becoming scarcer owing to the advent of digital liquid crystal displays, which will not provide such evidence.

> **TIP**
>
> The fire scene can provide some time indicators.

Although not used much except as travel alarm clocks, mechanical clocks can provide a time of failure as well. Some homes may even have antique or reproduction mechanical mantel or grandfather clocks that can be of assistance in establishing the timeline. When subjected to excessive heat, the mechanics of these clocks will fail, providing a time when that occurred. Because these clocks need winding, it will be necessary to ascertain from the owner if the clock was working and if it had recently been rewound.

The 2008 edition of NFPA 921, *Guide for Fire and Explosion Investigations*, cautions about using the depth of char for establishing a timeline.[6] Laboratory test results with pine indicated that this wood has a burn ratio of 1 in. of char in 45 minutes. However, so many variables and unknowns exist that using this ratio in an actual burn scene would be unreliable. Sometimes the investigator can make a generalization based on experience, especially if there is a gross disparity in actual times and alleged times, such as when the homeowner says he left only 5 minutes ago, but exposed 2 × 4's at the area of origin are almost consumed.

When the timeline is complete, the fire has been determined to be incendiary in nature, and a suspect has been identified, the investigator must then verify whether the suspect had the opportunity to set the fire based on known events and a thorough background investigation.

Incendiary/Arson Indicators

The world of fire investigation is not black and white. No single indicator can emphatically prove that something is this or that. Instead, the culmination of all the facts will eventually lead the investigator to construct an accurate hypothesis of the fire origin and cause. The indicators discussed in this section may be signs of an incendiary fire, but a thorough investigation of each situation is necessary to determine whether they are part of the facts that support the final determination of the cause of the fire and persons responsible.

> **TIP**
>
> The world of fire investigation is not black and white.

Location and Timing of the Fire

The arsonist wants the fire to burn as long as possible before it is discovered. Thus, the arsonist is likely to set the fire at a time when fewer people are around to discover the fire. After midnight and in the early morning hours, most people are in bed—so this would be the arsonist's preferred time frame. The location and time have a correlation as well. If the fire location is an industrial complex, such as an industrial park that has few people around on Saturdays or Sundays, then the weekend would offer an opportunity to set a fire and delay its discovery.

The location of the fire on the property also may be a key giveaway that the fire was intentionally set to allow fire growth. This location has to allow the arson fire to spread. Fire burns upward and outward, so setting the fire at the lowest point in the building provides a better opportunity to cause more damage. Note that this does not eliminate the attic of the building as a location. Attic spaces usually have no windows, which will delay the fire's discovery. This type of location is also difficult for the fire fighters to access, which will also add to the desired destruction.

Another telltale decision involves the available fuel. To make the fire look accidental, the fire may be set in the area with the most volatile fuel. The fuel load and heat release rate of the fuel may make this area the best bet to get the fire going and cause as much destruction as possible, regardless of who sees it and how fast the fire department gets to the scene.

> **TIP**
>
> When perpetrators are considering the location to set an intentional fire, they may choose a location that will allow the fire to spread rapidly.

The most obvious location issue arises with multiple set fires. If the fire appears to have two or more areas of origin that cannot be explained, the investigator must consider the possibility that the fire was deliberately set in more than one location.

When considering the location, also consider the ability of the fire to spread. Check whether any of the suppression systems (e.g., sprinkler systems) were shut off, if fire doors were blocked, and if detectors were disconnected. Arsonists typically disarm anything that could slow the fire or lead to its early detection.

Fuel, Trailers, and Ignition Source

Most arsonists try to make the fire look accidental while doing the most damage. One way to make this happen is to use the fuels that are normally found in the structure. The debris left behind is then what someone would expect to find in an accidental fire. Using some of these fuels to move the fire from one part of the structure to another takes creativity and may go undetected, depending on the amount of destruction and the amount of damage done by the use of suppression hose streams.

> **TIP**
>
> Most arsonists try to make the fire look accidental while doing the most damage.

An arsonist who is in a hurry, needs rapid damage, and does not care that it will look incendiary typically uses an ignitable liquid such as gasoline. Of course, some people may use gasoline without considering the risk of detection, not even thinking about being caught or being identified. Other individuals may use unique methods that involve mixing chemicals to allow for a delayed ignition. Some of these chemical mixtures have an extremely high heat-release rate that can ignite any common combustible.

Given these possibilities, the investigator needs to learn to recognize the residue of various items that could be possible heat sources. These items include ordinary items typically found in any residential or industrial occupancy. It is also beneficial to recognize where these items were found. Residue of a road flare would not be unusual in a garage, auto parts store, or the back of a vehicle, but would be unusual if found on a kitchen floor.

> **TIP**
>
> The investigator needs to learn to recognize the residue of various items that could be possible heat sources.

When considering this evidence, a strong background of working fires certainly comes in handy. The average fire fighter has learned to recognize different items in their burned state. To confirm or refresh their knowledge, it is always good practice for investigators to be involved in training sessions where items are set on fire in various scenarios; they can then do a scene examination. Another training scenario is to subject certain items to fire and examine the residue.

Cigarettes around the fire scene can often be found based on the filter residue, but the filter itself at the area of origin may disappear altogether with any disturbance of the debris. Residue of a match (either paper or wood) may be difficult to find, but it is not unusual to find residue of either. Again, a lot depends on the amount of debris at that exact site and how much it has been disturbed by fire streams or salvage.

The evidence of an electrical arc from a copper conductor could survive a typical residential fire, but aluminum will be more apt to melt. The condition of the wire may indicate an electrical short, but the investigator must determine whether the short caused the fire or the fire caused the short.

Some common materials, when mixed together, can create sufficient heat to act as a heat source. Depending on the mixture, they can act as a delay device as well. Each leaves a form of residue. These chemical mixtures are not usually listed in texts to prevent these books from becoming "how-to" texts instead of investigation tools. Specialized training venues, such as courses put on by law enforcement agencies ranging from the federal government to local police departments, can assure students are actual investigators and provide this specific information.

Some explosive items, such as black powder, merely burn when not confined so that the fire spreads rapidly. These trailers make a burn pattern across the surface on which they are poured. Fuses for model rockets will burn (like the fuse on a firecracker) and leave a scorch mark. In contrast, the safety fuses used in modern blasting leave a waxy residue where they are stretched across a surface.

What investigators find at the scene will dictate the additional field investigation required. If they detect traces of an ignitable liquid such as gasoline, as shown in the pattern in **FIGURE 15-4** and **FIGURE 15-5**, the investigators may want to visit local gas stations to see if anyone recently purchased fuel by filling containers. It would be even more beneficial to the investigation to find the persons who purchased the gas using credit cards. Videotapes from gas stations may be of assistance, but tend to be reused on a regular basis. Thus, it is beneficial to visit local stations as soon as is practical.

FIGURE 15-4 An ignitable liquid pour that went down a set of steps. The separate char on the left side of the lowest step is from a cardboard box that was left on the steps.
Courtesy of Russ Chandler.

FIGURE 15-5 Gasoline splashed onto walls from 1-gal milk jugs. Notice the rundown patterns in the center, as well as the wisping of smoke from the burning fuel on the pattern on the left of the photo.
Courtesy of Russ Chandler.

> **TIP**
>
> What investigators find at the scene will dictate the additional field investigation required.

If chemicals are used, then a visit to other local stores is recommended depending on the chemical involved. If the fire leaves a purple residue, the arsonist may have used potassium permanganate. This chemical can be bought in large containers from a local pharmacy. If the evidence suggests a model rocket motor was used, then a visit to the local hobby shop may prove interesting.

Black powder can be purchased at the local gun shop or sporting goods store. Pyrotechnic cord and safety fuses are available from various suppliers, but it may also be helpful to check for thefts or vandalism at local facilities such as explosive contractors or rock quarries. Although these sources are not guaranteed

FIGURE 15-6 A soft drink can that contained the residue of activated fiberglass resin. The use of the resin and the catalyst, methyl ethyl ketone peroxide, created an exothermic reaction that was the heat source for the fire.
Courtesy of Russ Chandler.

to provide evidence, failing to visit them may mean an opportunity lost for gathering potential clues.

Sometimes looking for the unusual is important. For example, while sifting through the debris in a trashcan located at the area of origin, the investigator noticed that a burned soda can was heavier than normal. Peeling back the metal revealed residue of a solid resin, typical of what is used in fiberglass work (**FIGURE 15-6**). The business that experienced the fire had nothing whatsoever to do with fiberglass. Notably, however, the owner of the building (the landlord) was the last person in the room of origin. Even more interesting, when the building owner showed up, he drove the truck for his second business, which advertised fiberglass repairs on the side of his truck. Resin when mixed with a catalyst produces an exothermic reaction that is capable of self-ignition as well as igniting nearby combustibles, such as a trashcan full of mostly paper products.

Challenge the Unusual

Investigators should make an extra effort to see what is actually present, rather than what they expect to see. They must be able to judge the situation and not compare it to the norms of their life. Both fire and police personnel should have no problem with this process. In their work, they have seen how many people live in many different ways. Bottom line: Keep an objective and open mind on what to look for in any fire situation.

> **TIP**
>
> Keep an objective and open mind on what to look for in any fire situation.

FIGURE 15-7 Burn pattern on the wall where a pot of gasoline ignited. Notice the V pattern from the smoke and the inverted V from the flames that burned the wall clean.
Courtesy of Russ Chandler.

One thing to look for is what is *not* there. Sometimes an absence can be obvious, as in **FIGURE 15-7**. In this case (which was discussed in Chapter 1, *Role of First Responders*), the smoke produced a great V pattern and the flame impact on the wall, which burned clean, created an inverted V. This pattern is typical in the beginning stages of a fire, and it was exactly what the engine company found upon arrival. It was alleged that the young man at this house had a metal cooking pot full of gasoline, and was ladling gasoline and pouring it into wine and beer bottles. Neighbors were watching, but no one bothered to report this strange behavior to the police. According to one neighbor, the man stopped his work, sat back and reached into his pocket, pulled out a cigarette, and lit it with his lighter. The neighbor thought for sure there would be a big explosion. Instead, there was a small flash and his right hand, which was holding the cigarette, caught on fire. In his shock and haste, the man started shaking his hand; it hit the top of the container of gasoline, igniting the fumes. The neighbor ran in her house and called 911.

> **TIP**
>
> One thing to look for is what is *not* there.

When the neighbor came back outside and looked across the street, the man was gone, along with the pot and the bottles. The fire department arrived and checked just inside the front door to assure there was no extension of fire. The engine company officer (first responder investigator) found a kitchen pot on the floor just inside the door. It was half full of a liquid that appeared to be gasoline, based on the odor. He was documenting his findings when the full-time investigator arrived. Also found in plain sight were several **Molotov cocktails**. After completing full documentation, photos, and diagrams, they moved the pot to the front porch, placing it on the small table: The burn patterns matched perfectly.

One of the fire fighters mentioned that he could hear a strange sound as he walked across the porch. Most responders on the scene were ready to clear out, but the mystery was too great to not investigate further. Someone finally looked under the porch—and there was the suspect, hiding in a space he could just barely fit in. When someone stepped on the porch, the boards depressed enough so that it pushed the air out of his lungs and caused him to make a small gasping noise. Surprisingly, he had only singed the hairs off his hand and arm and had slight first-degree burns. The fire investigator cuffed him and hauled him away for making explosive devices.

Further investigation led to additional witnesses, who confirmed that the suspect was part of a local gang. They were planning on taking action against a rival gang. Although the man pleaded not guilty, he was convicted for making explosive devices and sentenced to 4 years in prison.

It is also good to listen intently to the stories given by victims. In one case, the homeowner went to great lengths to describe how he just barely got out alive of the fire scene by running out the back door. But, as shown in **FIGURE 15-8**, the door had a key deadbolt, which was still secure, as was the lock on the doorknob. The structure had just one front door and one back door, and no one went out the back door during this fire. Such suspicious stories should prompt another line of questioning by the investigator.

FIGURE 15-8 Sooted back door with deadbolt. This door requires a key for entry from both sides; it was locked at the time of the fire.
Courtesy of Russ Chandler.

Contents

The contents of a burned structure can provide a wealth of information. A priority for investigators is to make sure that they are legally on the scene and have a valid reason for looking throughout the structure; just having a badge is not sufficient. In searching the structure, investigators may find some interesting facts that are not supported by the comments made by the occupants of the home.

TIP

Fire investigators need to make sure they are legally on the scene and have a valid reason for looking throughout the structure; just having a badge is not sufficient.

It is critical that all examinations—which are actually searches—are done within the confines of the law. The investigators have the right to search only the area of damage, as their role is to determine the area of origin and the cause of the fire. If a bedroom is not damaged from the fire, the fire service has no right to search that area. There is nothing wrong with checking to assure no extension of the fire occurred, but that area cannot be searched as part of the fire investigation. The U.S. Constitution guarantees the right to not be subject to unreasonable search and seizure. If no fire occurred in the basement or in a shed in the backyard, those areas cannot be searched for anything other than a reasonable expectation that the fire may have extended into them. If no extension of the fire took place, those areas cannot be searched. As repeatedly emphasized in this text, the recommended investigation process is to examine and search from the least damaged area to the most damaged area.

The only exception would occur when, while looking for the extension of the fire, the fire fighter finds evidence in plain sight. If a gasoline can was sitting in a bathroom, that evidence would be plain sight and can be secured, but a further search would require a search warrant. If evidence of any other crime is in plain sight, it can also be legally seized. However, if fire fighters see evidence not associated with the fire, such as the presence of drugs, they should turn that information over to local law enforcement and let them take action.

If the fire is suspected to be incendiary in nature, the full-time investigator must be summoned. That investigator may have to get a search warrant to make a complete search in and around the area.

False Insurance Claim

In one case, the homeowner kept expressing sorrow about the loss of her expensive wardrobe. All but one closet showed carbon char on the floor from what was believed to be the expensive clothing described by the homeowner. However, the closet with limited damage in the master bedroom still had clothing on hangers and most items were just sooted. Looking at the clothing, the lieutenant became suspicious. Tags on the shirts and dresses were from average or inexpensive chain stores. The lieutenant suspected that homeowner might be preparing for a large insurance payout by lying about the value of the clothing. If so, she might have set the fire to allow for that possibility. The lieutenant called for the full-time investigator.

The home suffered considerable damage; there were suspicious indicators but insufficient proof to prove the fire was intentionally set. As required, as there was no proof or an incendiary nature of the fire, it had to be labeled as accidental.

All information on this incident held by the fire department was provided to the insurance company fire investigator upon the insurance company's request. Two days later, that investigator contacted the fire department. The private investigator also could not find sufficient evidence to prove that the fire was set by the homeowner. However, the investigator did find evidence that the homeowner had gone to the local used clothing store and purchased almost 100 pieces of old clothing two days prior to the fire.

The homeowner had provided a signed "proof of loss" form to the insurance company validating her claim for damages to her home, including the clothing. The insurance company had more than sufficient information that the homeowner had lied on the proof of loss statement. Based on the clothing misrepresentation along with proof of other falsifications, her claim was denied based on providing a fraudulent claim.

Although arson would not be proved, this homeowner did not profit from the crime—in fact, she suffered quite a loss. She could have argued her case by suing the insurance company, but did not. There is no proof as to why. However, historically follow-up civil court cases against the insurance company in similar circumstances have allowed the discovery of further evidence that was sufficient to charge the homeowner with arson.

FIGURE 15-9 Empty cabinets in an occupied home are suspicious. The fact that all the doors were left open prior to the fire may suggest that the cabinets were emptied prior to the fire.
Courtesy of Russ Chandler.

FIGURE 15-10 Closet with clean areas showing that it had contents at the time of the fire but that they were removed after the fire.
Courtesy of Russ Chandler.

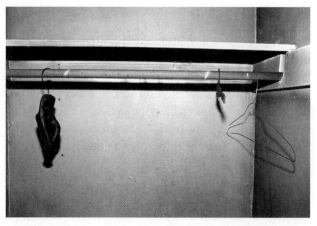

FIGURE 15-11 Closet rod showing that only three hangers were in the closet at the time of the fire. The hangers have been moved aside to show the protected area.
Courtesy of Russ Chandler.

FIGURE 15-12 Some of the stolen street signs found in the home of a fire victim.
Courtesy of Russ Chandler.

If a home is occupied and the cabinets are found open with no contents (**FIGURE 15-9**), the investigator must determine why. If the level of soot is the same inside of the cabinet as outside of it, then the doors were open during the fire and there were no contents. The investigator must use caution when making assumptions in such a case; the homeowner may be able to provide a logical reason for empty cabinets.

If the full-time investigator arrives on the scene the day after the initial fire, he or she may open a closet and find no contents. In such a case, the investigator should not jump to the conclusion that items were removed prior to the fire. The homeowner may have actually removed the contents after the fire to save them from further damage or to have them cleaned (**FIGURE 15-10**). Contents in closets that experienced fire damage may no longer be hanging but may have dropped to the floor. Always check the floor debris to determine the closet's contents. In closets with just smoke damage, it is a little easier to tell whether there were hanging contents at the time of the fire. The sooted bar will show areas that were protected by the hangers. In **FIGURE 15-11**, the three hangers have been moved aside to show the protected area. This finding by itself has no value in determining the nature of the fire, but is another piece of data that will be inserted into the final report.

The contents of the structure may indicate other activities that may be unrelated to the fire but show that an occupant has a propensity to commit other crimes. In one incident, a small fire occurred in a back room; by all accounts, it was accidental in nature. However, the contents of this residential back room included more than 50 stolen traffic and property signs (**FIGURE 15-12**). The theft of the street signs led to felony charges for the 20-year-old tenant.

TIP

The contents of the structure may indicate other activities that may be unrelated to the fire but show that an occupant has a propensity to commit other crimes.

FIGURE 15-13 Material on the desk covered with gasoline included a book titled *How to Beat the Bill Collector*. Having this book was not evidence by itself but did indicate further investigation might be necessary.
Courtesy of Russ Chandler.

Some of the contents may potentially tell a story in combination with the fire. One restaurant fire was clearly incendiary in nature. There were no food stocks for the following day, and even staff had questioned why no food deliveries were made, as they were needed to open the restaurant the next morning. At the scene, the office was clearly meant to burn because gasoline had been poured over the open drawers in the desk and file cabinet, but for some reason the office door was closed. On top of the desk, the material had been moved around a little because the sooting was uneven. In plain view was a book titled *How to Beat the Bill Collector* (**FIGURE 15-13**). This book by itself did not mean much, but several other papers were overdue notices. The investigator of this particular fire was a private fire investigator contracted by the insurance company. In this role, he was able to take photographs of all the contents of the structure. In addition to the rights of the insurance company coming into play, the owner had given permission to do the investigation.

To give some closure to this particular case, a great deal of evidence suggested that the owner had set his restaurant on fire. Even so, it was not strong enough to deny the claim or for local authorities to make an arrest. The key witness ultimately proved to be the 16-year-old wife of the owner. She later divorced him and contacted the county about 5 years after the fire, stating that she was willing to produce additional evidence and testify against her ex-husband. It was a very short deliberation by the jury, who found him guilty as charged.

Sifting a Scene

A complete scene sift is a task that is usually not necessary for forming the initial hypothesis about a fire's

origin and cause. Sifting a fire scene is a daunting task, but it can be handled effectively with sufficient time and a diligent, hard-working staff. The main reason why a sift is conducted is to help the fire department investigator find a suspected incendiary device. The private-sector fire investigator may also conduct a scene sift to prove insurance fraud when a proof-of-loss statement indicates inflated values and quantities of the contents.

TIP

Sifting a fire scene is a daunting task, but it can be handled effectively with sufficient time and a diligent, hard-working staff.

The first decision when contemplating a scene sift is the size of the openings of the screens that will be used to sift through debris. The best answer will completely depend on what the team expects to find. Screens can be commercially purchased, or they can be made on the site with basic supplies from a local hardware store. Intricate models can be made that will swing back and forth, whereas simple screens are more likely to be 4 ft^2 and set on two sawhorses. Each screen can be picked up and sifted by a team of two (**FIGURE 15-14**), or the sawhorses can be balanced on just two legs and rocked back and forth so that they take the weight of the sifting screen. A third person can shovel the debris onto

FIGURE 15-14 Sifting a scene: Two investigators handle the screen, while a third shovels in the debris. The fourth person documents, photographs, and packages any evidence found. Although this is not a task undertaken by the first responder, first responders are often drafted for this duty because of their basic knowledge and to provide training opportunities.
Courtesy of Russ Chandler.

the screen. The fourth person, usually the investigator in charge, documents and tags everything found in the process.

The full-time investigator may also sift the scene to look for evidence such as a weapon. The process is no different than that discussed in Chapter 4, *Science, Methodology, and Fire Behavior,* in terms of the insurance investigator's search for contents in a possible fraud case. In any case, the entire scene must be sifted to clearly indicate that the job was done properly.

During sifting, safety is the primary concern. Although this may be a cold scene, dust will rise as the debris is disturbed. Thus, it is critical that everyone on the scene is properly protected by respirators that have been properly fit tested as per Occupational Safety and Health Administration (OSHA) regulations. Crews must be dressed appropriately for the environment. The presence of hazardous materials will dictate whether specialized clothing is needed. For most residential structures, a pair of Tyvek coveralls will work well, along with a good pair of gloves.

Any time food is to be ingested at the site, the workers must be in a clean environment, and outer clothing should be taken off. Hands must be thoroughly cleaned. In addition, clothing exposed to the soot should not be washed in an ordinary washing machine, as it will become contaminated from the dust from the fire scene. Instead, scrubbing with a water hose and brush and washing clothing in a commercial turnout gear washer are safe ways to handle the clothing. Of course, used Tyvek coveralls should be disposed of, rather than washed or reused.

The final result will be a scene that has clear signs of the sifting process with various piles of finely sifted material. Adjacent piles will contain material that was examined and set aside. As an area is cleared off, the investigator examines the floor. The team then moves the screens forward and sifts the next area in a grid fashion. The process makes it possible to identify any items that survived the fire, such as firearms (**FIGURE 15-15**) or tools and other metal furnishings (**FIGURE 15-16**). Sometimes jewelry may be found in the debris (**FIGURE 15-17**), although not all jewelry will survive a fire. The outcome depends on the temperature to which the jewelry was exposed, whether the items were protected and insulated, and the melting temperature of the materials from which they are made.

FIGURE 15-15 Items such as firearms are still identifiable in the fire debris.
Courtesy of Russ Chandler.

FIGURE 15-16 Metal items recovered during sifting, such as tools and kitchen appliances and utensils.
Courtesy of Russ Chandler.

FIGURE 15-17 Ring found in the debris as a result of the sifting process.
Courtesy of Russ Chandler.

Wrap-Up

- Arson is a horrendous crime that affects the entire community and taxes the local emergency services. The crime of arson is defined by three elements: (1) something must have burned; (2) the burning must have been intentional; and (3) the element of malice must be present.

- Malice is the key indicator of whether a fire was an illegal act. For whatever reason, people often turn to using fire as a means to an end, even though they might never want to hurt anyone. Fire-setters can be anyone.

- Motives for setting a fire vary, but fall within six categories: vandalism, excitement, revenge, crime concealment, profit, and extremist. Even though the law may not require a motive to get a conviction, explaining the arsonist's motivation may help a jury convict the perpetrator of such a crime.

- Fires set by minors can be broken down into categories based on age. Child fire-setters are younger than age 7, and typically set fires out of curiosity or in the act of playing. Most courts do not allow any child younger than 7 to be charged with a crime. Juvenile fire-setters are generally 7 to 13 years old; adolescent fire-setters are 14 to 16 years old.

- Many indicators can give clues as to how a fire was started or why the fire occurred. These indicators aid the investigator in finding the evidence needed not only to determine that the fire was incendiary but also to identify who may have committed the crime.

KEY TERMS

Burglary The entry into a structure or part of a structure with the intent to commit a crime.

Collusion Agreement between two parties to conspire against a third party. In this case, it would be law enforcement and insurance industry using each others capabilities jointly to conspire against the insured.

Corpus delicti The body of the crime; the physical proof that a crime did occur.

Criminal act As stated in *Black's Law Dictionary*, an external manifestation of one's will that is prerequisite to criminal responsibility. There can be no crime without some act, affirmative or negative.

Deposition Obtaining testimony of witnesses that is recorded, authenticated, and reduced to writing and that can be used in future court testimony; used as a discovery device by either side in a civil or criminal trial.

Embezzlement Misappropriation of funds that were placed in someone's trust or belonging to the employer; theft.

Extortion Obtaining funds through threats or force.

Felony A crime punishable with prison time for a year or more, which is more serious than a misdemeanor. Felonies are usually violent crimes.

Fraud Deception intended for financial or personal gain; mostly criminal in nature and at the least wrongful.

Malice The intent to commit a wrongful act or inflict harm.

Misdemeanor A lesser crime or wrongdoing; when criminal in nature, the penalty can include prison time up to one year or a fine, or both.

Molotov cocktail A bottle filled with an ignitable liquid that is ignited by using a wick in the bottle, a reactive chemical coating, or wrapping of the bottle in a fuel-soaked rag.

Motive A reason for doing something; in this context, the reason for doing something wrong, which may not always be obvious.

Perpetrator The person who commits the criminal act.

Proof of loss A form provided by the insurance company to an insured party who experiences a loss. The insured fills it out with information about the loss, including the items lost and the amount of money that the insured expects; the form is signed and sworn by the insured. Most of the time, additional documentation is provided to support the claim.

Pyromania The tendency or impulse to set fires.

Robbery Feloniously taking any item of value from another directly or in their immediate presence and against their will while using force or fear.

Vandalism The act of deliberate destruction and damage of either private or public property.

REVIEW QUESTIONS

1. What are the three elements of arson?

2. Define *malice*.

3. What are the six motives for arson as identified by the National Center for the Analysis of Violent Crime? Give an example of each of these motives.

4. What are the three age groups distinguished among youth fire-setters?

5. Describe in detail five indicators that a fire may be incendiary in nature.

6. What are the differences between the terms *arson* and *incendiary*?

7. What is the difference between cooperation and collusion of the government investigator and the insurance investigator?

8. When the contents are missing in a structure fire, does their absence always mean it is an incendiary fire? Explain in detail.

9. What safety precautions should you take when sifting through the debris of a residential fire?

10. Why is pyromania no longer an accepted motive?

DISCUSSION QUESTIONS

1. Discuss the impact on the fire service when one of its own commits an act of arson. If convicted and after completing any punishment, should that person be allowed to come back as a fire fighter?

2. This chapter included a story about a grandmother who set a vehicle on fire. Should she be arrested? What good would it serve?

REFERENCES

1. *Black's law dictionary* (5th ed., p.102) (1979). St. Paul, MN: West Publishing Company.
2. *Black's law dictionary* (5th ed., p. 685) (1979). St. Paul, MN: West Publishing Company.
3. *Webster's new world college dictionary* (2nd ed., pp. 78, 709). (1984). New York: Simon & Shuster.
4. Gary, G. P., Huff, T. G., Icove, D. J., & Sapp, A. D. (n.d.). *A motive-based offender analysis of serial arsonists.* Washington, DC: Department of Justice, Federal Bureau of Investigation, and Federal Emergency Management Agency.
Retrieved from http://www.interfire.org/res_file/fuab_mb.asp
5. Council on Foreign Relations. (2010, March 16). *Provisional Irish republican army (IRA)*. Retrieved from https://www.cfr.org/backgrounder/provisional-irish-republican-army-ira-aka-pira-provos-oglaigh-na-heireann-uk
6. National Fire Protection Association. (2008). *NFPA 921: Guide for fire and explosion investigations* (Section 6.2.4.4, pp. 921–41). Quincy, MA: NFPA, 2008.

Appendix

FESHE Correlation Guide

Fire Investigation I (C0283) Correlation Guide

Course Outline	Fire Investigations for First Responders Chapter Correlation
Emergency Responder Responsibilities and Observations	Chapter 1, 8, 9, 11, 12, 15
■ Responsibilities of the Fire Department	Chapter 1, 8, 9, 12, 15
■ Responsibilities of the Firefighter	Chapter 1, 2, 8, 9, 11
■ Responsibilities of the Fire Officer	Chapter 1, 8, 9, 11, 12, 15
■ Observations When Approaching the Scene	Chapter 1
■ Observations upon Arrival	Chapter 1
■ Observations during Firefighting Operations	Chapter 1, 8, 9, 12
■ Identification of Incendiary Devices	Chapter 1, 2, 15
Constitutional Law	Chapter 3, 8, 9, 13, 15
■ Criminal Law	Chapter 3, 8, 9, 13, 15
■ Constitutional Amendments	Chapter 3, 9
Case Studies	Chapter 3, 8, 9, 12
■ *Michigan v. Tyler*	Chapter 3, 8, 9, 12
■ *Michigan v. Clifford*	Chapter 3, 8
■ Daubert Decision	Chapter 3
■ Benfield Decision	Chapter 3
■ Kuhmo/Carmichael Decision	Chapter 3

Course Outline	*Fire Investigations for First Responders* Chapter Correlation
Fire Investigations Terminology	Chapter 3, 4, 5, 6, 7, 8, 9, 10, 11, 12, 14, 15
▪ Terms as They Apply to Structural Fires	Chapter 5, 6, 7, 8, 9, 10, 11, 12
▪ Terms as They Apply to Vehicle Fires	Chapter 14
▪ Other Common Investigative Terms	Chapter 3, 4, 6, 7, 8, 9, 12, 15
Basic Elements of Fire Dynamics	Chapter 4, 10, 11, 13, 14
▪ Ignition	Chapter 4, 10, 13, 14
▪ Heat Transfer	Chapter 4, 10, 11, 14
▪ Flame Spread	Chapter 4, 10, 11, 13, 14
▪ Burning Rate	Chapter 4, 10, 11, 14
▪ Fire Plumes	Chapter 4, 11
▪ Fire Analysis	Chapter 4, 11
Building Construction	Chapter 2, 5, 10, 11, 12
▪ Types of Construction	Chapter 2, 5, 12
▪ Building Materials	Chapter 5, 10, 11, 12
▪ Building Components	Chapter 5, 10, 11, 12
Fire Protection Systems	Chapter 1, 6, 12
▪ Extinguishment Systems	Chapter 6
▪ Detection Systems	Chapter 1, 6
▪ Signaling Systems	Chapter 6
▪ Other Building Services	Chapter 6
Basic Principles of Electricity	Chapter 2, 4, 7, 10, 12, 14
▪ Basic Electricity	Chapter 2, 4, 7, 10, 12
▪ Wiring Systems	Chapter 7, 14
▪ Common Electrical Systems	Chapter 7

Course Outline	*Fire Investigations for First Responders* Chapter Correlation
Health and Safety	Chapter 2, 5, 7, 9, 12, 13, 14
■ Methods of Identification	Chapter 2, 12
■ Common Causes of Accidents	Chapter 2, 9, 12
■ Common Causes of Injuries	Chapter 2, 9, 12
Fire Scene Investigation	Chapter 1, 2, 3, 4, 5, 6, 7, 8, 9, 10, 11, 12, 13, 14, 15
■ Examining the Fire Scene	Chapter 1, 3, 4, 6, 7, 8, 9, 11, 12, 13, 14
■ Securing the Fire Scene	Chapter 1, 2, 3, 8, 9, 12, 13, 14
■ Documenting the Fire Scene	Chapter 1, 8, 9, 12, 13, 14
■ Evidence Collection and Preservation	Chapter 7, 8, 9, 10, 12, 13
■ Exterior Examination	Chapter 1, 8, 9, 12, 13, 14
Determining Point of Origin	Chapter 1, 8, 9, 10, 11, 12, 13, 14, 15
■ Interior Examination	Chapter 1, 8, 11, 12, 13, 14
■ Area of Origin	Chapter 10, 11, 12, 13, 15
■ Fire Patterns	Chapter 9, 10, 11, 12, 13, 14, 15
■ Other Indicators	Chapter 1, 9, 11, 12, 15
■ Scene Reconstruction	Chapter 8, 11, 12, 14, 15
■ Point of Origin	Chapter 11, 12
Types of Fire Causes	Chapter 1, 4, 7, 9, 10, 11, 13, 14, 15
■ Accidental	Chapter 1, 4, 9, 10, 13, 14, 15
■ Natural	Chapter 4, 9, 10, 13
■ Incendiary	Chapter 1, 4, 7, 9, 10, 11, 13, 14, 15
■ Undetermined	Chapter 4, 13, 14
Vehicle Fires	Chapter 7, 10, 14, 15
■ Examination of Scene	Chapter 14

Course Outline	*Fire Investigations for First Responders* Chapter Correlation
▪ Examination of Exterior	Chapter 14
▪ Examination of Driver and Passenger Areas	Chapter 10, 14
▪ Examination of Engine Compartment	Chapter 14
▪ Examination of Fuel System	Chapter 14
▪ Examination of Electrical System	Chapter 7, 14
Fire Setters	Chapter 1, 8, 9, 10, 11, 13, 15
▪ Characteristics of Arson	Chapter 1, 8, 9, 10, 11, 13, 15
▪ Common Motives	Chapter 9, 15

Glossary

A

Accelerant An item or substance used to ignite, spread, or increase the rate of fire growth.

Accelerator A component of a dry sprinkler system that is designed to induce air (from the dry pipe system) into a chamber of the dry pipe valve to reduce the amount of time it takes the system to activate allowing the flow of water into the system.

Accountability tag An identification tag with the holder's name and sometimes unit number or assigned company. More sophisticated tags also include the individual's medical history, including blood type and medical allergies. Used in the accountability system to track the location of all personnel at an emergency scene.

Administrative search warrant Warrant issued by a magistrate or judge that allows the investigator to be on the scene to determine the cause of the fire for the good of the people.

Air entrainment An event in which, upon the ignition of a fire, the flames stretch upward and the developing heat causes the air to rise. The process of the air rising up creates a void at the lowest area of the fire; air then moves into this area, becoming part of the ignition process.

Annealing The collapse of coil springs, loss of temper, as the result of heat.

Area of origin A structure, part of a structure, or general geographic location within a fire scene, in which the "point of origin" of a fire or explosion is reasonably believed to be located.

Arrow patterns As described in the NFPA 921, the patterns on wooden structural members that show a direction or path of fire travel.

Arson The crime of deliberate and willful setting of a fire with malice

Authority having jurisdiction (AHJ) A term used by code- and standard-writing organizations to indicate the organization, office, or individual who is responsible for approving equipment, installations, procedures, and other resources and policies; an entity responsible for developing, implementing, and maintaining a qualification process.

Autoignition temperature The minimum temperature at which a properly proportioned mixture of vapor and air will ignite with no external ignition source.

Awning window A window designed with a hinge at the top that opens outward at the bottom with the use of a hand crank; some designs use a motor to open and close the window. Units usually still have the hand crank in case the motor fails.

B

Backdraft An explosion resulting from the sudden introduction of air (oxygen) into a confined space containing oxygen-deficient superheated products of incomplete combustion.

Balloon construction A type of building construction in which wood studs are continuous from basement to the roof, creating a void from bottom to top. Should fire occur at lower levels, it can easily extend to the roof and other parts of the structure.

Barricade tape Wide, brightly colored tape with a clear message that will prevent entry onto the fire scene. The wording can be anything from *Fire Line—Do Not Cross* to *Crime Scene—Keep Out*.

Bead The melted end of a metal conductor that shows a globule of resolidified material with a sharp line of demarcation to the remainder of the wire.

Bernoulli effect As it applies to chimneys, a phenomenon in which wind blowing across the opening of a chimney decreases the pressure in the flue, increasing the updraft.

Bill of Rights The first 10 amendments to the U.S. Constitution, which include such rights as freedom of speech, due process, and a speedy public trial.

Biodiesel fuel Fuel derived from vegetable sources with added animal fat and alcohol that can be used in diesel engines with no modification of the engine.

Building official The person responsible for the enforcement of the local and state building codes in that jurisdiction. The building official's office is responsible for issuing construction permits and certificates of occupancy as a means of ensuring compliance with the code.

Building Officials Office An agency empowered by the authority having jurisdiction to enforce the provisions of a state or local building code, electrical code, plumbing code, and in some instances, the fire code.

Burglary The entry into a structure or part of a structure with the intent to commit a crime.

Burner assembly A unit that houses the electrodes in a furnace that ignite the oil pumped out of the jet nozzle; it also contains the fan, which blows air into the chamber, allowing for a bright and efficient flame.

C

Calcination The process of driving off the moisture in the gypsum (including the moisture chemically bonded within the gypsum), discoloring and softening the gypsum in the process.

Canister A manufactured metal or plastic cylindrical device filled with filtering material. Canisters are designed to be attached to a filter mask with an airtight seal, forcing all air entering the mask to pass through the filtering medium, which removes hazardous airborne particulates.

Carbon monoxide (CO) A colorless, odorless gas usually produced as a by-product of a fire. It can also be created from faulty appliances such as a furnace with a leaking exhaust flue.

Cartridge fuse An electrical device designed to interrupt electrical current when an overload occurs. It consists of a tube encasing the fusible link.

Case law Laws that are established by judicial decisions in the courtroom, rather than being promulgated by legislation.

Casement window A window hinged in the frame from one side, similar to the awning window; it operates with the use of a crank.

Casement windows Windows with hinges on the side that usually open with a rotating crank assembly.

Catalytic converter A device installed as part of the exhaust system of an automobile; designed to reduce emissions by burning off pollutants.

Ceiling jet A thin layer of buoyant gases that moves rapidly just under the ceiling in all directions away from the plume.

Chain of custody A means of documenting who had control of the evidence from the time of collection through the trial and up to its release or destruction.

Chimney chase A decorative hollow covering around a cement block or metal chimney, designed for aesthetic purposes; it is usually framed in wood and covered with the same material as the siding of the structure.

Chimney liner The covering inside the chimney that enables the flow of hot gases and smoke from the chimney and keeps them from seeping through the mortar and into the structure.

Circuit breaker A protection device within an electrical system that will automatically cut off the electricity should an excessive overload occur.

Class K Extinguishing agents for fires in cooking appliances involving cooking oils.

Clinker A solid glass-like object found as residue in a large hay fire.

CO/ALR A designation given to electrical devices that can safely use aluminum wiring. It is used only with those devices that meet a predetermined test accepted by both Underwriters Laboratory and the industry.

Cold scene The remains of a fire scene after the fire has been completely extinguished. Although the debris may still be warm, there is little or no chance of being reignited. Cold scenes may be days old.

Collusion Agreement between two parties to conspire against a third party. In this case, it would be law enforcement and insurance industry using each other's capabilities jointly to conspire against the insured.

Commercial evidence tape Tape with a sticky side used to seal evidence containers. Tape contains wording such as "evidence" or "do not tamper." When tape is removed, it leaves a visible indication that clearly shows the package or container has been opened.

Comparison sample An uncontaminated sample of what is being tested that helps examiners identify any contaminants.

Compressed natural gas (CNG) A natural lighter-than-air gas compressed for use as a fuel; it consists principally of methane in gaseous form plus naturally occurring mixtures of hydrocarbon gases.

Conduction The transfer of heat through a solid object.

Conductor Anything that is capable of allowing the flow of electricity; a term commonly used to describe the wires that provide electricity within a structure.

Conduit A tube or trough made from plastic or metal that is designed to provide protection for electrical conductors.

Consent Permission given by the person responsible for or controlling a property, which allows the investigator to search the property.

Constitutional law Laws that establish the relationship between the executive branch, the legislative branch, and the judiciary branch of government.

Contamination Anything introduced into the fire scene or into the evidence that makes test results unreliable or raises doubts about the results of the laboratory testing.

Convection The transfer of heat through air currents (or liquids).

Corpus delicti The body of the crime; the physical proof that a crime did occur.

Crankshaft A shaft that takes the power developed by the engine and transmits it through mechanical motion to other parts of the vehicle.

Craze (crazing) Multiple small, close fractures in glass created from rapid cooling; the opposite of fractures caused by rapid heating. (Contrast with tempered glass, which is designed to break into small fragments to limit injuries should it be accidently broken.)

Crime An illegal act that is an action or omission as defined in legislative laws, allowing for prosecution by the state or federal courts.

Criminal act As stated in *Black's Law Dictionary*, an external manifestation of one's will that is prerequisite to criminal responsibility. There can be no crime without some act, affirmative or negative.

Criminal search warrant A warrant issued by a magistrate or judge based on the sworn testimony (and written affidavit) of an investigator that probable cause exists that a crime has been committed and the person or place to search and the items being sought.

D

Deductive reasoning Taking the facts of the case and thoroughly and meticulously challenging all facts known, using logic.

Deluge sprinkler system Fire suppression method using open sprinkler heads; when a valve opens, it supplies to all heads simultaneously, immediately flowing water.

Demarcation A distinctive line between a burned area and an unburned or lesser-burned area; can also be used to describe other delineations such as smoke patterns.

Demonstrative evidence Evidence in the form of the representation of an object, in contrast to physical (real) evidence, documentary evidence, or testimony evidence.

Deposition Obtaining testimony of witnesses that is recorded, authenticated, and reduced to writing and that can be used in future court testimony; used as a discovery device by either side in a civil or criminal trial.

Diesel fuel A combustible petroleum distillate that is used in diesel engines.

Discovery A pretrial device used by both (all) sides in a case to obtain all the facts about the case to prepare for the trial.

Discovery sanctions Penalties for failing to comply with discovery rules.

Disposable butane lighter A device filled with a pressurized liquid ignitable gas; friction or an electrical arc is used to ignite the escaping gas as designed. Such a lighter is also capable of putting out the flame, cutting off the fuel and making the device safe enough to be carried by the user.

Documentary evidence Evidence in the form of media such as paper records, photographs, videos, and voice tape recordings.

Door assembly All of the component parts of the door from the framing which include the sill (bottom of the frame), the jamb (sides of the door frame) and the head. This will also include the door itself and its components from the hardware including the door knobs, hinges and strike plates. This assembly may also include additional components such as windows beside or above the door.

Double-hung window Two separate sashes installed in a window frame, with both the top and bottom sashes able to independently slide vertically up and down.

Drying oils Organic oils used in paints and varnishes that, when dried, leave a hard finish.

Drywall Also called sheetrock; a wallboard of gypsum with a coating of paper on each side.

Duct smoke detector Device that samples the air in the air distribution system for products of combustion. When activated, it closes dampers, thereby preventing smoke from traveling through the system.

E

Electric hybrid A vehicle that uses both an electric engine and a conventional engine (e.g., a gasoline gas). This type of vehicle provides an economical form of transportation compared to gasoline or diesel vehicles.

Electric matches Electrical devices that can act as an igniter from a remote distance.

Electronic fuel injection (EFI) A mechanical system, controlled electronically, that reduces a fuel to a fine spray and injects it into the cylinders of an internal combustion engine.

Electrons Negatively charged particles that are part of all atoms.

Embezzlement Misappropriation of funds that were placed in someone's trust or belonging to the employer; theft.

Emergency communications center (dispatch) Sometimes referred to as the E 911 center, the place where calls from the public for emergency assistance are received and whose staff (dispatchers) alert and send the appropriate emergency unit(s) to handle the situation.

Emergency communications dispatcher A person who works in an emergency communications center, whose duties are to receive calls from the public for emergency assistance, and then to send the appropriate emergency unit(s) to handle the situation. Commonly referred to as the dispatcher.

Emergency medical technicians (EMTs) Individuals with specialized medical skills, trained to offer basic first aid in the field under emergency circumstances, and then to assist in transporting the patient to a medical facility if necessary.

Empirical data Information collected that is based on observations, experiments, or experience.

Energy release rate (ERR) The amount of energy produced in a fire over a given period of time.

Ethanol A type of alcohol produced by natural fermentation of sugars; usually associated with liquors but can also be used as an additive to vehicle fuels. It is a colorless volatile flammable liquid.

Evidence transmittal form A form designed to identify the evidence to be submitted to the laboratory and the tests to be conducted on that evidence; it can also serve as the chain of custody of the evidence.

Exemplar A comparison sample that enables the investigator to compare the damaged item to an undamaged version of the product.

Exhauster An early action device installed on a dry sprinkler system. When a sprinkler head activates, this device will remove the air in the system, allowing a more rapid delivery of water to the sprinkler head.

Exigent circumstances An emergency; for the good of the people, without permission, public safety personnel can enter property on fire to save lives and control the fire in such a situation.

Exothermic reaction The release of heat from a chemical reaction when certain substances are combined.

Expert witness A person who is permitted to testify in court due to his or her special knowledge and expertise in a particular area relevant to the case.

Exterior finishes The outer finish of a building that provides protection against the elements.

Extortion Obtaining funds through threats or force.

F

Fact (or lay) witness A person with knowledge about what happened, what he or she saw, or what he or she heard that is firsthand and relevant to the case.

Felony A crime punishable with prison time for a year or more, which is more serious than a misdemeanor. Felonies are usually violent crimes.

Felony A crime that most likely involves violence and is more serious than misdemeanors. Punishment for such is usually more than a year in jail or even death.

Fire alarm control unit (FACU) A device that serves as the "brain" of the fire alarm system, monitoring and managing its operation. It serves as a link between initiating devices and notification devices. Most systems have primary and secondary power.

Fire door Any door with a fire resistance rating; it is used to limit the spread of fire and smoke to other areas of the building.

Fire point The lowest temperature at which a fuel, in an open container, gives off sufficient vapors to support combustion once ignited from an external source.

Fire tetrahedron A solid figure with four triangular faces; each face represents one of the four items necessary for a fire (heat, fuel, oxygen, and uninhibited chemical reaction).

Fire wall A wall or partition designed to limit the spread of fire. The specific construction will dictate how long it will prevent the spread of fire.

First responder Any public safety responder who may be or may have the potential of being first on the scene as the result of being sent there by the emergency communications center.

Fixed-temperature heat detector Device that activates at a predetermined temperature.

Flameover The point at which a flame propagates across the undersurface of a thermal layer; also called rollover.

Flash point The temperature at which a liquid gives off sufficient vapors that, when mixed with air in proper proportions, ignites from an exterior ignition source. Because of the limited supply of vapors, only a flash of fire across the surface occurs.

Flashover A transition phase of fire where the exposed surface of all combustibles within a compartment reach autoignition temperature and ignite nearly simultaneously.

Floating neutral A neutral that is not grounded. In a single-phase circuit, it can cause an unbalanced flow between each of the two current-carrying wires. This results in an unbalanced load, which can cause an overload and failure in appliances attached to the circuit.

Fraud Deception intended for financial or personal gain; mostly criminal in nature and at the least wrongful.

Fuses Devices used as screw-in plugs or a cartridges that contain a wire or thin strip of metal designed to melt at a specific temperature associated with certain overcurrent situations. The melting of the wire or metal strip stops the flow of electricity as designed.

G

Gas chromatography (GC) A laboratory test method that separates the recovered sample into its individual components; it provides a graphic representation of each component along with the amounts of each component present.

Gasoline A hydrocarbon fuel refined from crude oil to be used primarily in internal combustion engines.

Greenfield The manufacturer's name for a flexible conduit. This name has become a common term to describe any flexible conduit, either metal or plastic.

Ground fault circuit interrupter (GFCI) A circuit breaker that is more sensitive than the standard breaker, tripping at a slight ground fault with the intent of preventing electrocution.

Grounding block A mass of metal within an electrical panel, affixed with holes and screws to allow for the insertion of wires; the tip of the screw binds and holds the wire in place.

Guide A document that is advisory or informative in nature and that contains only nonmandatory provisions. It may contain mandatory statements, such as when a guide can be used, but the document as a whole is not suitable for adoption into law.

Gypsum board (sheetrock, dry wall, plaster board) A wall covering, usually with a gypsum core, bonded between two layers of paper that serves as the first and basic finish on walls and ceilings.

H

Hearsay Information that is heard or received from other people than the originating party. This information cannot be substantiated and most times is considered rumor.

Heat detectors A device designed to sense and respond to thermal energy (heat) at a predesignated rate and or temperature. There are two classifications, *rate-of-rise* which will respond to a given rate of temperature increase and a *fixed temperature* device that will react at a specific temperature.

Heat release rate (HRR) The rate at which heat is generated from a burning fuel.

Hydrocarbon detector An electronic device intended to be used in the field to identify the location of ignitable liquids; it is used to improve the probability of obtaining the best sample to submit to the laboratory for testing.

I

Imminent danger to life and health (IDLH) An atmosphere in which anyone entering endangers themselves unless proper protection is taken, which could include full turnout gear with self-contained breathing apparatus.

Incendiary fire A fire set willfully and intentionally, with malice.

Incident Command System (ICS) An on-scene, standardized, all hazard management process.

Insulator A nonconductor of electricity; commonly thought of as a material such as glass or porcelain, but today includes plastic and rubber (as in the coverings of electrical wire).

Interior finishes The exposed surfaces of walls and ceilings; can be wood, paint, or wallpaper applied over gypsum or sheetrock.

International Association of Arson Investigators (IAAI) A private, nonprofit organization dedicated to the professional development of fire and explosion investigators through training and research of new technology

Ionization detector Initiator that uses a small amount of radioactive material to ionize air between two metal plates, creating a constant current between two plates. When smoke enters the chamber, it reduces the current, setting off the alarm.

J

Jalousie window A window in which glass, plastic, acrylic, or wood louvred panes slightly overlap and join, with a track operating each of the slats in unison; the panes tilt outward from the bottom edge with the use of a crank handle.

Jet nozzle In an oil furnace, a nozzle with an extremely small opening. The oil is pumped up to this nozzle, where it squirts out and is ignited by the electrodes, providing a proper flame for the burn chamber. Nozzles come in varying sizes.

Junk science Findings that are untested or unproven theories that are inappropriately presented as scientifically based.

K

K-9 accelerant dog A dog specifically trained by a reputable organization to identify the presence of an accelerant and give the appropriate signal of its location.

Kinetic energy Energy as the result of a body in motion.

Knob and tube An early method of running electrical wiring through a structure. Wires were supported by porcelain knobs, which isolated the wire from the structure. Holes were drilled into beams and studs, where a porcelain tube was inserted and acted as an insulator, allowing the wire to safely pass through. The porcelain knobs and tubes acted as insulators that kept the electricity from going to ground.

L

Liquid propane-driven vehicle There are two types liquid propane driven vehicles; dedicated which run only on propane or bi-fuel propane vehicles which can run on propane or gasoline. A propane vehicle has similar acceleration and cruising speed as those run on gasoline fueled vehicles.

Litigant Someone who is involved in a lawsuit.

Load-bearing wall Any wall that bears structural weight (load) resting on it, transferring that weight to the foundation below.

Lockout A situation in which an electrical disconnect box is in the open position with no current flowing and a lock is in place to keep the circuit from being energized. Some lockouts provide space for multiple locks to provide safety for all personnel working on the scene.

Lower explosive limit (LEL) The lower limit of flammability of a gas or gas mixture at ambient temperature and pressure. It is the threshold just before the fuel becomes too lean, but still can ignite.

M

Malice A wrongful act that includes the intentional desire to cause harm.

Manifold (1) The engine component that sends the fuel–air mixture to the cylinders (intake manifold). (2) An engine component that attaches to the cylinder head, with the exhaust pipes then attaching to the manifold (exhaust manifold).

Mass spectrometry (MS) A test used to identify the presence of an ignitable liquid in a submitted sample; it indicates the composition of a physical sample by generating a spectrum representing the masses of sample components.

Meter base The receptacle for the electrical meter installed by the power company.

Methane gas Colorless, odorless, flammable gas created naturally by decomposing vegetation; it can also be created artificially.

Methanol A flammable liquid (alcohol) that is formed from the distillation of wood but can also be manufactured synthetically; used as antifreeze in vehicles and solvents, among other application.

Methenamine pill test Also known as the "Standard Method for Ignition Characteristics of Finished Textile Floor Covering Materials." A standard test of carpets to determine whether they meet product safety standards, in which a pill is ignited and dropped on the tested material to see if it will ignite the material and cause a spread of the fire beyond a limited measurable mark.

Methyl ethyl ketone (MEK) A water-soluble flammable liquid used as a solvent; also referred to as methyl ethyl ketone peroxide. It is similar to acetone.

Micro torch A miniature device that uses butane as a fuel to create flame; typically used by hobbyists. The mechanism can inject air into the chamber, allowing for a high heat output.

Misdemeanor A lesser crime or wrongdoing; when criminal in nature, the penalty can include prison time up to one year or a fine, or both.

Molotov cocktail A bottle filled with an ignitable liquid that is ignited by using a wick in the bottle, a reactive chemical coating, or wrapping of the bottle in a fuel-soaked rag.

Motive A reason for doing something; in this context, the reason for doing something wrong, which may not always be obvious.

Multi-gas detector A portable device for reading the levels of oxygen and hazardous gases in the atmosphere.

Multi-meter An electronic device used to measure AC/DC voltage, current, resistance, capacitance, and frequency. Also referred to as a multi-tester.

N

National Fire Incident Reporting System (NFIRS) A national computer-based reporting system that fire departments use to report information on fires and other incidents to which they respond. This uniform system collects data for both local and national use.

National Fire Protection Association (NFPA) A private, non-profit organization dedicated to reducing the occurrence of fires and other hazards.

NFPA 1033, *Standard for Professional Qualifications for Fire Investigator* Performance requirements and the qualifications to meet the requirements for the fire investigator.

NFPA 921, *Guide for Fire and Explosion Investigations* A document designed to assist individuals investigate fire and explosion incidents in a systematic and efficient manner.

NM Stands for *nonmetallic*; refers to a covering over wire conductors intended for use within a structure to deliver electricity to various points within the structure.

Non-load-bearing wall A wall that supports only itself.

Non-vented goggles Eyewear that makes an airtight seal with the face, preventing the introduction of any airborne particulates.

O

Occupational Safety and Health Act Also referred to as the OSH Act. Federal law that governs the safety and health practices for all industries, both government and the private sector. The purpose is to guide and require employers to provide a safe working environment for their employees.

Occupational Safety and Health Administration (OSHA) A federal agency created by the Occupational Safety and Health Act whose function is to ensure safety in the workplace.

Off-gassing The release of gas or vapors in the process of aging or decomposing. Also, vapors or gases that were absorbed by fabric, carpet, and so forth during a fire and then released after the fire is extinguished.

Officer in charge (OIC) The fire officer who has ultimate charge and control of the overall fire or emergency scene. Usually but not always the senior officer on the scene.

Ohm's law Mathematical equation that describes the relationship between the voltage (V), current (I), and resistance (R); it can be expressed in three ways, depending on what needs to be solved: $I = V \div R; V = I \times R; R = V \div I$

Outside stem and yoke (OS&Y) Sprinkler control valve with a visible stem that allows an investigator to identify whether the valve is open or closed by observing the length of the stem (screw) exposed. Occasionally referred to as outside screw and yoke.

Overhaul Process carried out by fire suppression personnel to examine and uncover any remaining heat source that could rekindle the fire. This often includes the term *salvage,* where the fire fighters take action to protect contents of the structure from further damage by removing or covering with tarps referred to as salvage covers.

Oxidize The result of heating a metal surface; the heat consumes any covering, resulting in the rusting (oxidation) of the metal surface.

Oxidizer Any material that forms from a fuel and supports combustion.

P

Perpetrator The person who commits the criminal act.

Photoelectric detector Detector using a light beam and a receiver. Visible particles of smoke entering the chamber either scatter the light or block the light, causing the alarm to activate.

Physical evidence policy A locality's policy that guides and directs the proper collection, storage, handling, use, and disposal of all evidence.

Piezoelectric ignition Certain crystals that generate voltage when subjected to pressure (or impact).

Plaster and lathe A wall and ceiling covering created by thin strips of wood closely mounted on studs where plaster is spread on the boards, allowing it to adhere, leaving a smooth plaster finish.

Plug fuse A threaded fuse that screws into a socket in an electrical panel; also known as a fuse plug.

Plume The column of smoke, hot gases, and flames that rises above a fire.

Point of origin The exact physical location within the area of origin where a heat source and a fuel first interact, resulting in a fire or explosion.

Pooling patterns An accumulation of a flammable liquid, a puddle. Identified by staining of the floor surface or if the flammable liquid burned, it will show a sharp line of demarcation between the burned and unburned area.

Post indicator valve (PIV) Sprinkler control valve with an indicator that will read open or shut.

Potential evidence Something that may, or could, possibly be used to make something else evident; findings not yet used or that have yet to be used to prove a point or issue or support a hypothesis.

Pro bono Work done free of charge; at no cost.

Proof of loss A form provided by the insurance company to an insured party who experiences a loss. The insured fills it out with information about the loss, including the items lost and the amount of money that the insured expects; the form is signed and sworn by the insured. Most of the time, additional documentation is provided to support the claim.

Pyromania The tendency or impulse to set fires.

R

Radiation Transmission of energy through electromagnetic waves. It travels at the speed of light, but only in straight lines.

Rate-of-rise detector A device that responds when heat exceeds a predetermined rise in temperature.

Resistance heating The heat produced when electrical current passes through a material of high resistance. This converts the electrical energy into heat.

Respirator A mask (full- or half-face) designed to cover the mouth and nose, allowing the wearer to breathe through filters attached to the mask and thereby prevent the inhalation of dangerous substances, usually in the form of dust or airborne particulates.

Robbery Feloniously taking any item of value from another directly or in their immediate presence and against their will while using force or fear.

Rocket motors Hobby propulsion kits that provide thrust through the burning of an ignitable mixture.

Rollover The point at which a flame will propagate across the undersurface of a thermal layer.

Rotor The part that rotates in an electric motor.

S

Safety match A match designed to ignite only when scraped across a chemically impregnated strip.

Salvage A suppression activity used to protect the contents of a property from smoke or water damage by removing objects from the structure, covering materials within the structure with tarps, and/or removing water with squeegees, pumps, water vacuums, and so forth.

Scale A flaking of metal as the result of oxidation.

Scientific method The systematic pursuit of knowledge involving the recognition and formulation of a problem, the collection of data through observation and experiment, and the formulation and testing of a hypothesis.

Self-contained breathing apparatus (SCBA) designed equipment that fire fighters wear when entering an unsafe atmosphere. The bottle in the apparatus is filled with compressed air.

Service drop Overhead wiring from the electrical company that is attached to the weather head to deliver electricity to the structure.

Service lateral Underground wiring that comes from the electrical company and enters the meter base.

Service panel The destination for electrical wire that travels from the meter base; it provides a means for overcurrent protection and the ability to distribute electricity throughout the structure through branch circuits.

Sheetrock Often called drywall, gypsum, or wallboard; a crumbly material called gypsum sandwiched between two layers of thick paper. Gypsum is fire resistive.

Single-hung window A window with two separate sashes installed in a window frame, with the top sash mounted as a stationary fixture and the bottom sash installed to slide vertically up and down.

Sleeving The slight separation of the insulation from around the electrical conductor(s), producing an effect that allows the insulation to loosely slide back and forth on the conductors.

Smoke detectors Automatic initiating devices that detect visible particles of smoke and send a signal to initiate an alarm.

Spalling A condition in which the surface of concrete pops off as a result of water in the concrete reaching boiling temperature and turning to steam. When this happens, the steam expands 1700 times in volume, creating the energy required to pop the concrete off from the surface.

Spark A glowing bit of molten metal debris.

Spoliation The destruction of evidence; the destruction or the significant and meaningful alteration of a document or instrument. It constitutes an obstruction of justice.

Spontaneous combustion A process in which a chemical reaction takes place internally, creating an exothermic reaction that builds until it reaches the ignition temperature of the material involved.

Spontaneous heating A process in which a material increases in temperature without drawing heat from the surrounding area.

Spontaneous ignition Initiation of combustion from within a chemical or biological reaction that produces enough heat to ignite the material.

Static electricity The buildup of a charge, negative or positive, as the result of items coming in contact and then breaking that contact, either taking away or leaving electrons, and resulting in the change of the static charge of the items involved.

Stator The part of an electric motor that stays stationary, housing the rotor.

Statutory laws Written laws usually enacted by a legislative body.

Steel-sole boots Boots with a lightweight steel plate that provides a barrier between the foot and the surface being walked on. The intent is to keep the wearer safe from punctures through the boot.

Steel-toe boots Work boots designed with a steel protective cap covering the toes to prevent crushing blows that would otherwise injure the wearer's toes.

Strike-anywhere match A match where the head is chemically impregnated to ignite when scraped across any rough surface.

Subpoena The command of a court to have a person appear at a certain place at a specific time to give testimony.

Subrogation A legal process usually associated with insurance companies in which the insurance pays the insured for his or her loss, then seeks reimbursement from the responsible party. This process makes obtaining a settlement under an insurance policy go smoothly.

Suspicious Upon examination of the scene there are items, issues or indications that are questionable, either the facts surrounding the fire scene or the actions of an individual that may indicate that the fire cause to be other than accidental.

Suspicious Upon examination of the scene there are items, issues or indications that are questionable, either the facts surrounding the fire scene or the actions of an individual that may indicate that the fire cause to be other than accidental

Systematic search A method or process used to search an area or a building that allows for complete coverage of the area, and that is done consistently at each fire scene or incident.

T

Tagged The action of placing a tag on an electrical disconnect switch warning others not to turn the handle to the on position due to a safety concern.

Tandem breaker Two single 15 (or 20 amp) breakers manufactured together, allowing both breakers to fit in a single slot in the electrical panel. Used to save space in the panel. They cannot be used for 208/240 volts because they are both on the same bus bar in the panel. These units are sometimes referred to as piggybacks, cheaters, or slimline breakers.

Technical rescue teams Assigned teams who have the application of special knowledge, skills, and equipment to safely resolve unique and/or complex rescue situations. Also referred to as heavy technical rescue teams, structural collapse teams.

Testimonial evidence Testimony offered to prove truth of the issue at hand, usually from a witness.

Thermal runaway A situation in which heat production increases during a fire; the exothermic reaction creates heat that increases the reaction rate, in turn creating more heat, which then increases the exothermic reaction.

Tool mark impressions Potential identifiable marks from tools used by the perpetrator of the crime.

Trailer A deliberate use of a flammable liquid or other combustible items to intentionally move the fire from one point to another.

Trailer The use of flammable liquids or easily ignitable combustibles (e.g., paper, cardboard, ropes) to enable a fire to travel from one area to another or to reach a specific target. Combustibles can be used in conjunction with the flammable liquids. Trailers can be both outside as well as in a structure.

Turbo charger A device that uses the energy of the vehicle exhaust gases to compress (increase) the air going into the engine, which in turn increases the power of the engine.

Tyvek coveralls Coveralls made out of Tyvek, which the manufacturer DuPont describes as lightweight, strong, vapor-permeable, water-resistant, and chemical-resistant material that resists tears, punctures, and abrasions. These coveralls are essential to protect the investigator working the extinguished fire scene against carbon and other contaminants and should be properly disposed after each incident.

U

U.S. Supreme Court The highest court in the United States.

UF The type of insulation on the electrical wiring that is used as an underground feeder. The insulation tends to be thicker and solid up to the insulated conductor.

Upper explosive limit (UEL) The threshold where the vapor concentration can be ignited, just before the concentration becomes too rich.

V

Vandalism The act of deliberate destruction and damage of either private or public property.

Vapor density Density of a gas or vapor in relation to air, with air having a designed vapor density of 1.

Vehicle identification number (VIN) A unique set of numbers and letters that serves as the serial number of a vehicle. These numbers, when decoded, can provide information about the vehicle, such as where the vehicle was built, the manufacturer, the vehicle brand, the engine size and type, the vehicle model year, and the plant where vehicle was assembled.

Voltage detector A tester or probe that either glows or emits a sound when placed in proximity of an energized electric circuit (electrical wires or cords).

Voltage Electromotive force or pressure, difference in electrical potential; measured as a volt (V), which is ability to make 1 ampere flow through a resistance of 1 ohm.

W

Watt A unit of power equal to 1 joule per second.

Weather head The weatherproof device where the electrical service from the overhead (or underground) power enters the building; sometimes referred to as a weather cap.

Index

© Jones & Bartlett Learning. Photographed by Glen E. Ellman.